Springer Monographs in Mathematics

A. Nepomnyashchy   I. Simanovskii
J.C. Legros

Interfacial Convection in Multilayer Systems

With 187 Illustrations

Springer

A. Nepomnyashchy
Department of Mathematics
Technion–Israel Institute of Technology
Haifa 32000
Israel
nepom@math.technion.ac.il

I. Simanovskii
Department of Mathematics
Technion–Israel Institute of Technology
Haifa 32000
Israel
yuri11@inter.net.il

J. C. Legros
Microgravity Research Centre
Universite Libre de Bruxelles, Belgium
50 av. F.D. Roosevelt
Bruxelles 1050
Belgium
jclegros@ulb.ac.be

Mathematics Subject Classification (2000): 76E17, 76E15

Library of Congress Control Number: 200692270

ISBN-10: 0-387-22194-8
ISBN-13: 978-0387-22194-6

Printed on acid-free paper.

Printed in the United States of America. (SPI/EB)

9 8 7 6 5 4 3 2 1

springer.com

Preface

The interfacial convection in multilayer systems is a widespread phenomenon that is of great importance in numerous branches of technology, including chemical engineering, space technologies, coating, etc.

The most well-known modern engineering technique that requires an investigation of the interfacial convection in systems with many interfaces is the liquid encapsulation crystal growth technique (Johnson, 1975; Géoris and Legros, 1996) used in space lab missions. It is known that time-dependent thermocapillary convection leads to solute segregation and, hence, to striation in crystals (Eyer *et al.*, 1985). The liquid encapsulation technique allows one to reduce significantly the convection and to grow high-quality, striation-free crystals by putting the melt between the fluid layers (Eyer and Leiste, 1985).

Another important problem that needs a multilayer approach for its self-consistent description is the coalescence between droplets or bubbles (Leshansky, 2001; Yeo *et al.*, 2001; Yeo and Matar, 2003) or between droplets and a bulk liquid (Savino *et al.*, 2003) under conditions of a heat/mass transfer. The influence of the interfacial convection on the coalescence rate, as well as the droplet migration under the action of an applied temperature/concentration gradient (Subramanian, 1981; Balasubramaniam and Subramanian, 2000), are significant for many engineering processes, including various extraction processes (Groothuis and Zuiderweg, 1960), steel refining (Mukai *et al.*, 2003), mixing (Fan and Zhang, 2001), and phase separation in alloys (Lu *et al.*, 2001).

One can mention also multilayer coating techniques used in the production of photographic films and multilayer fibers for optoelectronic devices (Yarin, 1995) and emulsified liquid membrane separation techniques widely used now in various extraction processes and industrial wastewater treatment (Noble and Douglas Way, 1987), where the interfacial convection appearing in a spherical liquid membrane layer can highly enhance mass transfer rates and sufficiently influence the stability of emulsified liquid membranes.

A scientific interest in such systems is due to the fact that the interfacial convection is characterized by a variety of physical mechanisms and types of instability, with the characteristic wave numbers ranging in a wide domain. The understanding of the underlying physical processes that can be achieved through the exploration of the mutual influence and the interaction between different interfaces is necessary for a successful application of this phenomenon. The problem of the convection in systems with many interfaces is important from the point of view of hydrodynamic stability theory, and of the theory of heat and mass transfer. Simultaneous interaction of interfaces with their bulk phases and with each other can lead to a much more complex dynamics and unexpected effects.

At present, the vast theoretical and experimental stuff related to the interfacial convection in multilayer systems is dispersed among numerous papers, many of which are hardly available. There are no monographs devoted to the interfacial convection in multilayer systems.

The present book, which contains a systematic investigation of convection in systems with interfaces, fills in this gap. For the first time, a classification of all known types of convective instability in such systems has been done, and the peculiarities of multilayer systems are discussed. The book provides an overview of the wide variety of steady and oscillatory patterns, waves, and other dynamic phenomena characteristic for multilayer fluid systems. Various physical effects, including heat and mass transfer, thermal and mechanical couplings on the interfaces, interfacial deformability, and the influence of surfactants on different types of convective motions are investigated using the multilayer approach; that is, the physical phenomena are studied in all fluids. This approach has its roots in the pioneering works of Sternling and Scriven (1959), Smith (1966), Zeren and Reynolds (1972), Imaishi and Fujinawa (1974a, 1974b), Gumerman and Homsy (1974a, 1974b), Gershuni and Zhukhovitsky (1982), Ferm and Wollkind (1982), Renardy and Joseph (1985), Renardy (1986), and Wahal and Bose (1988).

The book consists of seven chapters. In Chapter 1 (Introduction), the main mathematical models, which are used in subsequent parts of the book, are formulated. Chapter 2 contains the description of numerous types of convective instability in different systems with a single interface. In addition to the well-known monotonic Rayleigh-Bénard and Bénard-Marangoni instabilities, the systems with an interface are subject to many kinds of oscillatory instability generated by buoyancy, by the thermocapillary effect, by the competition between buoyancy and thermocapillarity, by the mode mixing of interfacial and internal waves, by an interplay between a thermal gradient and a surfactant distribution, and so on. Also, a specific non-Rayleigh mechanism of instability can produce a buoyancy instability by heating from above (anticonvection). All of these phenomena are studied both in the framework of a linear instability theory and by a nonlinear analysis.

The manifestations of the interfacial convective instabilities in multilayer systems are studied in Chapters 3 and 4. In Chapter 3, the situation in which the deformations of interfaces are not significant is considered. The main part of this chapter is devoted to the investigation of the Marangoni convection in three-layer systems, which is studied theoretically and by means of experiments (on the Earth and in space). Also, the peculiarities of the Rayleigh convection, mixed Rayleigh-Marangoni convection, and anticonvection are described. Situations that are impossible in the case of a single interface are analyzed.

The types of instability, which are essentially connected with interfacial deformations, are explored in Chapter 4. The progress in understanding of the nonlinear aspects of the problem is achieved by means of the long-wave asymptotic approach.

Chapter 5 presents the results on the stability of convective flows in systems with interfaces. The problems studied in this chapter include different types of instability of thermocapillary and buoyancy-thermocapillary flows, with and without interfacial deformations.

A specific class of interfacial phenomena related to the dynamics of very thin films is considered in Chapter 6. Chapter 7 is devoted to the discussion of the new directions for the scientific analysis and applications.

In our opinion, this book will be useful for experts in fluid mechanics, heat and mass transfer theory, nonlinear dynamics, and applied mathematics as well as for physicists and chemical engineers interested in the investigation of the interfacial physico-chemical processes and in their applications. The book can be used also by graduate students.

The scientific results included into this book have been obtained in collaboration with our co-authors and friends: Th. Boeck, L. M. Braverman, K. Eckert, Ph. Géoris, A. Yu. Gilev, A. A. Golovin, M. Hennenberg, I. L. Kliakhandler, A. Oron, A. Thess, S. Van Vaerenbergh, A. Viviani, I. I. Wertgeim, and M. A. Zaks. This book could not be written without fruitful discussions and scientific advice of our colleagues: I. Aranson, G. I. Burde, P. Colinet, S. H. Davis, C. S. Iorio, E. Istasse, D. Kessler, G. Lebon, B. Matkovsky, D. Melnikov, R. Narayanan, L. M. Pismen, A. Ye. Rednikov, B. Scheid, D. Schwabe, V. Shevtsova, L. Shtilman, G. Sivashinsky, and M. G. Velarde.

We dedicate this book to the memory of our late teachers and friends E. M. Zhukhovitsky and G. Z. Gershuni.

Our work on the subject of the book was sponsored by grants of the German-Israeli Foundation and of the Israel Science Foundation. The preparation of the book has been carried out in the framework of the ICOPAC (Interfacial Convection and Phase Change) network sponsored by the European Union.

Contents

1

Introduction

In the present chapter, we discuss the role of the interfacial convection for natural and engineering processes and present the main physical mechanisms that produce the convective instability, the thermocapillary effect, and the buoyancy. Then we consider the mathematical models used for the description of the abovementioned physical effects. In the framework of the simplest model (the one-layer model), the motion of the gas is neglected, whereas the heat/mass transfer in the gas is described in a phenomenological way by means of the Biot number. A more accurate approach (the two-layer model) takes into account the physical processes in fluids on both sides of the interface. Also, the three-layer model is formulated.

1.1 Motivation of the Problem

Interfacial convection is a striking phenomenon that takes place any time the interfacial tension is not uniform along the interface between two fluids. The role of this phenomenon is especially large on small scales, where the bulk effects like buoyancy are negligible. Specifically, this type of convection is crucial in microfluidic systems, where it provides a reliable transport mechanism (Sammarco and Burns, 1999; Kataoka and Troian, 1999; Darhuber *et al.*, 2003) and a mechanism for mixing (Frenkel and Halpern, 2002; Halpern and Frenkel, 2003). Interfacial convection is also a basic mechanism of fluid motion under microgravity conditions (Ostrach, 1982; Walter, 1987; Ratke *et al.*, 1989). Among the numerous modern engineering processes where the interfacial convection plays an important role one can mention laser welding (Fuhrich *et al.*, 2001; Dowden, 2002), fabrication of microporous polymer films (Maruyama *et al.*, 1998; Pitois and François, 1999a, 1999b; Srinivasao *et al.*, 2001), and ordering of nanoparticles (Maillard *et al.*, 2000; Truskett and Stebe, 2003). Traditional fields of application of the interfacial convection are chemical engineering (Levich and Krylov, 1969; Sawistowski, 1971) and materials processing (Szekely, 1979; Hondros *et al.*, 1998). Interfacial phenomena are significant for various biological processes (e.g., flows in the respiratory system) (Zelig and Haber, 2002; Grotberg and Jensen, 2004).

The numerous applications of the interfacial convection cause a need for adequate and reliable physical and mathematical models for its description.

An interface always separates two different physical media. Nevertheless, the most traditional and simple approach to the description of the interfacial convection is based on the *one-layer* model. In the framework of that model, only the processes in the liquid phase are considered in a complete way, whereas the influence of the gas is treated phenomenologically. The one-layer approach is not sufficient for the description of many real phenomena, which can be explained only when the *multilayer* approach is used; that is, the fluid motion and the heat/mass transfer are studied in all fluids. The application of the latter approach to the interfacial convection is the subject of the present book.

The multilayer approach has been applied for the investigation of several modern engineering processes. As the first example, let us mention the liquid encapsulation technique of crystal growth (Johnson, 1975; Géoris and Legros, 1996) used in space labs missions, which allows one to reach a high quality of the growing crystals by putting the melt between the fluid layers. The multilayer approach is needed for a self-consistent description of the droplet-droplet coalescence, where the Marangoni convection in an interdroplet film can considerably effect the coalescence time during extraction (Groothuis and Zuiderweg, 1960; Rother and Davis, 1999; Berejnov *et al.*, 2002). This approach is adequate for studying the interfacial convection in the presence of chemical reactions (see Bratsun and De Wit, 2004 and references therein) and phase transitions (Merkt and Bestehorn, 2003). As the potential applications of the multilayer approach, let us mention the multilayer coating techniques used in the production of photographic films and multilayer fibers for optoelectronic devices (Yarin, 1995) and emulsified liquid membrane separation techniques widely used now in various extraction processes and industrial wastewater treatment (Noble and Douglas Way, 1987).

Scientific interest in such systems results from the fact that different kinds of instability to disturbances with the wave numbers ranging in a wide domain can emerge from the motionless conductive state. These instabilities can be driven by different interfaces or can be induced by the interaction of disturbances generated on both interfaces. Simultaneous interaction of interfaces with their bulk phases and with each other can lead to a much more complex dynamics and unexpected effects.

1.2 Mathematical Models

1.2.1 Sharp-Interface Approach

The interface between two immiscible fluids has a thickness of microscopic size. Therefore, from the point of view of the macroscopic theory, the interface has to be considered as a *two-dimensional* physical system ("surface phase") characterized by *surface thermodynamic functions* (see, e.g., Rowlinson and Widom, 1982, Ch. 2). This point of view is called the *sharp-interface* approach, in a contradistinction to the *diffuse-interface* approach, which will be discussed below. For the construction of the mathematical model of the interfacial convection, the most important thermodynamic function is the *interfacial tension* σ, which is an intensive thermodynamic variable defined as the derivative of the interfacial internal energy U with respect to the area A of the interface at a fixed value of the interfacial entropy S, $\sigma = (\partial U/\partial A)_S$. The interfacial tension is a function of other intensive thermodynamic variables like temperature T, concentration C (in the case of a binary mixture), etc. Therefore, the interfacial tension generally varies along the interface.

In the framework of a sharp-interface formulation of the interfacial phenomena, the interfacial tension contributes into the stress balance condition at the interface (Landau and Lifshitz, 1987). If fluid 1 is located in the region $z > h(x, y, t)$, fluid 2 is located in the region $z < h(x, y, t)$, and the normal vector $\mathbf{n}$ is directed into fluid 1, the standard balance condition for stresses at $z = h(x, y, t)$ can be written as

$$\left(\sigma_{1,ik} - \sigma_{2,ik}\right) n_k = -\sigma\left(\frac{1}{R_1} + \frac{1}{R_2}\right) n_i + \frac{\partial \sigma}{\partial x_i}, \tag{1.1}$$

where $\sigma_{m,ik}$, $m = 1, 2$, is the stress tensor of the mth fluid, R_1 and R_2 are the curvature radii of the interface. For incompressible Newtonian viscous fluids,

$$\sigma_{m,ik} = p_m \delta_{ik} - \eta_m \left(\frac{\partial v_{m,i}}{\partial x_k} + \frac{\partial v_{m,k}}{\partial x_i}\right).$$

The first term on the right-hand side of equation (1.1) corresponds to the Laplace formula, which determines the static pressure jump across a curved interface, and the second term determines an additional tangential force caused by the variation of the interfacial tension.

Some more boundary conditions are valid on the two-dimensional ("sharp") interface $z = h(x, y, t)$. In the case of Newtonian viscous fluids, it is postulated that the velocities of both fluids are equal at the interface:

$$\mathbf{v}_1 = \mathbf{v}_2. \tag{1.2}$$

The motion of the interface is governed by the kinematic condition

$$\frac{\partial h}{\partial t} + v_{1,x}\frac{\partial h}{\partial x} + v_{1,y}\frac{\partial h}{\partial y} = v_{1,z}. \tag{1.3}$$

In the case of a nonisotermal fluid, the temperature field is continuous at the interface $z = h(x, y, t)$,

$$T_1 = T_2, \tag{1.4}$$

and the balance of normal heat fluxes reads

$$\left(\kappa_1 \frac{\partial T_1}{\partial x_i} - \kappa_2 \frac{\partial T_2}{\partial x_i}\right) n_i = 0, \tag{1.5}$$

where κ_m is the heat conductivity of the mth fluid.

In the present book, we adopt the sharp-interface approach formulated above and use the boundary conditions (1.1) – (1.5) as the basis for the analysis of the interfacial convection. We have to mention, however, some different approaches.

First, the boundary conditions formulated above take into account the rheology of the bulk fluids but ignore a possible additional dissipation *inside* the two-dimensional interface, the *surface viscosity* (Goodrich, 1981; Edwards *et al.*, 1991). We do not include the surface viscosity in our model for two reasons. First, experimental data concerning this phenomenon are scarce. Also, in some cases, the surface viscosity (as well as the effective surface elasticity) turns out to be a secondary phenomenon caused by the presence of surfactants on the interface (Levich, 1981) rather than a primary property of the interface itself.

The energy balance equation (1.5) ignores the contribution of the energy spent for the interface deformation (Napolitano, 1978). We neglect the latter effect because it is small under real conditions of the experiment (Pukhnachev, 1987).

Let us note also that the sharp-interface approach described above ignores completely the internal structure of the interface. The latter can be studied by means of the microscopic statistical physics (Rowlinson and Widom, 1982) or in the framework of a phenomenological *diffuse-interface* approach (for a review, see Anderson *et al.*, 1998). The description of the diffuse interface includes usually the evolution of an artificial *order parameter* governed by a modification of the Cahn-Hilliard equation (Hohenberg and Halperin, 1977). A more physically self-consistent approach uses density as the order parameter (Pismen and Pomeau, 2000). The applications of the diffuse-interface approach to studying the interfacial convection are briefly described in Chapter 6.

1.2.2 One-Layer Model

Let us start the systematic description of the models of the interfacial convection with the simplest, one-layer, model, which considers the interface as a "free surface". In reality, the "free surface" is a simplified description of an interface between a liquid and a gas. When the one-layer approach is applied, the full problem for the fluid motion and for the heat/mass transfer is formulated only in the liquid phase, whereas the influence of the gas phase is described in a phenomenological way by means of the Biot number.

Equations and Boundary Conditions

For a description of the bulk motions of the fluid, we will use the Boussinesq approximation (see Gershuni and Zhukhovitskii, 1976). This approximation is valid in the case where the density variations caused by temperature inhomogeneities are relatively small, and the characteristic temperature gradient is much larger than the adiabatic temperature gradient $g\beta T/c_p$, where g is the gravity acceleration, β is the thermal expansion coefficient, T is the absolute temperature, and c_p is the specific heat at constant pressure (Landau and Lifshitz, 1987). In the framework of this approximation, the dependence of the density ρ on the pressure is ignored. Also, the dependence of the density on the temperature is ignored in the continuity equation, whereas, in the equation of motion it is taken into account. The equations of motion are identical to ones of an incompressible fluid with the addition of a volume *buoyancy force* $-\mathbf{g}\beta T$. In the heat transfer equation, the compressibility effects and the viscous heat generation are neglected. Also, the values of dynamic viscosity, kinematic viscosity, heat conductivity, and thermal diffusivity (η, ν, κ, χ, respectively) are assumed to be constant.

We can write the equations in the following form:

$$\frac{\partial \mathbf{v}}{\partial t} + (\mathbf{v} \cdot \nabla)\mathbf{v} = -\frac{1}{\rho}\nabla p + \nu \nabla^2 \mathbf{v} + g\beta T \boldsymbol{\gamma}, \tag{1.6}$$

$$\frac{\partial T}{\partial t} + (\mathbf{v} \cdot \nabla)T = \chi \nabla^2 T, \tag{1.7}$$

$$\nabla \cdot \mathbf{v} = 0. \tag{1.8}$$

Here, $\boldsymbol{\gamma}$ is a unit vector directed upward, and p is the difference between the *hydrostatic* pressure in the liquid and the pressure p_g in the ambient gas (the total pressure is equal to $p - \rho g z + p_g$, where z is the vertical coordinate).

Now, we have to write down the boundary conditions on the free surface. For the sake of simplicity, we will assume that the interface is described by the equation

$$z = h(x, y, t). \tag{1.9}$$

Taking into account both Laplace's surface pressure and the thermocapillary stresses, we can write the following boundary condition (Landau and Lifshitz, 1987) describing the balance of stresses on the interface:

$$(p - \rho g h) - \sigma \left(\frac{1}{R_1} + \frac{1}{R_2} \right) = \sigma'_{ik} n_i n_k, \tag{1.10}$$

$$\sigma'_{ik} \tau_i^{(l)} n_k - \frac{\partial \sigma}{\partial x_i} \tau_i^{(l)} = 0, \quad l = 1, 2, \tag{1.11}$$

where

$$\sigma'_{ik} = \eta \left(\frac{\partial v_i}{\partial x_k} + \frac{\partial v_k}{\partial x_i} \right)$$

is the viscous stress tensor, $\mathbf{n}$ is the normal vector directed into the gas phase, $\boldsymbol{\tau}^{(l)}$ ($l = 1, 2$) are the tangent vectors to the surface, and R_1 and R_2 are the curvature radii of the surface. The motion of the interface itself is governed by the following kinematic condition:

$$\frac{\partial h}{\partial t} + v_x \frac{\partial h}{\partial x} + v_y \frac{\partial h}{\partial y} = v_z. \tag{1.12}$$

For temperatures and heat fluxes, one uses an empirical condition, e.g.,

$$\kappa \frac{\partial T}{\partial x_i} n_i = -K(T - T_g), \tag{1.13}$$

where K is the heat exchange coefficient, which might depend on local curvature of the surface, local temperature, etc., and T_g is a characteristic temperature of the ambient gas.

The assumption that the heat exchange coefficient is constant is a weak point of the one-layer approach. Some improvements of the boundary condition (1.13) that take into account the influence of the surface deformation on the temperature field in gas has been suggested in the literature (VanHook *et al.*, 1997; Zeytunian, 1998). The physically self-consistent solution of this problem is using the multi-layer approach described below. However, the one-layer approach is sufficient for a qualitative description of the physical origin of the convective instability.

The Origin of Convective Instability

In order to explain the origin of convective instability, we consider a horizontal liquid layer between a hot rigid plate and a cold free surface; the total temperature drop across the layer is θ.

When the temperature difference θ is sufficiently small, there is no convection in the fluid. The system keeps the *conductive regime* (mechanical equilibrium state), which is characterized by a constant vertical temperature gradient. When θ is larger than a certain critical value, this regime becomes *unstable* and the convective motion appears in the fluid.

There are two basic physical effects that produce the convective instability in a layer with a free surface: thermocapillary effect and buoyancy.

The *thermocapillary effect* is the dependence of the surface tension σ on the temperature T. Typically, the surface tension decreases when the temperature grows (the *normal* thermocapillary effect; $\alpha = -d\sigma/dT > 0$). The *anomalous* thermocapillary effect ($\alpha < 0$) was observed in aqueous alcohol solutions, nematic liquid crystals, binary metallic alloys, etc. (see, e.g., Legros, 1986 and references therein) as well as in some liquid-liquid systems such as 10 cS silicone oil / ethylene glycol (Braverman *et al.*, 2000). The surface tension also decreases with growth of the surface concentration of *surface active agents* (surfactants) adsorbed at the interface. The spatial inhomogeneity of thermodynamic variables leads to a surface tension inhomogeneity, which generates a thermally induced tangential surface stress $\mathbf{f}_t = \nabla\sigma$.

Let us explain the origin of the thermocapillary (Marangoni) instability in a layer heated from below in the case of the normal thermocapillary effect. Assume that we create a small positive disturbance of the surface temperature near a certain point ("a hot spot"). A thermocapillary tangential stress appears that generates a motion of the fluid near the surface *away* from this hot spot (see Fig. 1.1(a)). To compensate the spreading of the fluid near the surface, new portions of the *hot* fluid will rise from below. Hence, the temperature disturbance generates a motion that strengthens this disturbance (*a positive feedback*). The dissipative effects (viscosity and heat diffusion) prevent the growth of velocity and temperature disturbances; therefore, the convection arises only if the equilibrium vertical temperature gradient is sufficiently large.

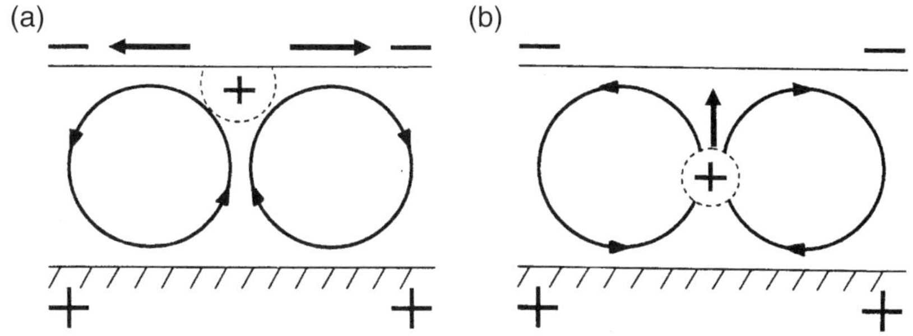

FIGURE 1.1. Schematic description of (a) Marangoni instability and (b) Rayleigh instability.

The *buoyancy* is caused by the dependence of the density on the temperature or on the concentration of an admixture. The density of the fluid typically decreases when the temperature grows (the water in the temperature interval between 0°C and 4°C is the most well-known counterexample). The explanation of the buoyancy (Rayleigh) instability is similar to that of the Marangoni instability (see Fig. 1.1(b)). The only difference is the fact that the motion is caused by the bulk buoyancy force rather than by the surface thermocapillary force.

Nondimensional Parameters

Now, we return to the quantitative description of the interfacial convection. Let us rewrite the system (1.6) – (1.13) in a non-dimensional form. Assume that a is a characteristic spatial size of the region filled by the fluid and θ is a characteristic temperature difference across this region. We will choose a, a^2/ν, ν/a, $\rho\nu^2/a^2$, and θ as units for length, time, velocity, pressure, and temperature, respectively. The system of equations (1.6) – (1.13) takes the following form (we use the same letters denoting nondimensional variables):

$$\frac{\partial \mathbf{v}}{\partial t} + (\mathbf{v} \cdot \nabla)\mathbf{v} = -\nabla p + \nabla^2 \mathbf{v} + GT\boldsymbol{\gamma}, \tag{1.14}$$

$$\frac{\partial T}{\partial t} + (\mathbf{v} \cdot \nabla)T = \frac{1}{P}\nabla^2 T, \tag{1.15}$$

$$\nabla \cdot \mathbf{v} = 0, \tag{1.16}$$

where P is the *Prandtl number* and G is the *Grashof number*, which governs the buoyancy convection. The product $R = GP$ is called the *Rayleigh number*.

Assume that the dependence of the surface tension σ on the temperature is linear: $\sigma = \sigma_0 - \alpha T$. The boundary condition on the interface $z = h$ is

$$p - Gah - W_0(1 - \delta_\alpha T)\left(\frac{1}{R_1} + \frac{1}{R_2}\right) = \sigma'_{ik} n_i n_k, \tag{1.17}$$

$$\sigma'_{ik}\tau_i^{(l)}n_k + \frac{M}{P}\frac{\partial T}{\partial x_i}\tau_i^{(l)} = 0, \quad l = 1, 2, \tag{1.18}$$

$$\frac{\partial h}{\partial t} + v_x\frac{\partial h}{\partial x} + v_y\frac{\partial h}{\partial y} = v_z, \tag{1.19}$$

$$\frac{\partial T}{\partial x_i}n_i = -Bi(T - \bar{T}_g), \tag{1.20}$$

where $Ga = ga^3/\nu^2$ is the *Galileo number*, $W_0 = \sigma_0 a/\eta\nu$, $M = \alpha\theta a/\rho\nu\chi$ is the *Marangoni number*, $Bi = Ka/\kappa$ is the *Biot number*, $\delta_\alpha = \alpha\theta/\sigma_0$, and $\bar{T}_g = T_g/\theta$. Let us mention here also the (static) *Bond number* $Bo = \rho g a^2/\sigma_0 = Ga/W_0$, which is used often in the literature.

The ratio of the Rayleigh number to the Marangoni number can be written as $R/M = (a/a_c)^2$, where $a_c^2 = \alpha/g\beta_1\rho_1$. In "thick" layers, $a \gg a_c$, the buoyancy mechanism of instability prevails and the thermocapillary effect can be neglected. In "thin" layers, $a \ll a_c$, the thermocapillary effect plays the dominant role and the buoyancy is not important.

Influence of Convection on the Deformation of the Interface

It is necessary to emphasize that the formulated boundary value problem is *incorrect* from the physical point of view if both the Grashof number G and the Galileo number Ga are taken of the same order. The Boussinesq approximation is based on the assumption of small relative deviations of density: $\delta_\beta = \beta\theta = G/Ga \ll 1$ (see Gershuni and Zhukhovitskii, 1976). If the latter condition is violated, the consideration of non-Boussinesq corrections in the equation of motion and in the continuity equation are mandatory (Velarde *et al.*, 2001). Also, such corrections are necessary for a self-consistent description of effects caused by small but nonzero δ_β. Otherwise, some artifacts can be produced by using the system (1.14) – (1.20).

Thus, when considering the buoyancy convection ($G = O(1)$), we should assume that $Ga \gg 1$ if we are in the framework of the Boussinesq approximation. Typically, on the Earth, Ga and W_0 are large, whereas the Bond number $Bo = Ga/W_0$ is $O(1)$. In the limit $Ga \to \infty$, the terms on the right-hand side of the boundary condition (1.17) can be omitted in the leading order, for they are $O(1)$. Also, the quantity p describing the difference of pressures on both sides of the interface can be considered as a constant, because the gradient of pressure is $O(1)$. Thus, the shape of the interface is determined by the balance of normal stresses, which are of hydrostatic and capillary origin (Landau and Lifshitz, 1987):

$$h + Bo^{-1}(1 - \delta_\alpha T)\left(\frac{1}{R_1} + \frac{1}{R_2}\right) = C, \tag{1.21}$$

where C is a constant. In the case where the motion takes place in a closed cavity, equation (1.21) is solved with a boundary condition on a lateral boundary corresponding to a certain contact angle. The influence of the lateral boundary vanishes on the distances larger than $Bo^{-1/2}$.

For an "infinite" layer ($L \gg Bo^{-1/2}$, L is the horizontal size of the system), one can assume $h = 0$. In this case, the boundary condition (1.17) for the normal stress is omitted, whereas the boundary conditions (1.18) – (1.20) can be written as

$$\frac{\partial v_x}{\partial z} + \frac{M}{P}\frac{\partial T}{\partial x} = 0, \quad \frac{\partial v_y}{\partial z} + \frac{M}{P}\frac{\partial T}{\partial y} = 0, \tag{1.22}$$

$$v_z = 0, \tag{1.23}$$

$$\frac{\partial T}{\partial z} = -Bi(T - \bar{T}_g). \tag{1.24}$$

In the opposite limit (for instance, under the reduced gravity), the second term on the left hand side of (1.21) prevails over the first term. If δ_α is small, the interface has a constant curvature.

Equation (1.21) should be considered as a zeroth approximation for the full problem in the limit of small δ_β. The fields of variables (velocity, pressure, temperature) found in the region with the shape governed by equation (1.21) should be used for the calculation of the next-order correction to the interface shape. For instance, in the case of an infinite layer, the shape of the interface $h = O(Ga^{-1})$ in the presence of buoyancy convection and in the absence of the temperature dependence of the surface tension should be obtained from the equation

$$h - Bo^{-1}\left(\frac{\partial^2 h}{\partial x^2} + \frac{\partial^2 h}{\partial y^2}\right) = Ga^{-1}\left[p - 2\frac{\partial v_z}{\partial z}\right]\Bigg|_{z=0}. \tag{1.25}$$

The next step (computation of the influence of the surface deformation on the convective motion) cannot be done within the Boussinesq approximation, because this influence is of the same order in δ_β as some terms omitted in this approximation.

Let us discuss now the case of finite values of the modified Galileo number. Within the Boussinesq approximation, $\delta_\beta \to 0$ and thus $R \to 0$, so only the thermocapillary convection should appear. This situation should take place in thin layers or under reduced gravity conditions. If W_0 is large, while δ_α is small, the shape of the surface is not essentially influenced by the thermocapillary motion. The zeroth-order solution should be constructed for a fixed shape of the surface; then the corrections to the surface shape should be calculated, etc. However, in the case of the long-wave convection on the background of a flat surface, the capillary term in the boundary condition for normal stresses does not prevail, and the full problem should be considered.

Convection in the Presence of a Mass Transfer

In the previous subsection, we considered the inhomogeneity of the temperature as the only source of the local density and surface tension variations. In the case of a solution, two additional physical phenomena should be taken into account. First, the density ρ depends on the solute concentration C. If the fluid system is subject to the mass transfer rather than to the heat transfer, the buoyancy term on the right-hand side of equation (1.6) is replaced by $g\beta_c C\boldsymbol{\gamma}$, where β_c is the solutal expansion coefficient, which can be positive or negative, depending on the kind of solute. Also, the surface tension usually depends on the solute concentration near the free surface (the *solutocapillary effect*). As the matter of fact, it is necessary to distinguish between the *volume* solute concentration C and the *surface* concentration Γ of the surfactant adsorbed at the interface.

If the adsorption kinetics is disregarded and it is assumed that the surface concentration of the surfactant is just proportional to the volume concentration near the surface, the solutocapillary effect can be described by the quantity $\alpha_c = d\sigma/dC$. Typically, α_c is positive for inorganic solutes and negative for organic solutes. The mathematical formulation of the problem is quite similar to that in the case of the heat transfer, with the obvious replacements $T \to C$, $\beta \to \beta_c$, $\alpha \to -\alpha_c$, $\chi \to D$, and $\kappa \to D$, where D is the volume diffusion coefficient of the solute. The surface absorption violates this similarity. An additional equation for the surface density of the surfactant should be written (see Sec. 2.5.4).

1.2.3 Two-Layer Model

The one-layer approach is unsufficient for the description of many phenomena caused by processes in fluids on both sides of the interface. These phenomena cannot be understood without an analysis of the interfacial hydrodynamic and thermal interaction between both fluids.

As the first example, let us mention the onset of Marangoni convection in a liquid-gas system. The one-layer approach predicts the monotonic Marangoni instability only for heating from the side of the liquid (Pearson, 1958). The two-layer approach reveals the appearance of the monotonic Marangoni instability for both ways of heating, depending on the ratio of layers thicknesses (Smith, 1966); see Sec. 2.4.1.

The two-layer approach can be unavoidable also in the case of the buoyancy convection. It can be found in textbooks (Landau and Lifshitz, 1987; Gershuni and Zhukhovitsky, 1976) that the buoyancy-driven convective instability of a mechanical equilibrium state appears only when the heating is from below (in the case of fluids with positive heat expansion coefficients). Strange as it may be, it is not correct in the presence of an interface between two fluids. An amazing phenomenon, which is called "anticonvection", has been predicted for some specific two-fluid systems heated from above (Welander, 1964; Gershuni and Zhukhovitsky, 1980; Simanovskii, 1980); see Sec. 2.3.

It is known that the stability problem for the mechanical equilibrium state in a system with an interface is not self-adjoint (see, e.g., Simanovskii and Nepomnyashchy, 1993); thus, an oscillatory instability is possible. However, the one-layer approach is unable to reveal several oscillatory instabilities in systems with a nondeformable interface. The first example of the Marangoni oscillatory instability was found by Sternling and Scriven (1959) in a two-fluid system (see also Reichenbach and Linde, 1981; Nepomnyashchy and Simanovskii, 1983b;

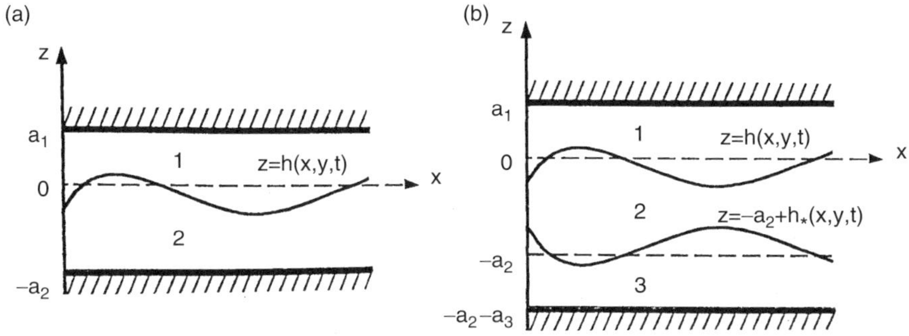

FIGURE 1.2. Basic configurations: (a) two-layer system; (b) three-layer system.

and Sec. 2.5.1). The hydrodynamic and thermal interaction between convective motions on both sides of the interface can produce oscillations also in the case of the Rayleigh convection (Gershuni and Zhukhovitskii, 1982; Gilev *et al.*, 1987c; Rasenat *et al.*, 1989); see Sec. 2.2. A specific kind of oscillations caused by the competition of buoyancy and thermocapillary effect can appear in a two-layer system in the case when the buoyancy convection is excited in the top layer (Gilev *et al.*, 1987a; Sec. 2.5.2).

We will present here the full mathematical model governing convection in a system of two horizontal layers of immiscible fluids with different physical properties (see Fig. 1.2(a)). The system is bounded from above and from below by two rigid plates kept at constant different temperatures (the total temperature drop is θ). The variables corresponding to the top layer are marked by subscript 1 and the variables corresponding to the bottom layer are marked by subscript 2. Density, kinematic and dynamic viscosity, heat conductivity, thermal diffusivity, and the heat expansion coefficient of the mth fluid are respectively ρ_m, ν_m, η_m, κ_m, χ_m, and β_m; a_m is the thickness of the mth layer ($m = 1, 2$). The surface tension coefficient σ is a linear function of the temperature T: $\sigma = \sigma_0 - \alpha T$.

Let us introduce the following notation:

$$\rho = \rho_1/\rho_2,\ \nu = \nu_1/\nu_2,\ \eta = \eta_1/\eta_2,\ \kappa = \kappa_1/\kappa_2,\ \chi = \chi_1/\chi_2,$$
$$\beta = \beta_1/\beta_2,\ a = a_2/a_1.$$

As the units of length, time, velocity, pressure, and temperature, we use the parameters of the top layer: a_1, a_1^2/ν_1, ν_1/a_1, $\rho_1\nu_1^2/a_1^2$, and θ, respectively.

The nonlinear equations of convection *in the framework of the Boussinesq approximation* have the following form (Simanovskii and Nepomnyashchy, 1993):

$$\frac{\partial \mathbf{v}_1}{\partial t} + (\mathbf{v}_1 \cdot \nabla)\,\mathbf{v}_1 = -\nabla p_1 + \nabla^2 \mathbf{v}_1 + G T_1 \boldsymbol{\gamma},$$
$$\frac{\partial T_1}{\partial t} + \mathbf{v}_1 \cdot \nabla T_1 = \frac{1}{P}\nabla^2 T_1, \tag{1.26}$$
$$\nabla \cdot \mathbf{v}_1 = 0;$$

$$
\begin{aligned}
\frac{\partial \mathbf{v}_2}{\partial t} + (\mathbf{v}_2 \cdot \nabla)\,\mathbf{v}_2 &= -\rho \nabla p_2 + \frac{1}{\nu}\nabla^2 \mathbf{v}_2 + \frac{G}{\beta} T_2 \boldsymbol{\gamma}, \\
\frac{\partial T_2}{\partial t} + \mathbf{v}_2 \cdot \nabla T_2 &= \frac{1}{\chi P}\nabla^2 T_2, \\
\nabla \cdot \mathbf{v}_2 &= 0.
\end{aligned}
\qquad (1.27)
$$

Here, $G = g\beta_1 \Theta a_1^3/\nu_1^2$ is the Grashof number (g is the gravity acceleration), $P = \nu_1/\chi_1$ is the Prandtl number of the fluid in the top layer, and $\boldsymbol{\gamma}$ is a unit vector directed upward.

The boundary conditions at the rigid boundaries are

$$\mathbf{v}_1 = 0, \; T_1 = 0 \;\text{ at }\; z = 1, \qquad (1.28)$$

$$\mathbf{v}_2 = 0, \; T_2 = s \;\text{ at }\; z = -a, \qquad (1.29)$$

with $s = 1$ for heating from below and $s = -1$ for heating from above.

The boundary conditions at the deformable interface $z = h$ include the following: the balance of normal stresses:

$$
\begin{aligned}
& p_1 - p_2 + W_0(1 - \delta_\alpha T_1)\left(\frac{1}{R_1} + \frac{1}{R_2}\right) + Ga\,\delta h \\
&= \left[\left(\frac{\partial v_{1i}}{\partial x_k} + \frac{\partial v_{1k}}{\partial x_i}\right) - \eta^{-1}\left(\frac{\partial v_{2i}}{\partial x_k} + \frac{\partial v_{2k}}{\partial x_i}\right)\right] n_i n_k;
\end{aligned}
\qquad (1.30)
$$

the balance of tangential stresses:

$$\left[\left(\frac{\partial v_{1i}}{\partial x_k} + \frac{\partial v_{1k}}{\partial x_i}\right) - \eta^{-1}\left(\frac{\partial v_{2i}}{\partial x_k} + \frac{\partial v_{2k}}{\partial x_i}\right)\right] \tau_i^{(l)} n_k - \frac{M}{P}\tau_i^{(l)}\frac{\partial T_1}{\partial x_i} = 0, \; l = 1, 2; \qquad (1.31)$$

the continuity of the velocity field:

$$\mathbf{v}_1 = \mathbf{v}_2; \qquad (1.32)$$

the kinematic equation for the evolution of the interface shape:

$$\frac{\partial h}{\partial t} + v_{1x}\frac{\partial h}{\partial x} + v_{1y}\frac{\partial h}{\partial y} = v_{1z}; \qquad (1.33)$$

the continuity of the temperature field:

$$T_1 = T_2; \qquad (1.34)$$

and the continuity of the heat flux normal components:

$$\left(\frac{\partial T_1}{\partial x_i} - \kappa^{-1}\frac{\partial T_2}{\partial x_i}\right) n_i = 0. \qquad (1.35)$$

Here, $Wo = \sigma_0 a_1/\eta_1 \nu_1$, $Ga = g a_1^3/\nu_1^2$ is the Galileo number, $M = \alpha\theta a_1/\rho_1 \nu_1 \chi_1$ is the Marangoni number, $\delta_\alpha = \alpha\theta/\sigma_0$, and $\delta = \rho^{-1} - 1$; $\mathbf{n}$ is the normal vector, $\boldsymbol{\tau}^{(l)}$ $(l = 1, 2)$ are the tangent vectors to the interface, and p_1, p_2 are the differences between the overall pressure and the hydrostatic pressure.

As it was explained in Sec. 1.2.2, it is necessary to be cautious when considering the mutual influence of the convection and the interfacial deformation. In the framework of the Boussinesq approximation ($G = O(1)$, $Ga \gg 1$), the deformation of the interface should be neglected as a *non-Boussinesq effect*, except the case of a system with very close densities of both fluids, $Ga \cdot \delta = O(1)$. If the densities of fluids are not close, the shape of the interface is determined in the zeroth order by the balance of hydrostatic and capillary stresses:

$$\delta h + Bo^{-1}(1 - \delta_\alpha T)\left(\frac{1}{R_1} + \frac{1}{R_2}\right) = C, \tag{1.36}$$

where $Bo = Ga/Wo = \rho_1 g a_1^2/\sigma_0$ is the Bond number and the constant C corresponds to the pressure difference across the interface. The zeroth-order boundary condition (1.36) should replace equation (1.30) by solving equations (1.26) and (1.27), because these equations are written in the zeroth order in δ_β.

In the case of a closed cavity, the contact angles on the lateral solid walls are taken into account, and, generally, $h \neq 0$ and $C \neq 0$. If the layers are infinite in the horizontal direction, the *zeroth-order* shape of the interface is just $h = 0$ and $C = 0$. The boundary conditions on the interface $z = 0$ are

$$\frac{\partial v_{1x}}{\partial z} - \eta^{-1}\frac{\partial v_{2x}}{\partial z} - \frac{M}{P}\frac{\partial T_1}{\partial x} = 0, \quad \frac{\partial v_{1y}}{\partial z} - \eta^{-1}\frac{\partial v_{2y}}{\partial z} - \frac{M}{P}\frac{\partial T_1}{\partial y} = 0, \tag{1.37}$$

$$\mathbf{v}_1 = \mathbf{v}_2, \; T_1 = T_2, \; \frac{\partial T_1}{\partial z} - \kappa^{-1}\frac{\partial T_2}{\partial z} = 0. \tag{1.38}$$

After the calculation of the velocity, pressure, and temperature fields in the framework of the Boussinesq approximation (boundary value problem (1.26), (1.27), (1.37), (1.38)), one can find the *correction* of the first order in δ_β to the shape of the interface $h(x, y, t)$ by solving the following equation:

$$\delta h + r_c^2(1 - \delta_\alpha T|_{z=0})\left(\frac{\partial^2 h}{\partial x^2} + \frac{\partial^2 h}{\partial y^2}\right) = Ga^{-1}\left[p_2 - p_1 + 2\left(\frac{\partial v_{1,z}}{\partial z} - \eta^{-1}\frac{\partial v_{2,z}}{\partial z}\right)\right]\Bigg|_{z=0}. \tag{1.39}$$

Recall that the influence of the interface deformation $h \sim \delta_\beta$ on the convective motion is a non-Boussinesq effect, and it cannot be explored without a consideration of other non-Boussinesq effects that have the same order of smallness.

In the case of the thermocapillary convection, using the system of equations (1.26), (1.27) (with $G = 0$) together with the boundary conditions (1.30) – (1.35) corresponding to a deformable interface is not forbidden. However, if the relative change δ_α of the surface tension coefficient is small, then the coefficient

$W_0 = M/\delta_\alpha P$ is large. As one can see from the boundary condition (1.30), the deformation of the interface produced by the thermocapillary convection has a *small curvature* in the latter case. If the wavelength of the interfacial deformation is not large, the deformation is suppressed by the surface tension, and the boundary conditions (1.37) and (1.38) corresponding to a flat interface can be used. However, the *long-wave* deformations are not suppressed by the surface tension, and they should be considered using the original boundary conditions (1.30) – (1.35).

The Case of the Mass Transfer

As it was explained in Sec. 1.2.2, the problem of the mass transfer is similar to that of the heat transfer, if the adsorption kinetics is neglected. When the two-layer approach is used, it is necessary to take into account that while the normal component of the mass flux is continuous across the interface,

$$\left(\frac{\partial C_1}{\partial x_i} - D^{-1}\frac{\partial C_2}{\partial x_i}\right) n_i = 0,$$

where $D = D_1/D_2$ is the ratio of the diffusion coefficients, the concentration field *is not continuous* across the interface,

$$k^L C_1 = C_2,$$

where $k^L = k_1^L/k_2^L$ and k_m^L is the Langmuir adsorption coefficient in the mth fluid. However, the formal transformation $T_1 = k^L C_1$, $T_2 = C_2$ restores the boundary conditions (1.34) and (1.35) with $\kappa = Dk^L$.

1.2.4 Three-layer Model

We will consider also a model for a *three-layer* fluid system (see Fig. 1.2(b)). The third layer of the mean thickness a_3 is added. The parameters of the bottom fluid are denoted by subscript 3. The surface tension coefficient of the lower interface is $\sigma_* = \sigma_{*0} - \alpha_* T$. We define

$$\rho_* = \frac{\rho_1}{\rho_3}, \quad \nu_* = \frac{\nu_1}{\nu_3}, \quad \eta_* = \frac{\eta_1}{\eta_3} = \rho_*\nu_*, \quad \chi_* = \frac{\chi_1}{\chi_3},$$

$$\kappa_* = \frac{\kappa_1}{\kappa_3}, \quad \beta_* = \frac{\beta_1}{\beta_3}, \quad a_* = \frac{a_3}{a_1}, \bar{\alpha} = \frac{\alpha_*}{\alpha}.$$

The boundary value problem for the description of a three-layer system is an extension of the system of equations (1.26) – (1.35), which contains the additional equations

$$\frac{\partial \mathbf{v}_3}{\partial t} + (\mathbf{v}_3 \cdot \nabla)\,\mathbf{v}_3 = -\rho_* \nabla p_3 + \frac{1}{\nu_*}\nabla^2 \mathbf{v}_3 + \frac{G}{\beta_*} T_3 \boldsymbol{\gamma},$$

$$\frac{\partial T_3}{\partial t} + \mathbf{v}_3 \cdot \nabla T_3 = \frac{1}{\chi P}\nabla^2 T_3, \tag{1.40}$$
$$\nabla \cdot \mathbf{v}_3 = 0$$

and the additional boundary conditions,

$$\mathbf{v}_3 = 0, \;\; T_3 = s \;\text{ at }\; z = -a - a_*. \tag{1.41}$$

The boundary conditions (1.29) are replaced by the following set of boundary conditions at $z = -a + h_*$:

$$p_2 - p_3 + W_{*0}K_*(1 - \delta_{\alpha *}T_1) + Ga\,\delta_* h_*$$
$$= \left[\eta^{-1}\left(\frac{\partial v_{2i}}{\partial x_k} + \frac{\partial v_{2k}}{\partial x_i}\right) - \eta_*^{-1}\left(\frac{\partial v_{3i}}{\partial x_k} + \frac{\partial v_{3k}}{\partial x_i}\right)\right] n_{*i}n_{*k}, \tag{1.42}$$

$$\left[\eta^{-1}\left(\frac{\partial v_{2i}}{\partial x_k} + \frac{\partial v_{2k}}{\partial x_i}\right) - \eta_*^{-1}\left(\frac{\partial v_{3i}}{\partial x_k} + \frac{\partial v_{3k}}{\partial x_i}\right)\right] \tau_{*i}^{(l)} n_{*k}$$
$$-\frac{\bar{a}M}{P}\tau_{*i}^{(l)}\frac{\partial T_3}{\partial x_i} = 0, \tag{1.43}$$
$$l = 1, 2,$$
$$\mathbf{v}_2 = \mathbf{v}_3, \tag{1.44}$$
$$\frac{\partial h_*}{\partial t} + v_{3x}\frac{\partial h_*}{\partial x} + v_{3y}\frac{\partial h_*}{\partial y} = v_{3z}, \tag{1.45}$$
$$T_2 = T_3, \tag{1.46}$$
$$\left(\kappa^{-1}\frac{\partial T_2}{\partial x_i} - \kappa_*^{-1}\frac{\partial T_3}{\partial x_i}\right) n_{*i} = 0, \tag{1.47}$$

where $W_{*0} = \sigma_{*0}a_1/\eta_1\nu_1$, $\delta_{\alpha_*} = \alpha_*\theta/\sigma_{*0}$, $\delta_* = \rho_*^{-1} - \rho^{-1}$, K_* is the curvature of the lower interface, $\mathbf{n}_*$ is the normal vector, and $\boldsymbol{\tau}_*^{(l)}$ are the tangential vectors of the lower interface.

As was discussed in the previous subsection, the conditions of the deformability of interfaces, which read now as $Ga\delta = O(1)$ and $Ga\delta_* = O(1)$, are incompatible with the assumption $G \neq 1$ in the framework of the Boussinesq approximation, unless $\delta \ll 1$ and $\delta_* \ll 1$. Therefore, the boundary conditions (1.42) – (1.47) can be used only if $G = 0$ or $\delta \ll 1$, $\delta_* \ll 1$. If the quantities $Ga\delta$ and $Ga\delta_*$ are large, the deformations of interfaces are small ($|h| \ll 1$, $|h_*| \ll 1$), and their influence on the flow and temperature distribution can be ignored. In this case, the boundary conditions (1.42) and (1.45) can be omitted and other boundary conditions can be simplified. Finally, we obtain the following system of boundary conditions at $z = -a$:

$$\eta^{-1}\frac{\partial v_{2x}}{\partial z} - \eta_*^{-1}\frac{\partial v_{3x}}{\partial z} - \frac{\bar{a}M}{P}\frac{\partial T_3}{\partial x} = 0, \;\; \eta^{-1}\frac{\partial v_{2y}}{\partial z} - \eta_*^{-1}\frac{\partial v_{3y}}{\partial z} - \frac{\bar{a}M}{P}\frac{\partial T_3}{\partial y} = 0, \tag{1.48}$$
$$\mathbf{v}_2 = \mathbf{v}_3, \;\; T_2 = T_3, \;\; \kappa^{-1}\frac{\partial T_2}{\partial z} - \kappa_*^{-1}\frac{\partial T_3}{\partial z} = 0. \tag{1.49}$$

2

Types of Convective Instability in Systems with an Interface

In this chapter, we present a brief description of the linear stability approach and discuss the problem of nonlinear pattern selection for stationary and wavy patterns. Then we discuss different types of convective instability characteristic for systems with an interface. As the basic model, we use the two-layer model (see Sec. 1.2.3).

We start with the consideration of a pure buoyancy instability. Special attention is paid to the truly two-layer phenomena like convective oscillations and anticonvection. Also, we discuss the influence of the interfacial deformation on the buoyancy instability.

The next sections are devoted to the investigation of the convection generated by the thermocapillary effect (the Marangoni convection) and by the combined action of the buoyancy and the thermocapillary effect. We discuss the appearance of stationary interfacial patterns under different conditions. Also, we study various mechanisms leading to the development of Marangoni waves, including mode mixing of interfacial and internal waves, influence of the thermocapillary/thermosolutary effect on the gravity-capillary waves, and the oscillations generated by the presence of a surfactant.

2.1 The Problem of Stability

The problem (1.26) – (1.35) for any choice of parameters has the solution

$$\mathbf{v}_1^0 = \mathbf{v}_2^0 = 0,\; h^0 = 0,\; T_1^0 = -s\frac{z-1}{1+\kappa a},\; T_2^0 = -s\frac{\kappa z - 1}{1+\kappa a},$$

$$p_1^0 = -\frac{G}{1+\kappa a}\left(\frac{z^2}{2} - z\right),\; p_2^0 = -\frac{G}{\rho\beta(1+\kappa a)}\left(\frac{\kappa z^2}{2} - z\right), \tag{2.1}$$

which corresponds to the mechanical equilibrium state. Its stability can be investigated in the framework of the linear stability theory. The boundary value problem

(1.26) – (1.35) is linearized near the solution (2.1). The solutions of the linearized problem are presented as a superposition of normal modes characterized by a wave vector $\mathbf{k} = (k_x, k_y)$ and a complex growth rate $\lambda = \lambda_r + i\lambda_i$:

$$(\tilde{\mathbf{v}}_1(z), \tilde{p}_1(z), \tilde{T}_1(z), \tilde{\mathbf{v}}_2(z), \tilde{p}_2(z), \tilde{T}_2(z)) \exp(ik_x x + ik_y y + \lambda t); \tag{2.2}$$

in what follows, the "tilde" will be omitted.

Since the problem is isotropic, the growth rate λ depends only on the wave vector modulus $k = |\mathbf{k}|$ but not on its direction. That is why it is sufficient to consider only two-dimensional disturbances with $\mathbf{k} = (k, 0)$, which do not depend on the coordinate y. Introducing the stream function disturbances

$$v_{mx} = \psi_m', \quad v_{mz} = -ik\psi_m \ (m = 1, 2),$$

where the prime stands for d/dz, and eliminating pressure disturbances in the usual way, we obtain the following boundary eigenvalue problem:

$$-\lambda D\psi_m = -c_m D^2\psi_m + ikGb_m T_m, \tag{2.3}$$

$$\lambda T_m - ikA_m\psi_m = \frac{d_m}{P} DT_m, \ (m = 1, 2); \tag{2.4}$$

$D = d^2/dz^2 - k^2$, $c_1 = b_1 = d_1 = e_1 = 1$, $c_2 = 1/\nu$, $b_2 = 1/\beta$, $d_2 = 1/\chi$, and $e_2 = \rho$; $A_1 = dT_1^0/dz = -s/(1+\kappa a)$ and $A_2 = dT_2^0/dz = -s\kappa/(1+\kappa a)$ are the dimensionless temperature gradients;

$$z = 1: \ \psi_1 = \psi_1' = T_1 = 0, \tag{2.5}$$

$$z = -a: \ \psi_2 = \psi_2' = T_2 = 0, \tag{2.6}$$

$$z = 0: \ \psi_1''' - \eta^{-1}\psi_2''' + [\lambda\delta - 3k^2(1 - \eta^{-1})]\psi_1'$$

$$+ik\{Ga[\delta + \delta_\beta s(1 - \rho^{-1}\beta^{-1})/(1+\kappa a)] + Wk^2\}h = 0, \tag{2.7}$$

$$\eta(\psi_1'' + k^2\psi_1) - (\psi_2'' + k^2\psi_2) - \frac{\eta M}{P}\left(T_1 - \frac{s}{1+\kappa a}h\right) = 0, \tag{2.8}$$

$$\psi_1' = \psi_2', \ \psi_1 = \psi_2 = i\frac{\lambda}{k}h, \tag{2.9}$$

$$T_1 - T_2 = \frac{s(1-\kappa)}{1+\kappa a}h, \ \kappa T_1' - T_2' = 0; \tag{2.10}$$

here, $W = W_0(1 - \delta_a s/(1+\kappa a))$.

The eigenvalue problem (2.3) – (2.10) determines the spectrum of the growth rates. Because this problem is not self-adjoint, the eigenvalues can be complex: $\lambda = \lambda_r + i\lambda_i$. The condition $\lambda = 0$ corresponds to the boundary of a monotonic instability, whereas the conditions $\lambda_r = 0$, $\lambda_i \neq 0$ correspond to the boundary of an oscillatory instability.

If the deformation of the interface h can be neglected, the system of boundary conditions at the interface is simplified:

$$z = 0: \ \eta\psi_1'' - \psi_2'' - \frac{ik\eta M}{P}T_1 = 0, \ \psi_1' = \psi_2', \ \psi_1 = \psi_2 = 0, \tag{2.11}$$

$$T_1 = T_2, \ \kappa T_1' = T_2'. \tag{2.12}$$

2.2 Rayleigh-Bénard Convection

We will start the discussion of different types of instability with the consideration of the pure *buoyancy* instability in a two-layer system with an interface. Because the buoyancy is a *bulk* effect, one can expect that the appearance of the buoyancy convection in each layer is determined mainly by its *local* Rayleigh number. The interaction between the flows in both layers through the boundary conditions on the interface (which is an *interfacial* effect) and the influence of the Rayleigh number of another layer seem to be of a minor importance.

However, there exist some *truly two-layer* phenomena caused by hydrodynamic and thermal interactions of convective flows through the interface. First, this interaction can produce a specific type of *convective oscillations*. Formally, the oscillatory instability is possible in a two-layer system, because the eigenvalue problem that determines the spectrum of growth rates is non-self-adjoint. Actually, the oscillatory instability is found only in the case where the properties of fluids are essentially different.

Another phenomenon that takes place only due to the interaction through the interface is the "anticonvection" (i.e., the buoyancy convection by heating *from above*). This specific type of instability occurs when either the fluid properties or temperature gradients in both layers differ considerably.

The above-mentioned phenomena are considered in Secs. 2.2 and 2.3.

2.2.1 *Linear Stability*

As was explained in Sec. 1.2.2, in the framework of the Boussinesq approximation ($G = O(1)$, $Ga \gg 1$), the deformation of the interface should be neglected as a non-Boussinesq effect, except the case of close densities of both fluids, $Ga\,\delta = O(1)$. Thus, the linear stability problem is governed by the system of equations (2.3), (2.4) with the boundary conditions (2.5), (2.6), (2.11) (with $M = 0$), and (2.12). The spectrum of eigenvalues $\lambda(k)$ depends on eight parameters: G, P, η, ν, κ, χ, β, a. It is known that the problem is self-adjoint only if

$$\frac{\eta\beta\chi}{\nu} = 1 \tag{2.13}$$

(see Simanovskii and Nepomnyashchy, 1993).

The number of independent parameters decreases in the case of a monotonic instability boundary ($\lambda(k) = 0$). By means of the transformation $T_m(z) = P\theta_m(z)$, $m = 1, 2$, the system of equations (2.3), (2.4) is reduced to the following one:

$$D^2\psi_1 - ikR\theta_1 = 0, \;\; D\theta_1 + ikA_1\psi_1 = 0, \tag{2.14}$$

$$D^2\psi_2 - ikR\zeta\theta_2 = 0, \;\; D\theta_2 + ikA_2\chi\psi_2 = 0. \tag{2.15}$$

Recall that in the case of heating from below ($s = 1$), $A_1 < 0$, $A_2 < 0$. Here, $R = GP$ is the Rayleigh number, $\zeta = \nu/\beta$. Thus, the monotonic neutral curve $R = R(k)$ is determined by five parameters: η, κ, χ, ζ, a.

Let us define the *local* Rayleigh numbers

$$R_m = \frac{g\beta_m|\tilde{A}_m|a_m^4}{\nu_m\chi_m}, \quad m = 1, 2, \tag{2.16}$$

which are constructed using the parameters of the corresponding fluids ($\tilde{A}_m$ is the dimensional temperature gradient in the mth fluid). The values of R_m are connected with the nondimensional parameters defined in Chapter 1 in the following way:

$$R_1 = \frac{GP}{1+\kappa a}, \quad R_2 = \frac{GP\kappa}{1+\kappa a}\frac{\nu\chi a^4}{\beta}.$$

The ratio of local Rayleigh numbers

$$\frac{R_2}{R_1} = \frac{\kappa\nu\chi a^4}{\beta} \tag{2.17}$$

depends only on physical properties of both fluids and on the ratio of thicknesses a; it is not changed with the change of the heating intensity.

In the generic case, when the local Rayleigh numbers are essentially different from each other ($R_2 \gg R_1$ or $R_1 \gg R_2$), one can easily distinguish between neutral stability curves, which correspond to the onset of convection in different layers. The eigenfunction corresponding to the lowest neutral curve is localized mainly in a layer where the local Rayleigh number is higher. In another layer, a weak motion is induced.

As an example, let us consider the onset of the Rayleigh-Bénard convection in the 47v2 silicone oil / water system (system 1 in Table 2.1). This system was used in experiments on the convection in two-layer systems carried out by Degen *et al.*, 1998. Substituting the physical parameters of the system into equation (2.17), we find that $R_2/R_1 = 0.05a^4$.

First, let the ratio of thicknesses a be equal to 1.2, so that $R_2/R_1 = 0.104$. The corresponding neutral curves are shown in Fig. 2.1. Line 1 corresponds to the onset of convection in the top layer, where the local Rayleigh number is much larger than that in the bottom layer. A weak motion is induced in the bottom layer due to the tangential viscous stresses on the interface. Line 2, which consists of two fragments 2a and 2b, is the excitation boundary of the convection in the bottom layer. Line 3 (fragments 3a and 3b) is the instability boundary with respect to a higher mode, which is localized mainly in the top layer. It has, however, a

TABLE 2.1. Parameters of two-layer systems.

System	ν	η	κ	χ	β	P
1	2.0	1.7375	0.184	0.778	5.66	25.7
2	0.6493	0.549	0.6194	1.096	1.4516	94
3	1.14	1.02	0.698	0.934	0.963	7.84
4	15.1	0.0182	0.0396	138	17.7	0.758
5	1.116	0.915	0.169	0.472	7.1595	6.28
6	15.4	11.1	0.41	0.714	0.672	306

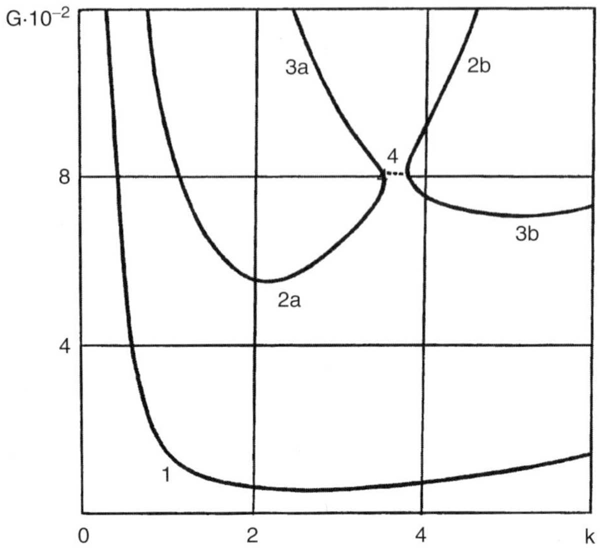

FIGURE 2.1. The neutral curves for the 47v2 silicone oil/water system (system 1 of Table 2.1); $a = 1.2$; $R_2/R_1 = 0.104$.

more complicated dependence on the vertical coordinate than the basic instability mode corresponding to line 1. Note that a simple crossing of neutral curves never takes place. Instead, an *oscillatory instability* (line 4) is developed in the interval of the wave numbers where the monotonic neutral curves 2 and 3 approach each other. The oscillatory instability is possible because the eigenvalue problem (2.3) – (2.6), (2.11), (2.12) is not self-adjoint. Recall that this eigenvalue problem is self-adjoint only if $kM = 0$, $\eta\beta\chi/\nu = 1$ (Simanovskii and Nepomnyashchy, 1993). For system 1, the value of the parameter $\eta\beta\chi/\nu = 3.83$ is rather far from 1, which is favorable for the appearance of an oscillatory instability. Here and later on, the monotonic (oscillatory) neutral curves are shown as solid (dashed) lines.

With the growth of a, the characteristic scale of the disturbances developed in the bottom layer and the local Rayleigh number R_2 grow. Therefore, the neutral curve 2 corresponding to the bottom-layer instability modes moves toward smaller values of the wave number k and of the Grashof number G. The neutral curve 1 is not significantly changed near its minimum: The critical wave number is about 2.7 and the critical Grashof number is changed from 60 to 71 when a is changed from 1.2 to 2.4. When monotonic neutral curves corresponding to convection in different layers approach each other, they produce oscillatory neutral curves corresponding to the excitation of correlated motions in both layers. For example, merging of the monotonic neutral curves 1 and 2a described above leads to the appearance of a long-wave oscillatory neutral curve 5 (see Fig. 2.2). The further increase of a leads to splitting of the monotonic neutral curves shown in Fig. 2.3. The short-wave minimum (on line 1a) corresponds to the convection onset in the relatively thin top layer, whereas the long-wave minimum (on line 2c) corresponds to the convection onset in the relatively thick bottom layer. For even larger

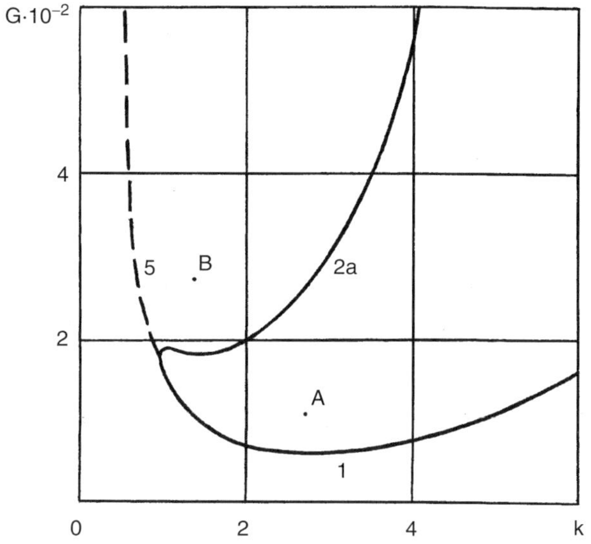

FIGURE 2.2. The neutral curves for the 47v2 silicone oil / water system (system 1 of Table 2.1); $a = 1.6$; $R_2/R_1 = 0.328$.

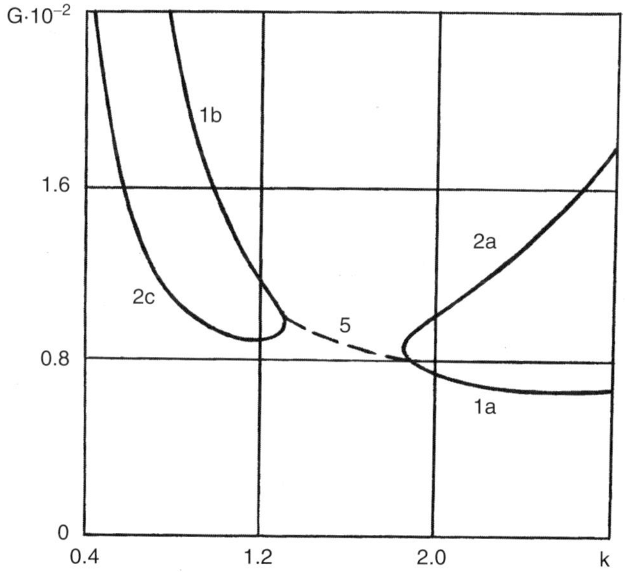

FIGURE 2.3. The neutral curves for the 47v2 silicone oil / water system (system 1 of Table 2.1); $a = 2$; $R_2/R_1 = 0.8$.

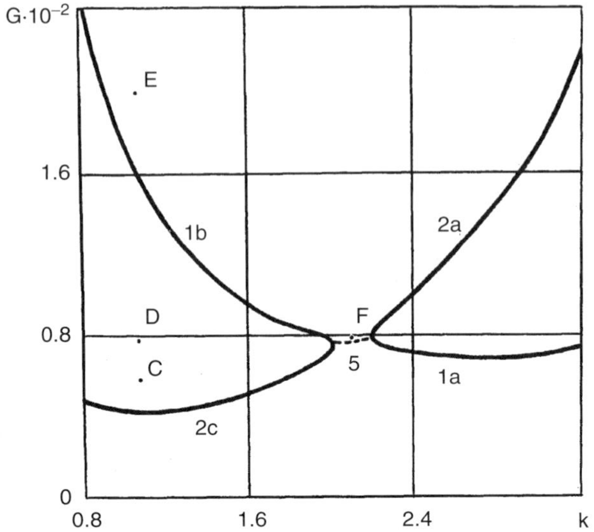

FIGURE 2.4. The neutral curves for the 47v2 silicone oil / water system (system 1 of Table 2.1); $a = 2.4$; $R_2/R_1 = 1.66$.

values of a, the critical Grashof number for the convection onset in the bottom layer becomes smaller than that for the convection onset in the top layer Fig. 2.4.

Note that the minimum values of the Grashof number for system 1 always correspond to the monotonic instability. Actually, no real two-layer systems are known with oscillatory instability more "dangerous" than the monotonic one. Some model systems have been constructed with the minimum value of the Grashof number at the oscillatory neutral curve (Gilev *et al.*, 1987c; Rasenat *et al.*, 1989; Renardy, 1996).

The conditions for the appearance of an oscillatory instability can be simplified in the presence of a constant, spatially uniform heat source, or sink at the interface.

Assume that a constant heat release rate Q (Q may be positive or negative) is set at the interface $z = 0$. Define $G_Q = g\beta_1 Q a_1^4/\nu_1^2\kappa_1$. Dimensionless temperature gradients in the equilibrium state are

$$A_1 = -\frac{sG + a\kappa G_Q}{G(1+\kappa a)}, \quad A_2 = -\frac{\kappa(sG - G_Q)}{G(1+\kappa a)}. \tag{2.18}$$

As was explained above, the oscillatory instability is developed when two monotonic neutral curves merge, forming a *gap* in a certain interval of wave numbers. Thus, for the prediction of the interval where the oscillatory instability is expected, it is sufficient to consider the shape of monotonic neutral curves.

Let us consider the monotonic stability problem in the case of the interfacial heat release. Using the transformation $\psi_m(z) = \Phi_m(z)/A_1$, $m = 1, 2$, we can rewrite equations (2.14) and (2.15) in the following form:

$$D^2\Phi_1 - ikRA_1\theta_1 = 0; \quad D\theta_1 + ik\Phi_1 = 0;$$

$$D^2\Phi_2 - ikRA_1\frac{\nu}{\beta}\theta_2 = 0; \quad D\theta_2 + ik\frac{A_2}{A_1}\chi\Phi_2 = 0.$$

The parameters characterizing the appearance of the monotonic instability are actually RA_1 and $\chi A_2/A_1$. One can introduce an artificial system without a heat release at the interface (i.e., with temperature gradients $\tilde{A}_1 = -s/(1+\kappa a)$, $\tilde{A}_2 = -s\kappa/(1+\kappa a)$), but with a renormalized value of the thermal diffusivity ratio $\tilde{\chi}$ satisfying the relation

$$\frac{A_2}{A_1}\chi = \frac{\tilde{A}_2}{\tilde{A}_1}\tilde{\chi}; \tag{2.19}$$

that is,

$$\tilde{\chi} = \chi \frac{sG - G_Q}{sG + a\kappa G_Q}. \tag{2.20}$$

The critical Rayleigh number for the original system with a heat release at the interface that produces temperature gradients A_1 and A_2 is connected with the critical Rayleigh number for that artificial system by the relation

$$A_1 R(A_1, A_2, \chi) = \tilde{A}_1 R(\tilde{A}_1, \tilde{A}_2, \tilde{\chi}). \tag{2.21}$$

The change of the ratio A_2/A_1 due to the interfacial heat release is equivalent to the change of the thermal diffusivity, hence to the change of the parameter $\eta\beta\chi/\nu$.

Specifically, for system 1 from Table 1 considered above, $\eta\beta\tilde{\chi}/\nu > \eta\beta\chi/\nu$ when $G_Q < 0$. Taking into account that the presence of the heat sink at the interface enhances the temperature gradient in the bottom layer and diminishes the temperature gradient in the top layer, one can expect that the oscillations will be generated in the case $R_2 < R_1$.

The neutral curves corresponding to the case $a = 1.6$, $G_Q = -150$ are shown in Fig. 2.5. One can see that the minimum value of the Grashof number on the oscillatory branch (line 5) of the neutral curve is lower than that on the monotonic branch (line 1). A pair of complexconjugate eigenvalues λ with a positive real part exists in the region between the neutral curve 5 and the *discriminant curve* 6. On line 6, the imaginary parts of the above-mentioned eigenvalues become equal to zero. Between lines 6 and 2a, there are two real eigenvalues with positive real parts.

2.2.2 Nonlinear Flow Regimes

A weakly nonlinear theory of oscillations has been described by Colinet and Legros (1994).

Here, we present results of fully nonlinear simulations of two-dimensional flows ($v_{my} = 0$ ($m = 1, 2$), the fields of physical variables do not depend on y). In this case, we can introduce the stream function ψ:

$$v_{mx} = \frac{\partial \psi_m}{\partial z}, \quad v_{mz} = -\frac{\partial \psi_m}{\partial x} \quad (m = 1, 2).$$

Eliminating the pressure and defining the vorticity

$$\phi_m = \frac{\partial v_{mz}}{\partial x} - \frac{\partial v_{mx}}{\partial z},$$

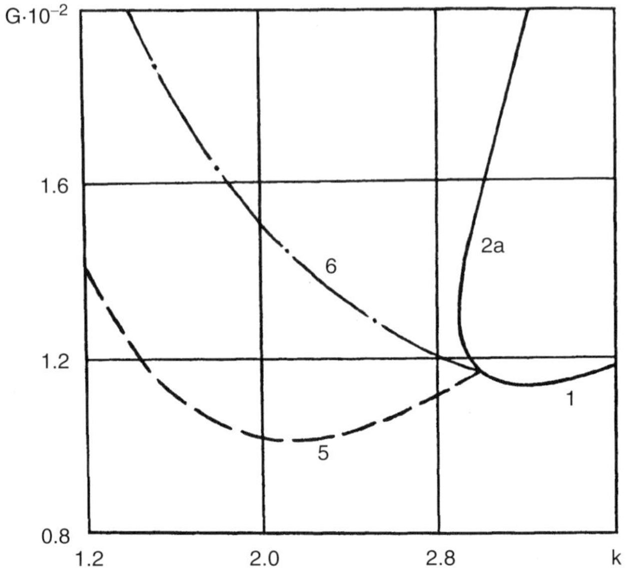

FIGURE 2.5. The neutral curves for the 47v2 silicone oil/water system (system 1 of Table 2.1); $a = 1.6$; $G_Q = -150$.

we can rewrite the boundary value problem (1.26) – (1.35), (1.37), (1.38) in the following form:

$$\frac{\partial \phi_m}{\partial t} + \frac{\partial \psi_m}{\partial z} \cdot \frac{\partial \phi_m}{\partial x} - \frac{\partial \psi_m}{\partial x} \cdot \frac{\partial \phi_m}{\partial z} = c_m \nabla^2 \phi_m + b_m G \frac{\partial T_m}{\partial x}, \tag{2.22}$$

$$\nabla^2 \psi_m = -\phi_m, \tag{2.23}$$

$$\frac{\partial T_m}{\partial t} + \frac{\partial \psi_m}{\partial z} \cdot \frac{\partial T_m}{\partial x} - \frac{\partial \psi_m}{\partial x} \cdot \frac{\partial T_m}{\partial z} = \frac{d_m}{P} \nabla^2 T_m \tag{2.24}$$

$$(m = 1, 2).$$

$$z = 1 : \psi_1 = \frac{\partial \psi_1}{\partial z} = 0; \quad T_1 = 0; \tag{2.25}$$

$$z = -a : \psi_2 = \frac{\partial \psi_2}{\partial z} = 0; \quad T_2 = s; \tag{2.26}$$

$$z = 0 : \psi_1 = \psi_2 = 0, \quad \frac{\partial \psi_1}{\partial z} = \frac{\partial \psi_2}{\partial z}, \quad \phi_2 = \eta \phi_1 + \frac{M\eta}{P} \frac{\partial T_1}{\partial x}; \tag{2.27}$$

$$T_1 = T_2, \quad \kappa \frac{\partial T_1}{\partial z} = \frac{\partial T_2}{\partial z}. \tag{2.28}$$

The coefficients b_m, c_m, and d_m have been defined in Sec. 2.1.

The calculations were performed in a finite region $-L/2 \leq x \leq L/2$, $-a \leq z \leq 1$ with the following types of boundary conditions: (a) periodic boundary conditions:

$$\psi_m(x+L) = \psi_m(x),\ \phi_m(x+L) = \phi_m(x),\ T_m(x+L) = T_m(x);\ m = 1, 2; \tag{2.29}$$

(b) free heat-insulated boundaries:

$$x = 0, L : \psi_m = \phi_m = \frac{\partial T_m}{\partial x} = 0;\ m = 1, 2; \tag{2.30}$$

(c) rigid heat-insulated boundaries:

$$x = 0, L : \psi_m = \frac{\partial \psi_m}{\partial x} = \frac{\partial T_m}{\partial x} = 0;\ m = 1, 2. \tag{2.31}$$

The boundary conditions (a) correspond to spatially periodic structures in a laterally infinite two-layer system and are used for the comparison of numerical results with those of the linear theory developed for the infinite system. The boundary conditions (b) are used for the calculation of symmetric stationary patterns and standing waves in a laterally infinite system. The boundary conditions (c) correspond to a closed cavity.

The computations have been done by means of the finite-difference method (for details, see Simanovskii, 1991; Simanovskii and Nepomnyashchy, 1993) for the system 47v2 silicone oil / water (system 1 of Table 2.1).

Stationary Flow Regimes

In the present subsection, we consider the case $G_Q = 0$ and we use the boundary conditions (2.30). In that case, the lowest minimum of the neutral curve corresponds to the monotonic instability and stationary flow regimes are developed in the system.

First, consider the case when $R_2 < R_1$. The stream lines of the typical convective flow, which corresponds to point A in Fig. 2.2, are shown in Fig. 2.6(a). The monotonic instability of the mechanical equilibrium state generates an intensive convective motion in the top layer and a relatively weak motion in the bottom layer caused mainly by the viscous stresses that act at the interface (*viscous* or *mechanical coupling*).

Note that the corresponding vortices in the top and bottom layers rotate in the opposite directions and the isotherms in both layers are distorted in the opposite way (see Fig. 2.6(b)). With the growth of G, beyond the boundary with respect to the bottom-layer instability (line 2a in Fig. 2.2), the general structure of the motion is kept (see Fig. 2.7, which corresponds to point B in Fig. 2.2). However, one can see that the vortices in the bottom layer rotating in the same direction have slightly different intensities: The vortices near the center of the computational region are more intensive than those near the lateral boundaries. In other words, the spatial period of the motion is *doubled* due to the influence of the relatively long-wave bottom-layer instability. The modulation of the vortices intensity strengthens with the growth of G (see Fig. 2.8).

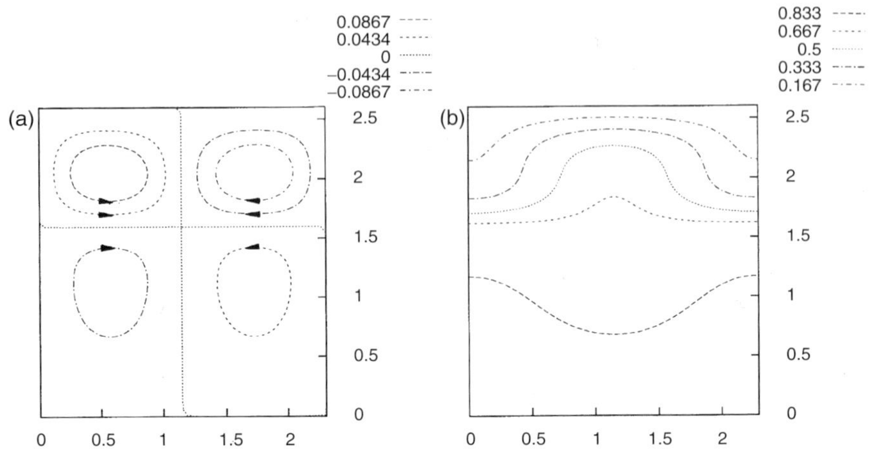

FIGURE 2.6. (a) Stream lines and (b) isotherms of the convective flow for $a = 1.6$; $L = 2.3$; $G = 109.4$.

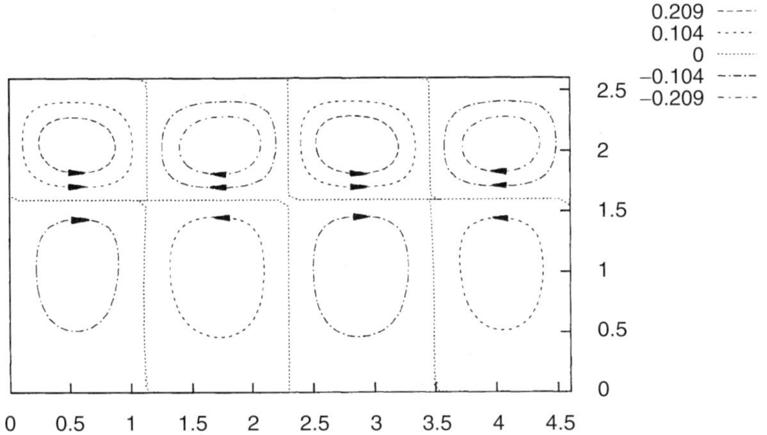

FIGURE 2.7. Stream lines of the convective flow for $a = 1.6$; $L = 4.6$; $G = 273.5$.

In the case $R_1 < R_2$, the convective motion is generated first in the bottom layer, whereas a relatively weak motion is induced in the top layer. The stream lines and isotherms of the convective flow corresponding to point C in Fig. 2.4 are shown in Fig. 2.9. Note that in part of the top layer, the direction of the vortices' rotation *coincides* with the rotation direction of the corresponding vortices in the bottom layer (see Fig. 2.9(a)), whereas the isotherms in both layers are distorted in a similar way (see Fig. 2.9(b)). Thus, the motion in the top layer is induced by both the *viscous coupling* and the *thermal coupling*. With the increase of G,

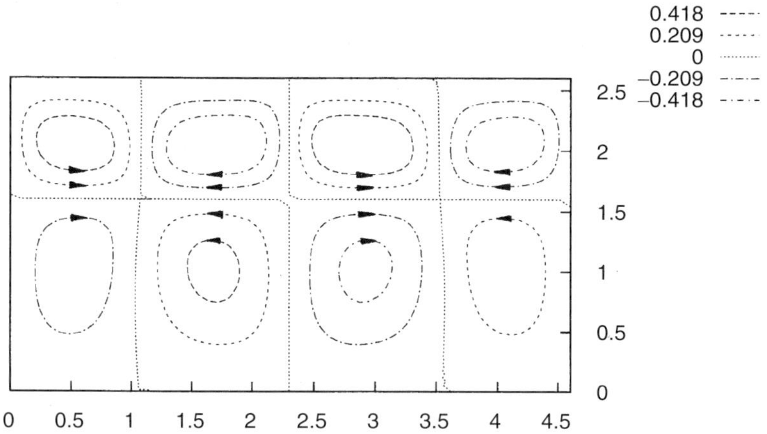

FIGURE 2.8. Stream lines of the convective flow for $a = 1.6$; $L = 4.6$; $G = 820.5$.

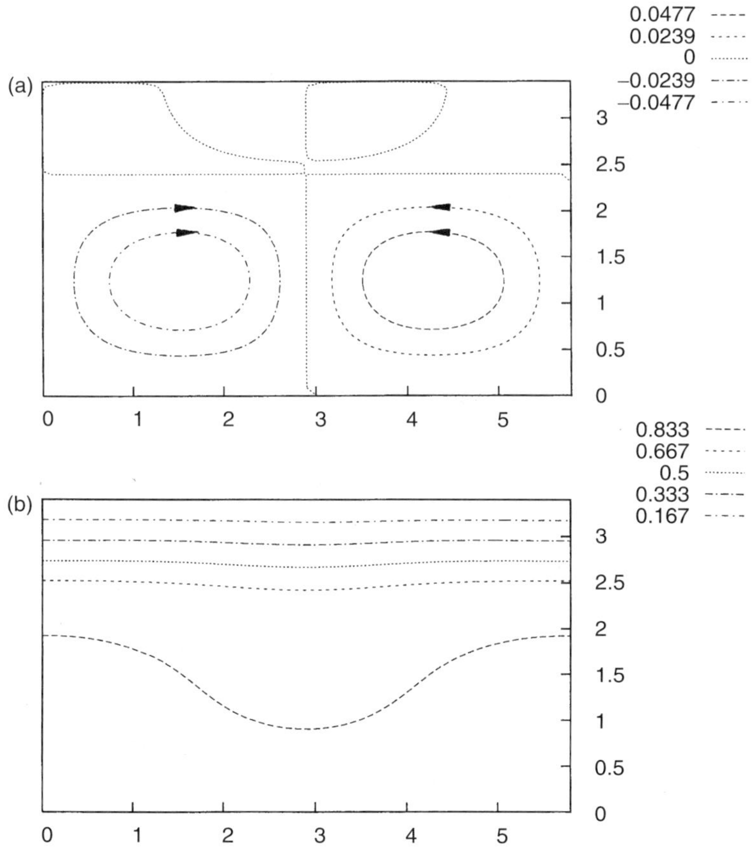

FIGURE 2.9. (a) Stream lines and (b) isotherms of the convective flow for $a = 2.4$; $L = 5.8$; $G = 54.8$.

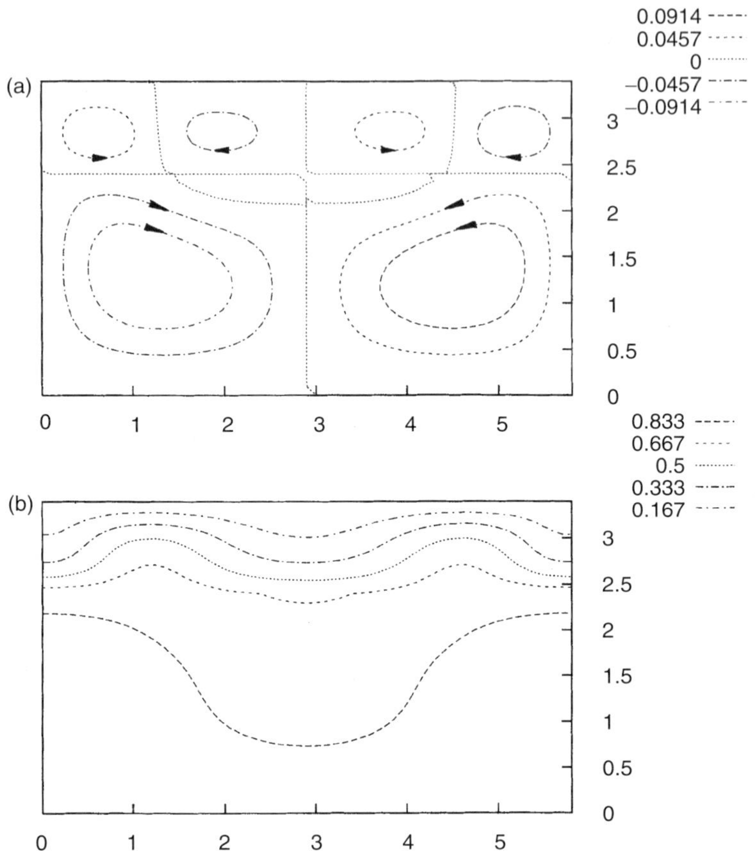

FIGURE 2.10. (**a**) Stream lines and (**b**) isotherms of the convective flow for $a = 2.4$; $L = 5.8$; $G = 79.2$.

the intensity of the motion in the top layer grows (Fig. 2.10; point D in Fig. 2.4) and becomes comparable with the intensity of the motion in the bottom layer (Fig. 2.11; point E in Fig. 2.4). Note that the structure that is developed in the relatively thin top layer has a smaller characteristic wavelength. The vortices in the both layers rotating in the same direction (due to the thermal coupling) are separated by weak vortices induced by the interfacial viscous stresses.

Thus, either a viscous coupling or a thermal coupling may prevail in the interaction between convective flows in the top layer and in the bottom layer. In the former case, the corresponding vortices in both layers have opposite signs, whereas in the latter case, they have one and the same sign. Also, the vortices in both layers may have different horizontal scales. All of these types of structure were first found in the numerical simulations by Simanovskii (1979a, 1979b); see also Simanovskii and Nepomnyashchy, 1993.

The first experimental observation of the thermal coupling between two-layer convective flows was presented by Nataf *et al.* (1988), who measured the velocity

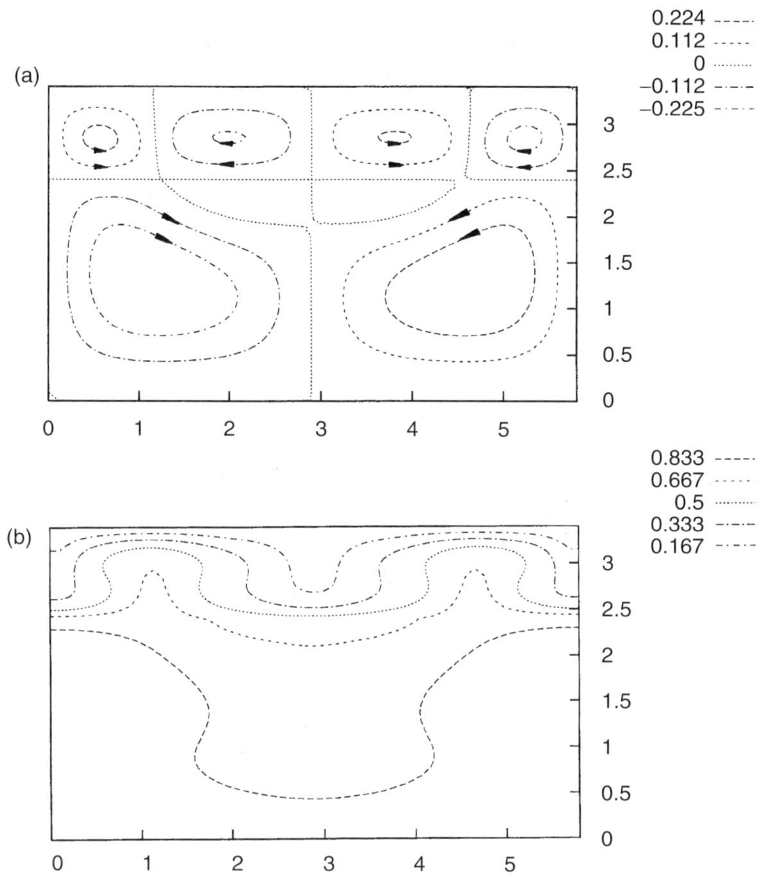

FIGURE 2.11. (a) Stream lines and (b) isotherms of the convective flow for $a = 2.4$; $L = 5.8$; $G = 243.6$.

field for the system silicon oil Rhodorsil / glycerol in the case of equal layers thicknesses (see also Cardin *et al.*, 1991). For the same fluid system but with non-equal layers thicknesses, Cardin and Nataf (1991) observed convective structures similar to those shown in Fig. 2.10(a), where the vortices in different layers had different characteristic horizontal scales. Rasenat *et al.* (1989) found both kinds of coupling in their experiments in oil / ethylene glycol and decane / ethylene glycol systems. An extensive experimental investigation of steady Rayleigh-Bénard convection in the system silicone oil / ethylene glycol was performed by Prakash and Koster (1996).

Oscillatory Flow Regimes

In order to find the oscillatory flow regimes, we consider the case corresponding to the neutral curve shown in Fig. 2.5, where the minimum on the oscillatory branch of the neutral curve is lower than that on the monotonic branch.

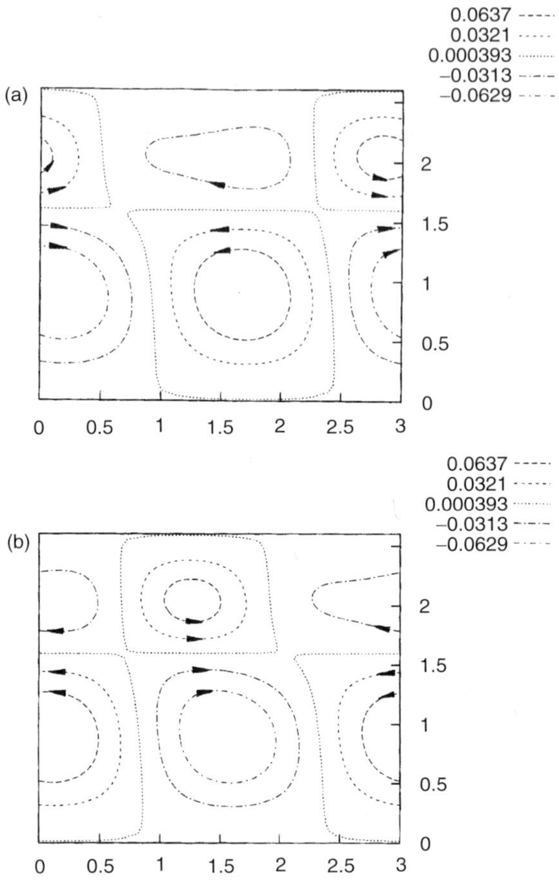

FIGURE 2.12. Snapshots of the stream function field for the traveling wave; $a = 1.6$; $L = 3$; $G = 111.5$; $G_Q = -150$. The wave moves from the right to the left.

The linear theory reveals the oscillatory instability, but it is unable to predict which kind of nonlinear regime, traveling wave or standing wave, appears in the supercritical region.

In order to answer that question, the simulations with periodic boundary conditions (2.29) have been performed. The traveling wave has been found near the threshold predicted by the linear stability theory (see Fig. 2.12).

The simulations have been carried out also for rigid heat-insulated boundary conditions (2.31). In order to describe the flow structure, we define the following integral variables characterizing the intensity of the motion in the left and in the right halves of the layers:

$$S_{l1}(t) = \int_0^{L/2} dx \int_0^1 dz \psi_1(x, z, t),\ S_{r1}(t) = \int_{L/2}^{L} dx \int_0^1 dz \psi_1(x, z, t), \tag{2.32}$$

$$S_{l2}(t) = \int_0^{L/2} dx \int_{-a}^{0} dz \psi_2(x, z, t), \; S_{r2}(t) = \int_{L/2}^{L} dx \int_{-a}^{0} dz \psi_2(x, z, t). \tag{2.33}$$

Perfectly symmetric standing waves, which satisfy the symmetry conditions

$$\psi_m(x, z, t) = -\psi_m(-x, z, t), \;\; T_m(x, z, t) = T_m(-x, z, t), \;\; m = 1, 2, \tag{2.34}$$

are developed near the instability threshold. Because of the above-mentioned symmetry, $S_{r1} = -S_{l1}$ (see Fig. 2.13(a)). The oscillations in the top layer and in the bottom layer are characterized by a phase shift (see Fig. 2.13(b)). The evolution of the stream function field during one-half of the oscillation period is shown in Fig. 2.14. Snapshots (a) – (c) of Fig. 2.14 correspond to the fragment of the

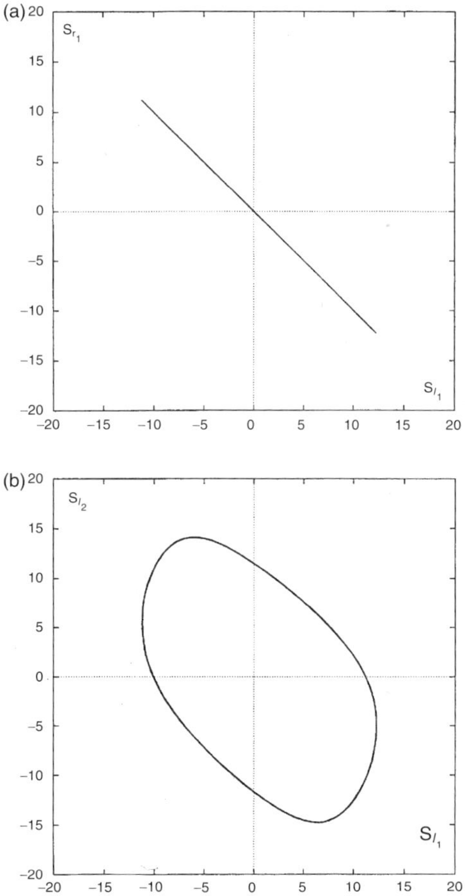

FIGURE 2.13. Phase trajectories for the oscillatory convective regime: (a) S_{r1} *versus* S_{l1}; (b) S_{l2} *versus* S_{l1}; $a = 1.6$; $L = 3$; $G = 116$; $G_Q = -150$.

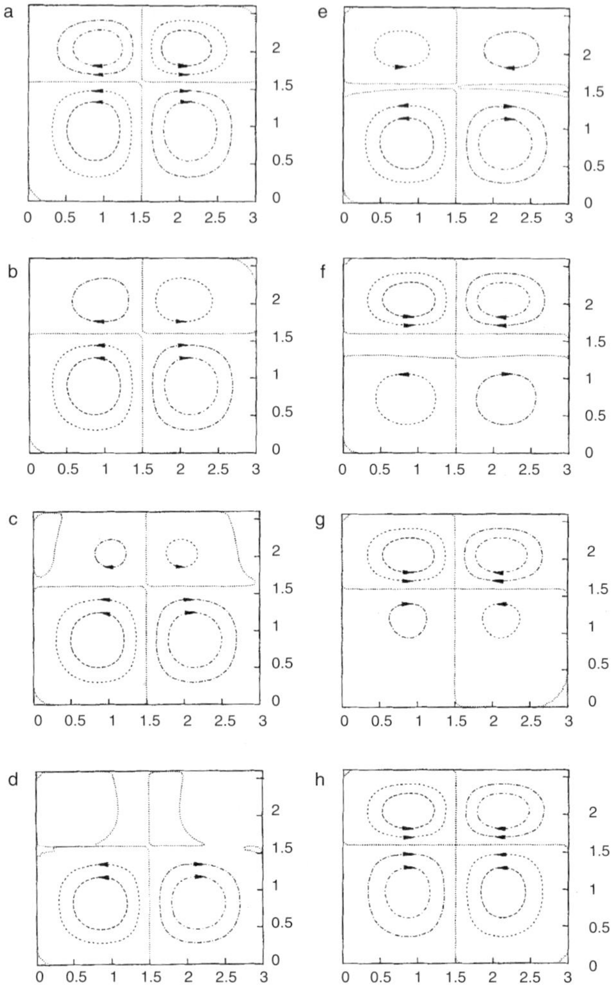

FIGURE 2.14. Snapshots of the stream function field during half of the period; $a = 1.6$; $L = 3$; $G = 116$; $G_Q = -150$.

phase trajectory in Fig. 2.13(b) located in the region $S_{l1} < 0$, $S_{l2} > 0$. Near the time instant when S_{l1} changes its sign, there are two vortices of different signs in each half of the top layer (see Fig. 2.14(d)). The boundary between both vortices is nearly vertical, and it moves in the horizontal direction. The sign of S_{l2} is changed in a different way. When $S_{l1} > 0$, $S_{l2} > 0$ (i.e., the rotation is in the same direction in both fluids), the flow in the bottom layer has a two-storied structure (see snapshots (e) and (f) of Fig. 2.14). The boundary between the vortices of different sign is almost horizontal, and it moves downwards until S_{l2} becomes negative (Fig. 2.14(f)). The stream function field in Fig. 2.14(h) is similar to that in Fig. 2.14(a), but the corresponding vortices have the opposite signs.

The analysis presented above is based on the assumption that the deformation of the interface is negligible (see Sec. 1.2.3). This assumption is not valid in the case where the densities of both fluids are close, $\delta = O(G^{-1}) \ll 1$. The latter case has been studied by Renardy *et al.* (for a comprehensive review, see Joseph and Renardy, 1993). The nonlinear simulation of the convection in a system with close densities has been carried out by Simanovskii (1983).

2.3 Anticonvection

In the present subsection, we will describe the amazing phenomenon of "anticonvection", which appears in two-layer systems when heating is from above ($s = -1$). We disregard the thermocapillary effect ($M = 0$) and assume that for both fluids, the density decreases with temperature ($G > 0$, $\beta > 0$). Because in both layers the density gradients are directed downward, one could expect that the buoyancy convection is impossible. Nevertheless, a specific kind of convective instability can occur in the system if the thermal diffusivity and the thermal expansion coefficient of the top fluid are much smaller than those of the bottom fluid ($\chi \ll 1$, $\beta \ll 1$) (Welander, 1964; Gershuni and Zhukhovitskii, 1980; Simanovskii, 1980; Gershuni *et al.*, 1981). The physical essence of this instability mechanism is as follows.

Let an element of the top fluid move down toward the interface. Due to the low thermal diffusivity of the top fluid, this element keeps its temperature higher than that of the surrounding fluid. Because the thermal expansion coefficient of the top fluid is small, this buoyancy force caused by that temperature difference cannot stop the downward motion of the top fluid element. This motion produces a hot spot at the interface, which generates an upward convective motion in the bottom, highly buoyant fluid. Because of the high thermal diffusivity of the bottom fluid, the latter motion does not change the temperature field, but it produces tangential stresses at the interface that support the initial velocity disturbance in the top layer. Similar arguments can be used for the explanation of the appearance of anticonvection in the case $\chi \gg 1$, $\beta \gg 1$.

Actually, the conditions for the excitation of the instability described above turned out to be rather restrictive, and in fact only one physical system satisfying these conditions was found (water / mercury). However, the appearance of anticonvection can be essentially simplified in the presence of a constant, spatially uniform heat source, or sink at the interface that is applied in such a way that the temperature gradient in one layer is much smaller than in the other one (Nepomnyashchy and Simanovskii, 1990a, 2001, 2002; Perestenko and Ingel, 1991, 1995; Nepomnyashchy *et al.*, 2000; Simanovskii *et al.*, 2002; Simanovskii, 2004). Assume that a constant heat release rate Q (Q may be positive or negative) is set at the interface $z = 0$. Dimensionless temperature gradients in the equilibrium state are determined by the following formulas:

$$A_1 = -\frac{sG + a\kappa G_Q}{G(1+\kappa a)}, \quad A_2 = -\frac{\kappa(sG - G_Q)}{G(1+\kappa a)} \tag{2.35}$$

($s = -1$ for heating from above). Thus, $A_1 \ll A_2$ if $G_Q \approx G/a\kappa$, and $A_1 \gg A_2$ if $G_Q \approx -G$. As is shown in Sec. 2.2.1, one can introduce an artificial system without a heat release at the interface (i.e., with temperature gradients $\tilde{A}_1 = 1/(1+\kappa a)$, $\tilde{A}_2 = \kappa/(1+\kappa a)$), but with a renormalized value of the thermal diffusivity ratio $\tilde{\chi}$,

$$\tilde{\chi} = \chi \frac{sG - G_Q}{sG + a\kappa G_Q}. \tag{2.36}$$

For any two-fluid system, the quantity $\tilde{\chi}$, which determines the existence of anticonvection, can be made very small if $A_2/A_1 \ll 1$, or very large if $A_2/A_1 \gg 1$ (Nepomnyashchy *et al.*, 2000). The first inequality holds as $G_Q > 0$, $sG < -a\kappa G_Q$ (recall that $s = -1$ for heating from above), and sG is close to $-a\kappa G_Q$. Similarly, the second inequality holds as $sG < G_Q < 0$ and sG is close to G_Q.

Let us describe some results obtained for the two-liquid system silicone oil 10 cS / ethylene glycol (system 2 in Table 2.1). The neutral curve corresponding to the case of heat sources is shown in Fig. 2.15 (line 1). In the minimum of this neutral curve, $sG \approx -3729$ is less than $-a\kappa G_Q \approx -3718$; thus, both gradients A_1 and A_2 are positive (see equations (2.35)), and the anticonvection is the only instability mechanism. In the point of minimum, the ratio $A_1/A_2 \approx 1.4 \times 10^{-3}$; that is, there is a strong heating from above in the bottom layer and nearly neutral

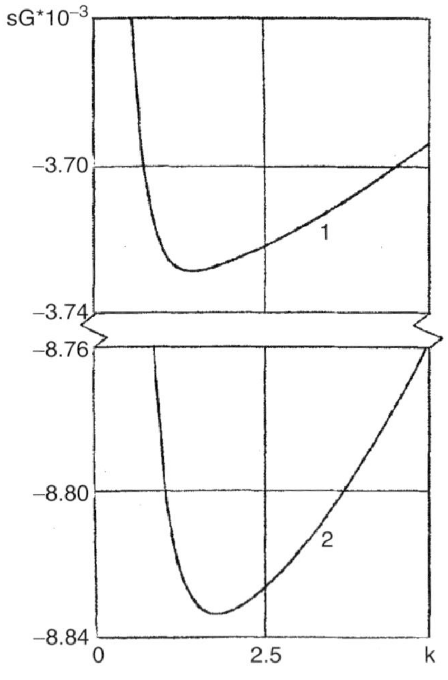

FIGURE 2.15. Neutral curves for the anticonvection at $G_Q = 6000$ (line 1) and $G_Q = -8835$ (line 2).

stratification in the top layer. Line 2 in Fig. 2.15 corresponds to the opposite case, $G_Q < 0$, $sG \approx G_Q$, characterized by a nearly vanishing temperature gradient in the bottom layer and a strong positive gradient in the top layer.

Nonlinear simulations confirm the existence of anticonvection in the system with heat sources and heat sinks (Nepomnyashchy and Simanovskii, 2001). A vertical crosssection of an anticonvective flow in the case $G_Q > 0$, $sG < -a\kappa G_Q$ $(0 < A_1 \ll A_2)$ is shown in Fig. 2.16. The stream lines are shown in Fig. 2.16(a), whereas Fig. 2.16(b) presents the fields of temperature deviations $T_m(x,z) - T_m^{(0)}(z)$, where $T_m^{(0)}(z)$ is the equilibrium temperature field corresponding to expressions (2.35). The intensity of the flow in the top layer, where the temperature gradient is relatively small, is much higher than that in the bottom layer. Note that the flow in the bottom layer has a "multistoried" structure.

Direct three-dimensional simulations of anticonvective flows show that the hexagonal pattern is the most typical one. However, in some cases, one observes a certain kind of the distorted square pattern (Nepomnyashchy *et al.*, 2003).

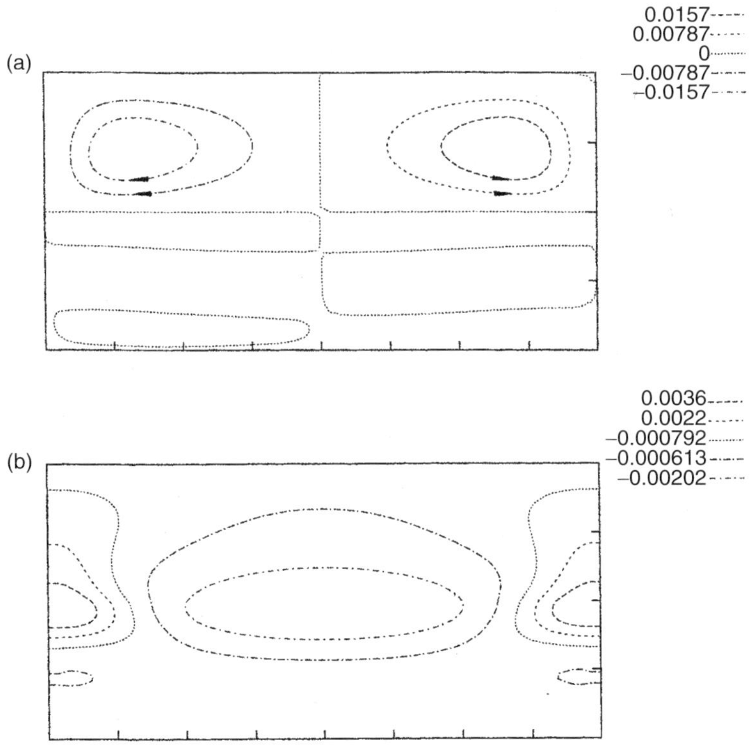

FIGURE 2.16. (a) Streamlines and (b) isolines of the temperature deviations for $G = 3717$; $G_Q = 6000$.

2.4 Stationary Marangoni Patterns

2.4.1 Exact Formulas

In the case of a pure Marangoni convection ($G = 0$, $M \neq 0$), the neutral curve of the stationary instability can be found analytically. Recall (see Sec. 1.2.2) that the influence of buoyancy can be ignored in the case when the layer thicknesses are much smaller than $a_c = (\alpha/g\beta_1\rho_1)^{1/2}$ (i.e., for thin layers or under microgravity conditions).

Solving the boundary value problem (2.3) – (2.10) with $\lambda = 0$, $G = 0$, one obtains the monotonic neutral curve determined by the following *exact* formula (Smith, 1966):

$$sM(k) = \frac{8Pk^2(1+\kappa a)(\kappa D_1 + D_2)(\eta B_1 + B_2)}{\kappa[\eta P(\chi E_2 - E_1) - 8k^5(D_1 + D_2)(\eta F_1 - F_2)(Ga\delta + Wk^2)^{-1}]}, \tag{2.37}$$

where

$$D_1 = \frac{C_1}{S_1}, \quad D_2 = \frac{C_2}{S_2}, \quad B_1 = \frac{S_1C_1 - k}{S_1^2 - k^2}, \quad B_2 = \frac{S_2C_2 - k}{S_2^2 - k^2a^2},$$

$$E_1 = \frac{S_1^3 - k^3C_1^3}{S_1(S_1^2 - k^2)}, \quad E_2 = \frac{S_2^3 - k^3a^3C_2^3}{S_2(S_2^2 - k^2a^2)}, \quad F_1 = \frac{1}{S_1^2 - k^2}, \quad F_2 = \frac{a^2}{S_2^2 - k^2a^2},$$

$$S_1 = \sinh k, \quad C_1 = \cosh k, \quad S_2 = \sinh ka, \quad C_2 = \cosh ka. \tag{2.38}$$

The denominator of expression (2.37) can change its sign when k is changed. In this case, the monotonic neutral curve is discontinuous; that is, the Marangoni convection can be excited for *both* ways of heating, from below ($s = 1$) and from above ($s = -1$), but in different intervals of wave numbers k. The term in the denominator that contains $Ga\delta$ and W is caused by the deformation of the interface. If $Ga\delta \gg 1$ and $W \gg 1$, which is typical under the normal gravity, this term can be dropped except for small k, so that

$$sM(k) = \frac{8k^2(1+\kappa a)(\kappa D_1 + D_2)(\eta B_1 + B_2)}{\kappa\eta(\chi E_2 - E_1)}. \tag{2.39}$$

In the long-wave limit,

$$sM(k) = -\frac{80P(1+\eta a)(1+\kappa a)^2 a}{\kappa[\eta a^3 P(1 - \chi a^2)k^2 + 120(1+a)(\eta a^2 - 1)(Ga\delta + Wk^2)^{-1}]}. \tag{2.40}$$

One can see that if $Ga\delta \gg 1$, the deformation of the interface becomes significant as $k \sim k_0 \sim (Ga\delta)^{-1/2}$. If $Ga\delta \sim 1$, $W \gg 1$, the deformation is significant as $k \sim k_0 \sim W^{-1/4}$.

Thus, one can distinguish between the *short-wave* Marangoni instability, which is insensitive to the interface deformation, and the *long-wave* instability, which is essentially associated with the deformations of the interface. In the long-wave limit, the critical Marangoni number

$$sM_c = sM(0) = \frac{2PGa\delta(1+\eta a)(1+\kappa a)^2 a}{3\kappa(1+a)(1-\eta a^2)}. \tag{2.41}$$

The deformational type of the monotonic instability is important in the case of very thin layers, close densities of fluids, or under microgravity conditions.

There exists another kind of long-wave Marangoni instability that appears when the fluids are situated between *poorly conducting* rigid boundaries. If we take

$$z = 1: \ \psi_1 = \psi_1' = T_1 = 0, \tag{2.42}$$

$$z = -a: \ \psi_2 = \psi_2' = T_2 = 0, \tag{2.43}$$

instead of equations (2.42) and (2.43), formula (2.21) is replaced by

$$sM(k) = \frac{8k^2(1+\kappa a)(\kappa H_1 + H_2)(\eta B_1 + B_2)}{\kappa\eta(\chi I_2 - I_1)}, \tag{2.44}$$

where

$$H_1 = \frac{S_1}{C_1}, \ H_2 = \frac{S_2}{C_2},$$

$$I_1 = \frac{S_1^2 C_1 - 2kS_1 + k^2 C_1 - k^3 S_1}{C_1(S_1^2 - k^2)}, \ I_2 = \frac{S_2^2 C_2 - 2kaS_2 + k^2a^2 C_2 - k^3a^3 S_2}{C_2(S_2^2 - k^2a^2)};$$

S_1, C_1, S_2, and C_2 are defined by formulas (2.38). The minimum of the neutral curve determined by equation (2.44) is achieved at $k = 0$,

$$sM(k) = -\frac{48(1+\kappa a)(a+\kappa)(1+\eta a)}{(1-\chi a^2)\kappa\eta a}. \tag{2.45}$$

We will discuss the above-mentioned types of monotonic Marangoni instability separately.

2.4.2 Short-Wave Marangoni Patterns

Linear Theory

In the "short-wave" region, $k = O(1)$, we can disregard the interface deformations and use equation (2.39) for the description of the monotonic neutral curve (in the case $M \neq 0$, $G = 0$).

Four types of monotonic neutral curve corresponding to equation (2.39) are shown in Fig. 2.17. For a fixed value of the wave number k, the monotonic instability of the equilibrium can appear only for a definite way of heating (either from below or from above), but not for both of them simultaneously. For the opposite

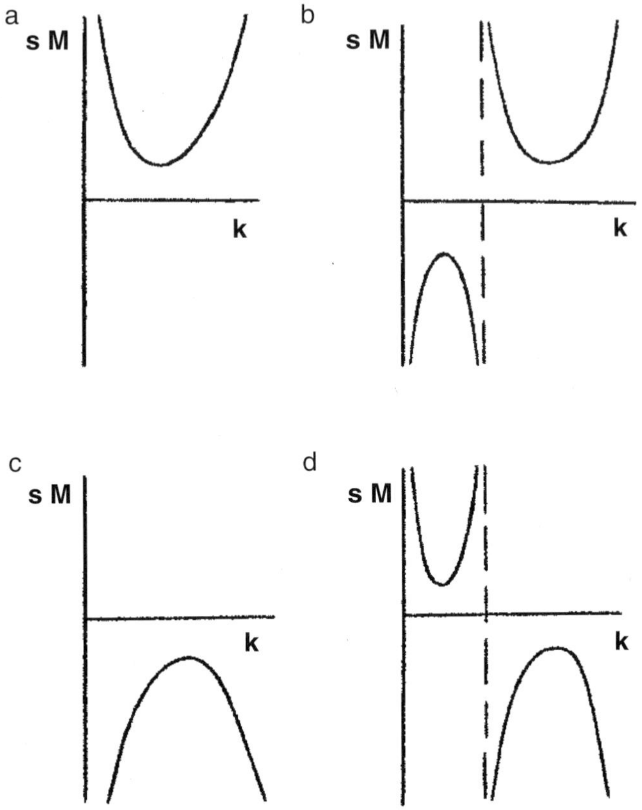

FIGURE 2.17. The types of monotonic neutral curve: (a) $\chi > 1$, $a > a_* = \chi^{-1/2}$; (b) $\chi > 1$, $a < a_*$; (c) $\chi < 1$, $a < a_*$; (d) $\chi < 1$, $a > a_*$.

way of heating, the equilibrium is stable with respect to monotonic disturbances for arbitrary M. This suppression of the monotonic instability is a circumstance favorable for the appearance of the oscillatory instability. Note that in the case $a = \chi = 1$, $\chi E_2 - E_1 = 0$; therefore, the monotonic instability disappears for any k.

As an example, let us consider the system n-octane / methanol (system 3 of Table 1). Because $\chi < 1$, this system belongs either to type c of Fig. 2.17 (if $a < a_* = 1/\sqrt{\chi} = 1.03$) or to type d (if $a > a_*$). In the latter case, an additional oscillatory neutral curve appears in the long-wave region in the case of heating from above. If $a < a^* = 1.38$, the minimum value of the Marangoni number for the monotonic neutral curve is lower than that of the oscillatory neutral curve (Colinet *et al.*, 1996); see Fig. 2.18. In this case, one can expect that stationary patterns appear near the instability threshold. The opposite case $a > a_*$, where the oscillations are developed, will be considered in Sec. 2.5.1.

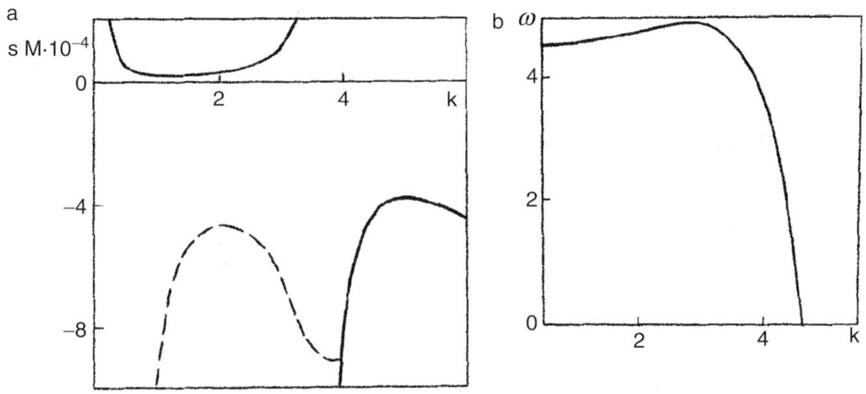

FIGURE 2.18. (a) Monotonic (solid lines) and oscillatory (dashed line) neutral curves for the system n-octane-methanol; $a = 1.3$. (b) Dependence of the frequency ω on the wave number k on the oscillatory neutral curve.

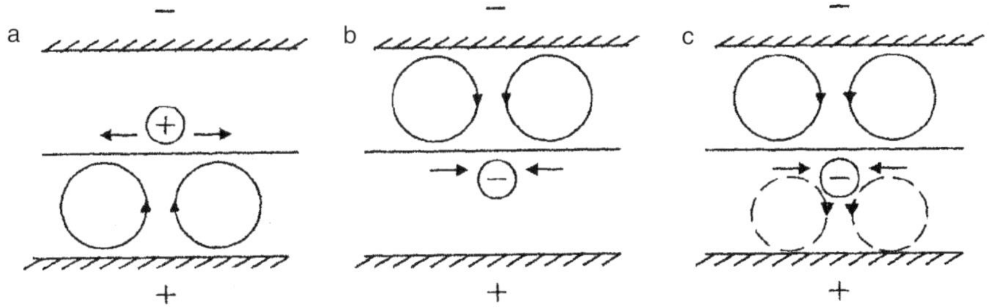

FIGURE 2.19. Interaction between buoyancy and thermocapillarity. (a) The buoyancy convection is in the bottom layer. (b) The buoyancy convection is in the top layer. (c) The origin of the oscillatory instability.

Let us discuss now the case of the combined action of the thermocapillarity and the buoyancy. Recall that the buoyancy effect in each layer may be characterized by the "local" Rayleigh numbers $R_m = g\beta_m|A_m|a_m^4/\nu_m\chi_m$, $m = 1, 2$, which are not independent (see Eq. (2.16)). If the ratio $R_2(G)/R_1(G)$ is essentially different from 1, the buoyancy convection appears first in the layer with the larger value of the "local" Rayleigh number. In another layer, a weak induced flow arises (see Sec. 2.2).

Let us consider the case $R_1(G) < R_2(G)$, where the buoyancy convection takes place mainly in the bottom layer. As one can see in Fig. 2.19(a), a temperature disturbance on the interface generates buoyancy volume forces and thermocapillary tangential stresses acting *in the same direction*. In this case, the action of the thermocapillary effect leads to a *decrease* of the minimized critical Grashof number (see Fig. 2.20).

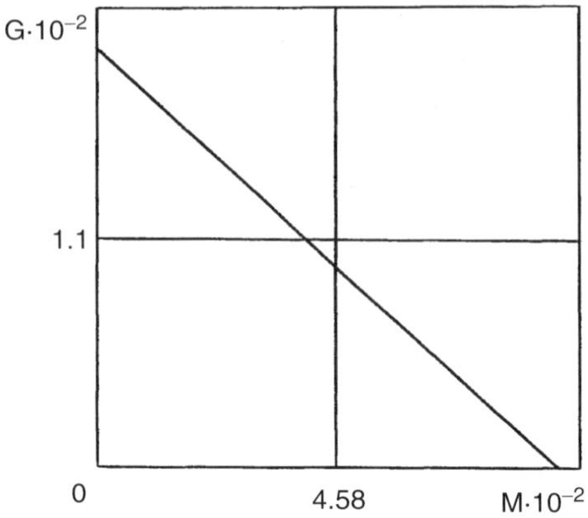

FIGURE 2.20. Dependence of G_c on M for the air / water system (system 4 in Table 1); $a = 1$.

In the opposite case, $R_1(G) > R_2(G)$, the buoyancy convection arises first in the upper layer (see Fig. 2.19(b)). In the latter case, the buoyancy volume forces and thermocapillary tangential stresses act *in the opposite way*. The asynchronic action of two factors working in opposite directions can produce an *overstability* that will be studied in Sec. 2.5.2.

Nonlinear Patterns

The typical stationary convective pattern that is produced by the pure short-wave monotonic Marangoni instability ($M \neq 0$, $G = 0$) is the hexagonal pattern (Scanlon and Segel, 1967; Koschmieder and Biggerstaff, 1986). The hexagonal pattern appears in a subcritical way, but the depth of the subcriticality (i.e., the minimum value of the parameter $\epsilon = (M - M_c)/M_c$ for which the pattern is first observed) has been found to be rather small, both in experiments (Schatz *et al.*. 1995) and in theory (Scanlon and Segel, 1967; Golovin *et al.*, 1997a). With the growth of the Marangoni number, a transition either to a square pattern or to a roll pattern is predicted, depending on the parameters of the two-layer system (Golovin *et al.*, 1997a).

In the case of the combined action of the thermocapillary effect and of the buoyancy ($M \neq 0$, $G \neq 0$), the hexagonal pattern and the roll pattern compete near the instability threshold (Parmentier *et al.*, 1996). The thermocapillary effect acts in favor of hexagons, whereas the buoyancy supports rolls. A detailed investigation of the competition among hexagonal, square, and roll patterns for different two-layer systems has been carried out by means of the bifurcation analysis in the paper of Engel and Swift (2000). It was found that although the hexagonal pattern is the preferred pattern near the threshold, a transition to other patterns may take place for rather small values of ϵ.

The latter prediction is in accordance with the experiments performed by Tokaruk *et al.* (2000). In the water / FC-75 system, the square pattern was observed already at $\epsilon = 0.05$, and no hexagonal pattern was found. For larger values of ϵ, the square pattern was replaced by curved rolls. Various transitions among hexagonal, square, and roll patterns have been observed in the experiments of Juel *et al.* (2000).

2.4.3 Long-Wave Marangoni Patterns; The Case of Poorly Conducting Boundaries

As was shown in Sec. 2.4.1, in the case of a fixed heat flux on the non-deformable boundaries (boundary conditions (2.42) and (2.43)), the minimum of the neutral curve is achieved at $k = 0$. This *long-wave* instability is connected with the conservation of the mean temperature. In this case, a *constant* ($k = 0$) temperature deviation uniform in the whole system is *neutral*: $\lambda(0) = 0$ (recall that $\lambda(k)$ is the growth rate of the disturbance with the wave number k). The instability appears when the sign of the second derivative $(\partial^2\lambda/\partial k^2)_{k=0}$ is changed rather than $\lambda(0)$ itself.

In the case of poorly conducting boundaries (low Bi on *both* rigid boundaries of the system), one can find that $k_c = O(Bi^{1/4})$. The flow generated by the instability is characterized by slow modulation of the temperature in horizontal directions. This circumstance can be used for a *reduction* of the original three-dimensional problem to a *two-dimensional* problem.

In the limit $P \to \infty$, the disturbances of velocity and pressure, as well as modes characterizing the inhomogeneity of the temperature in the vertical direction, are *slaved* by the two-dimensional field $\Theta(x, y)$ of the temperature averaged across the layer, which is the only *active* variable. In this limit, Knobloch (1990) derived (in the framework of a one-layer model) a closed equation for Θ that has the following structure (after a scaling):

$$\frac{\partial \Theta}{\partial t} = \alpha\Theta - \eta\Delta\Theta - \Delta^2\Theta + \nabla\cdot(|\nabla\Theta|^2\nabla\Theta) \tag{2.46}$$

$$+\beta\nabla\cdot(\Delta\Theta\nabla\Theta) + \delta\Delta|\nabla\Theta|^2,$$

where

$$\Delta \equiv \nabla^2.$$

The coefficient α is proportional to the Biot number, η is the scaled bifurcation parameter, and the coefficients β and δ are constants of order of unity. Earlier, a similar equation (but without quadratic nonlinear terms that appear because the top and the bottom boundary conditions are not identical) was obtained for the Rayleigh-Benard convection by Gertsberg and Sivashinsky (1981). The one-dimensional version of equation (2.46) was derived by Sivashinsky (1982) and Garcia-Ybarra *et al.* (1987a, 1987b).

The direct numerical simulation of equation (2.46) performed by Shtilman and Sivashinsky (1991) and Pontes *et al.* (1996) showed that the *hexagonal patterns* appear near the instability threshold. The direction of motion (flow up or down in the center of the hexagons) depends on the sign of the parameter $\beta - \delta$, in accordance with the theory of Knobloch (1990). Near the point $\beta = \delta$, a competition between hexagonal and square patterns is observed.

It is known (see Pismen 1986a) that at finite P, the three-dimensional convective motions may generate a large-scale horizontal flow. The stream function $\Psi(x, y)$ of such a flow satisfies the equation

$$\Delta\Psi = p^{-1}[\nabla\Delta\Theta, \nabla\Theta]_z,$$

where p is a coefficient proportional to P. An additional term, $[\nabla\Theta, \nabla\Psi]_z$, which describes the advection of the temperature by the horizontal flow, should be added to the left-hand side of equation (2.46). In the case of the Marangoni convection, the equations of this kind were derived by Shtilman and Sivashinsky (1991).

The influence of the interfacial deformation $h(x, y, t)$ on the long-wave instability discussed above has been investigated for a gas / liquid system (Golovin *et al.*, 1995). Because of the low heat conductivity of the gas, the temperature gradient in the gas phase is relatively large, and the disturbances of the heat flux caused by the interface deformations are generally not small. That is why the temperature disturbances in the gas cannot be eliminated and the analysis should be performed by a two-layer approach. Golovin *et al.* (1995) derived a coupled nonlinear system for variables Θ, Ψ, and h and studied the competition between different patterns by means of the amplitude equations approach. The numerical simulation of the above-mentioned system has been carried out by Golovin *et al.* (2003).

2.4.4 Long-Wave Deformational Instability

There is another kind of monotonic Marangoni instability essentialy connected to the surface deformation (see Eq. (2.40) and Eq. (2.41)). It can be explained qualitatively in the following way. Let us consider a layer heated from below with a local depression of the fluid level. The depressed area of the surface is closer to the heated lower boundary and, hence, is warmer. The corresponding thermocapillary tangential stress generates a motion away from the depression that strengthens this depression. In the case of long-wave surface deformations, the only factor that can control such an instability is gravity (because both capillarity and dissipative effects are negligible in the long-wave limit). Because of this, the long-wave limit of the critical Marangoni number $M_c = M(0)$ is proportional to the Galileo number Ga (see Eq. (2.41)). The deformational instability becomes important for very thin layers and under microgravity conditions.

As well as the instability described in the previous subsection, the long-wave deformational instability is connected with a certain *conservation law*. Now, the conserved quantity is *the volume of the fluid* (or the mean thickness d of the fluid layer) rather than the mean temperature. The *active* variable is the long-wave *deflection of the surface* $h(x, y)$.

The long-wave instability is typically nonsaturable; that is, it leads to a blow-up physically corresponding to the formation of dry spots (VanHook *et al.*, 1995). The absence of the saturation is inavoidable in the framework of the one-layer approach. Indeed, let us take a layer with the mean thickness corresponding to the critical value of the Marangoni number and consider a long-wave modulation of its local thickness $h(x, y)$. The local *critical* Marangoni number is proportional to the Galileo number and, hence, to $h^3(x, y)$. The *actual* local Marangoni number is proportional to $h(x, y)$. Thus, the instability is enhanced in the regions where the thickness decreases. The nonlinear evolution of the interfacial deformation is governed to the leading order by the *Funada equation* (Funada, 1987):

$$\frac{\partial h}{\partial t} = -\nabla_\perp^2 \left[A\nabla_\perp^2 h + B(M - M_c)h + Ch^2 \right], \tag{2.47}$$

where $A > 0$, $B > 0$, C are some constant real coefficients, and $\nabla_\perp^2 = \partial^2/\partial x^2 + \partial^2/\partial y^2$. It is well known that this equation, which is called also *Sivashinsky equation* (Sivashinsky, 1983), is subject to a blow-up in finite time (see, e.g., Bernoff and Bertozzi, 1995).

The weakly nonlinear equation (2.47) ceases to be valid for large deformations of the boundary. In the latter case, a more general equation derived by Davis (1987) may be used, which has the following structure:

$$\frac{\partial H}{\partial t} + \nabla \cdot [a(H)\nabla H + b(H)\nabla\Delta H] = 0, \tag{2.48}$$

where $H = d + h$ is the full thickness of the layer, the function $a(H) = \frac{1}{2}(MH^2 - M_c H^3)$ describes the destabilizing action of the thermocapillary effect and the stabilizing action of the gravity, and the function $b(H) = \frac{1}{2}B^{-1}M_c H^3$ corresponds to the stabilizing action of the surface tension. The expression for $a(H)$ was modified by Joo *et al.* (1991), who more exactly considered the heat transfer on the layer surface, and by Van Hook *et al.* (1997), who took into account the dependence of the Biot number on the gas and liquid layers' thicknesses ratio, thus incorporating some two-layer effects and getting a better agreement with experiments.

It should be noted that when considering the late stages of the rupture of the layer, it is necessary to take into account the additional interaction between the boundaries of the layer due to the *intermolecular forces* (e.g., *van der Waals forces*), which becomes much more important than gravity for thin films of fluids (Israelashvili, 1985; Derjaguin *et al.*, 1987). The intermolecular forces give a contribution into the expression of $a(H)$ in the form $(H^3/3\eta)\phi'(H)$, where η is viscosity and $\phi(H)$ is *potential of intermolecular forces* or *disjoining pressure* (see, e.g., Williams and Davis, 1982; Sharma and Ruckenstein, 1986a). In the case of van der Waals forces, this potential can be taken as $\phi = A/H^3$, where A is dimensionless *Hamaker constant* (Hamaker, 1937). The intermolecular forces may stabilize ($A > 0$) or destabilize ($A < 0$) the film. This problem is discussed in more detail in Sec. 6.2.

Numerical integration of equation (2.48) and its modifications taking into account additional physical effects like intermolecular forces, evaporation, etc. was performed by several authors (for a review, see Oron *et al.*, 1997). Near the rupture point, the derivation of the long-scale equation (2.48) is not well justified. Direct numerical investigation of the rupture process by means of the integration of the full nonlinear equations of convection with a moving free surface reveals an interesting nonmonotonicity of the film relief and cascade processes (Krishnamoorthy *et al.*, 1995, Boos and Thess, 1999).

The direct experimental investigation of the long-wave deformational instability was performed by VanHook *et al.* (1995). It was found that the long-wave instability generates a droplet trapped within a dry spot. In coincidence with the theory, the deformational mode competes with the short-wave pattern-forming mode and prevails, as G is small enough (for thin layers).

For intermediate values of the Galileo number, both instabilities can be observed simultaneously (VanHook *et al.*, 1995, 1997). The nonlinear interaction between both modes was studied theoretically by Golovin *et al.* (1994, 1997a, 1997b). It was shown that the appearance of short-wave convection patterns can prevent the rupture of the liquid layer.

The destruction of the fluid layer due to the nonsaturable long-wavelength deformational instability can be prevented by means of a control. This idea was suggested and implemented by Or *et al.* (1999). As the controlled variable, they used the temperature on the lower rigid boundary of the gas-layer system. The physical idea consists in enhancing the bottom heating beneath a spontaneously appearing interface elevation. The thermocapillary flow generated by such a temperature modulation diminishes the interface elevation. Using a special *nonlinear* feedback control strategy, the authors managed to completely suppress the subcritical instability, at least in the framework of a weakly nonlinear model.

When the two-layer approach is used, the dependence of the critical temperature difference on the thickness of the liquid layer may be *nonmonotonic* (Nepomnyashchy and Simanovskii, 1990b; Simanovskii and Nepomnyashchy, 1993). If the ratio of the layers thicknesses $a = a_c$ corresponds to the minimum value of the critical temperature difference, both the increase and the decrease of the local liquid layer thickness weaken the instability. In the latter case, the coefficient C in equation (2.47) vanishes, and the cubic nonlinearity has to be incorporated. If a is close to a_c, the quadratic and the cubic nonlinearities are of the same order, and the problem is governed by the *Cahn-Hilliard equation*,

$$\frac{\partial h}{\partial t} = -\nabla_{\perp}^{2}\left[A\nabla_{\perp}^{2}h + B(M - M_c)h + Ch^2 - Dh^3\right], \qquad (2.49)$$

which is saturable.

The general theory of nonlinear long-wave deformational instability in a two-layer system was developed by Badratinova (1985). The theory is rather complicated and includes, in addition to a nonlinear evolution equation governing the finite-amplitude long-wave deformations, two elliptic equations for pressure disturbances in both fluids.

2.5 Marangoni Waves in Systems with a Nondeformable Interface

2.5.1 *Oscillatory Marangoni Instability*

We will start our analysis of oscillatory instabilities in two-layer systems by the consideration of the case where the deformation of the interface is not essential. The instabilities of such kind can be called *longitudinal* instabilities, in contradistinction to *transverse* instabilities connected with interfacial deformations.

The existence of longitudinal, or dilational, waves was first discovered in the pioneering works by Lucassen (1968) and Lucassen-Reynders and Lucassen (1969). In order to understand the physical nature of this kind of waves, let us consider a fluid layer on a cooled rigid plate. When a liquid element rises to the free surface, it creates a cold spot. The surface tension gradient acts toward this spot, pushing the element to the bulk. Thus, some oscillations are generated. It can be shown that the dependence of the nondimensional frequency ω on the nondimensional wave number k in the limit of large M and k is determined by the relation

$$\omega^2 = \omega_{long}^2 = \frac{M}{P(\sqrt{P}+1)}k^2 \tag{2.50}$$

(see Levchenko and Chernyakov, 1981). However, in a one-layer system, the purely longitudinal waves *decay* for any values of M, and they do not produce an instability.

The longitudinal waves can exist also in two-layer systems. Indeed, the argumentation given above is kept (e.g., if we take into account the adjacent layer of a gas with a relatively small viscosity and high thermal diffusivity). In the general case, the two-layer system is characterized by four different timescales (viscous and heat diffusion time scales for each fluid). Because of this, the counteraction of convective heat fluxes generated by motions in the upper and lower layers has some *time delay*. Since the fields of velocity and temperature in two fluids do not change synchroneously, some *oscillations* can arise instead of a monotonic growth or decay of the disturbance. In the limit of high Mk^2, the frequency of oscillations can be calculated analytically (see Simanovskii and Nepomnyashchy, 1993; Rednikov *et al.*, 1998):

$$\omega^2 = \frac{sM(\sqrt{\nu}-\sqrt{\chi})\kappa\eta k^2}{P(1+\sqrt{P})(\sqrt{\nu}+\sqrt{\chi P})(\eta+\sqrt{\nu})(\kappa+\sqrt{\chi})(1+\kappa a)} \tag{2.51}$$

(all of the notations are the same as in the previous sections). Thus, the condition of the existence of longitudinal waves for arbitrary large M is

$$s(\sqrt{\nu}-\sqrt{\chi}) > 0. \tag{2.52}$$

That means that the oscillations appear by heating from below ($s = 1$) if $\nu/\chi < 1$ and by heating from above ($s = -1$) if $\nu/\chi > 1$. In other words, the oscillations can take place at large M if the heating is from the side of the fluid with the relatively small Prandtl number.

Moreover, if the temperature gradient is sufficiently large, the longitudinal waves can produce an instability. The best known example of longitudinal oscillatory instability in two-layer fluid systems was discovered by Sternling and Scriven (1959) almost simultaneously with the appearance of the Pearson's theory (1958) for cellular Marangoni patterns. The classical theory of Sternling and Scriven (1959) dealt with a system of two layers of *infinite thicknesses* ($d \to \infty$) but disturbances with finite wavelengths λ. Thus, it described actually the *short-wave asymptotics* of the neutral curve ($\lambda \ll d$). However, as well as in the case of the stationary instability, the most dangerous disturbances have the wavelength comparable with the thicknesses of the layers. In the case $\lambda \sim d$, the criteria for the instability can be essentially changed in comparison to the limit $\lambda \ll d$ (Reichenbach and Linde, 1981; Nepomnyashchy and Simanovskii, 1983a). However, the physical nature of the oscillatory instability is not changed: it is the competition between processes that take place in both fluids.

The system n-octane / methanol (system 3 in Table 2.1) heated from above ($s = -1$) is an example of the physical system where the two-layer longitudinal oscillations have been predicted (see Colinet et al., 1996). According to the theory presented in Sec. 2.4.2, the system belongs to type (d) of Fig. 2.17 if $a > a_* = \chi^{-1/2} = 1.03$. In this case, the monotonic instability takes place for $s > 0$, as $k < k_d$, and for $s < 0$ (by heating from above), as $k > k_d$. The absence of the monotonic instability in the long-wave region $k < k_d$ by heating from above is a condition favorable for the appearance of the oscillatory instability. Indeed, the linear stability analysis reveals the existence of an oscillatory neutral curve in the long-wave region.

For the ratio of the layer thicknesses $a = a_0 = 1.38$, the monotonic instability threshold (at $k = k_m = 5.10$) and the oscillatory instability threshold (at $k = k_o = 1.94$) coincide.

In the case $a > a_0$, the oscillatory instability is the most "dangerous" one. A typical neutral curve is shown in Fig. 2.21. For $a = 1.6$, the critical Marangoni number $M_c = 2.715 \times 10^4$ for the oscillatory instability corresponds to $k_c = 1.75$, whereas $k_d = 3.95$.

It is remarkable that there exists a codimension-2 point (k_*, sM_*) where the frequency of the oscillations tends to zero and the oscillatory neutral curve (dashed line) terminates in the monotonic one. The behavior of the real part of the growth rate as a function of the Marangoni number M for $k > k_*$ (the case of a monotonic instability) and for $k < k_*$ (the case of an oscillatory instability) is shown in Fig. 2.22 (dashed line corresponds to an oscillatory mode, solid lines correspond to monotonic modes). Note that on the monotonic neutral curve $sM = sM_m(k)$, where the growth rate $\lambda(M_m, k)$ is equal to zero,

$$\lambda'_M(k) \equiv \left(\frac{\partial \lambda}{\partial M}\right)_{M=M_m(k)}$$

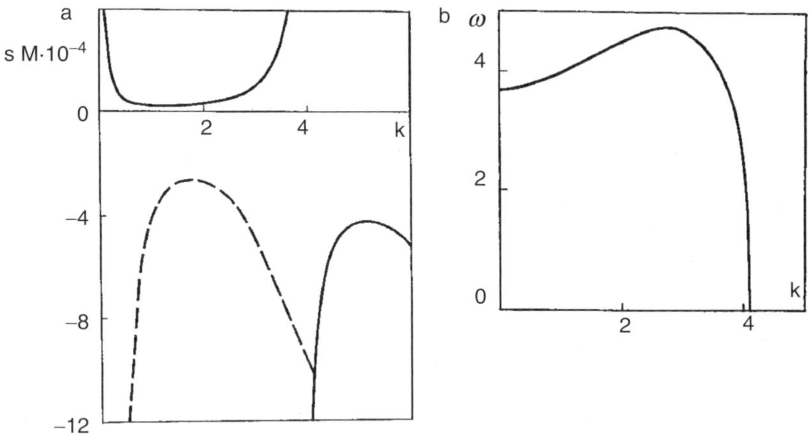

FIGURE 2.21. (a) Monotonic (solid lines) and oscillatory (dashed line) neutral curves for the system n-octane / methanol; $a = 1.6$. (b) Dependence of the frequency ω on the wave number k on the oscillatory neutral curve.

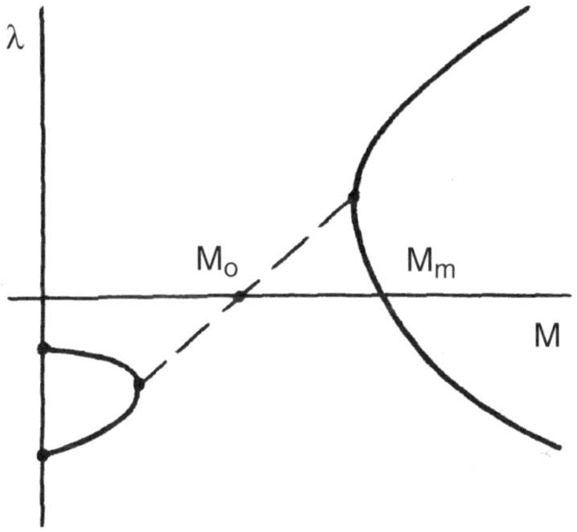

FIGURE 2.22. The dependence of the real part of the growth rate on the Marangoni number in the case of the oscillatory instability.

is positive in the case $k > k_*$ and negative in the case $k < k_*$. Thus, the role of the monotonic neutral curve is different in both cases. In the case $k > k_*$, it is indeed an instability boundary. In the case $k < k_*$, the monotonic neutral curve is the boundary of the *stabilization* for one of two monotonic modes. We can see that the inequality $\lambda'_M(k) > 0$ held on the monotonic neutral curve is the sign of a true monotonic instability and of the absence of the oscillatory instability.

Quite opposite, the inequality $\lambda'_M(k) < 0$ is the indication of the existence of an oscillatory instability for the same value of k and a smaller value of M.

The expression for the quantity $\lambda'_M(k)$ on the monotonic neutral curve can be obtained analytically, but it is rather cumbersome. It can be shown that $\lambda'_M(k)$ changes its sign at the value of $k = k_d$ corresponding to the discontinuity of the monotonic neutral curve. That means that *for any two-fluid system*, there exists an oscillatory instability near the point $k = k_d$ for a certain way of heating. The opposite way of heating does not produce an oscillatory instability. Recall that the discontinuity of the monotonic neutral curve at a certain $k = k_d$ takes place if $\chi > 1, a < a_* = \chi^{-1/2}$ or $\chi < 1, a > a_*$.

Some relatively simple criteria for the appearance of the oscillatory instability can be obtained in the limits of small k (long-wave disturbances), large k (short-wave disturbances), and large M.

Long-wave disturbances. A general expression for $\lambda'_M(k)$ in the limit of small k is presented in the book of Simanovskii and Nepomnyashchy (1993). The long-wave disturbances are especially important in the case where a is close to a_* (see Sec. 2.4). It can be shown that in this case, the condition $\lambda'_M(k) < 0$ of the oscillatory instability is fulfilled if $\nu > \chi, s = -1$ or $\nu < \chi, s = 1$. In other words, oscillations arise in a long-wave region if the heating is performed from the side of the fluid with larger value of the Prandtl number.

Short-wave disturbances. In the short-wave limit ($k \to \infty$), the analysis of the expression $\lambda'_M(k)$ shows that there is no short-wave oscillatory instability in the case $\chi < 1, s = -1$, whereas in the case $s = 1$, the short-wave oscillatory instability arises as $1 < \chi < \chi_{max}$, where

$$\chi_{max} = 1 + (1 + \nu\chi)\left(\frac{\eta + \nu}{1 + \eta} + 4P\right)^{-1}$$

(Simanovskii and Nepomnyashchy, 1993).

One more criterion can be obtained by the consideration of the short-wave asymptotics of the oscillatory neutral curve, which corresponds to the limit of semi-infinite fluid layers (Sternling and Scriven, 1959). The short-wave oscillatory instability appears if the quantities $sM_0 f_1$ and f_2 have opposite signs, where

$$sM_0 = \frac{8(1 + \kappa a)(1 + \kappa)(1 + \eta)}{\kappa\eta(\chi - 1)}$$

determines the short-wave asymptotics of the neutral curve in the leading order ($sM \sim sM_0 k^2$),

$$f_1 = \frac{P}{2}\left(\frac{\kappa + \chi}{1 + \kappa} + 1 + \chi\right) + \frac{\eta + \nu}{4(\eta + 1)} + \frac{\chi\nu - 1}{4(\chi - 1)},$$

$$f_2 = \frac{1 + \kappa a}{\kappa\sqrt{2}}\chi^{1/2}(\chi^{1/2} + \kappa)(\nu^{1/2} + \eta)(\nu^{1/2} + \chi^{1/2}P^{1/2})(1 + P^{1/2})$$

$$\times \frac{[(\eta+1)(\eta+\nu^{1/2})^{-1}+P^{-1/2}](\chi^{1/2}-\nu^{1/2})+(1-\chi^{1/2})(1+P^{-1/2})}{(\chi^{1/2}-\nu^{1/2})^2}.$$

Large Marangoni number. The next order correction to the expression (2.51) (Rednikov *et al.*, 1998) shows that in the limit of large M an oscillatory instability takes place if the condition (2.52) is satisfied and the wave number k of the disturbance satisfies the inequality

$$\frac{\tanh k - \nu^{1/2}\tanh ka}{\nu^{1/2}-\chi^{1/2}} > 0. \tag{2.53}$$

Let us emphasize that (2.52) and (2.53) are the conditions of the "survival" of the oscillatory instability for a fixed k in the limit $M \to \infty$. That means that the oscillatory instability can exist even when (2.52) is violated, but in that case, inside the instability region there should exist a discriminant curve on which the frequency of oscillations becomes equal to zero. For instance, the oscillatory instability was found numerically in the case $\eta = \nu = 0.5$, $\chi = \kappa = P = a = s = 1$ (Nepomnyashchy and Simanovskii, 1983a) where the criterion (2.52) is not satisfied. Note that in the case $\chi = 1, a = 1$, there is no monotonic instability at all (see Sec. 2.4). The case mentioned above was studied also by means of the full nonlinear simulations (Nepomnyashchy and Simanovskii, 1983b) in a closed cavity with aspect ratio $L = 2.5$. The instability appears as $M > M_1 \approx 3.14 \times 10^4$ and leads to the development of an oscillatory four-vortex flow with $S_{l1} = -S_{r1}$ (see Sec. 2.2). With the increase of the Marangoni number, the oscillations become essentially nonlinear and their period τ grows. In the region $4.5 \times 10^4 < M < M_2 \approx 5.0 \times 10^4$, the period of oscillations is well approximated by formula

$$\tau^{-2} = 0.31(M_2 - M).$$

As $M = M_2$, the limit cycle corresponding to the oscillatory motion becomes the separatrix of a saddle-node fixed point. In the region $M > M_2$, the limit cycle does not exist, and a steady convective flow arises. A more complicated sequence of nonlinear oscillations was observed in the case $L = 2$; it includes a period doubling bifurcation with the violation of the equality $S_{l1} = -S_{l2}$.

In the case of a closed cavity, the steady deformation of the interface caused by the contact angle on the lateral boundaries can essentially influence the flow regimes. This deformation generates a thermocapillary flow below the threshold of the instability, which suppresses the oscillatory instability mode. The numerical simulations fulfilled for the above-mentioned model system by Nepomnyashchy and Simanovskii (1984) showed that the oscillations are completely suppressed, as the contact angle is less than 84°.

As an example of nonlinear oscillations in a real physical system, let us consider the system n-octane / methanol (system 3 of Table 1).

The weakly nonlinear bifurcation analysis (Colinet et al., 1996) reveals a subcritical instability of the equilibrium state with respect to standing waves; that is, the instability is not saturated on the level of small-amplitude waves.

The nonlinear simulations of finite-amplitude oscillatory flow regimes have been performed by Nepomnyashchy and Simanovskii (2002) in a cell with the aspect ratio $L = 3.6$. The prediction of the weakly nonlinear theory has been justified. The oscillations have been observed in the subcritical region, when $M > M_{sub} = 1.77 \times 10^4$. In that region, sufficiently small disturbances decay on the background of the mechanical equilibrium state while some finite-amplitude disturbances generate a nondecaying oscillatory regime.

It was found that the oscillations were never time-periodic. In order to understand the nature of the periodicity violation, it is convenient to analyze the dependences $S_{l1}(t)$ and $S_{+1}(t) = S_{r1}(t) + S_{l1}(t)$ shown in Fig. 2.23. Typically, the quantity $S_{+1}(t)$, which characterizes the violation of the symmetry property $\psi(x, z) = -\psi(L - x, z)$, is relatively small, and the field of the stream function looks almost antisymmetric, whereas the temperature field is almost symmetric. The streamlines and isotherms corresponding to the "epoch" of weakly violated symmetry, are shown in Figs. 2.24 and 2.25. However, the regime of oscillations with the *antisymmetric* field of the stream function is unstable. One can see that the quantity $S_{+1}(t)$, which characterizes the amplitude of the mode with the *symmetric* stream-function field, grows on the background of the antisymmetric oscillatory flow. On the late stages of the symmetric mode growth, the flow patterns are obviously asymmetric (see Figs. 2.26 and 2.27). Finally, the growth of the symmetric mode leads to the complete destruction of the antisymmetric flow, but after some transient period, the almost antisymmetric stream-function field is restored, and the process is repeated. During a long period of time, the phase trajectory is close to the line $S_{l1} = -S_{r1}$, but the growth of the symmetric mode leads to the deviation of the phase trajectory from this line.

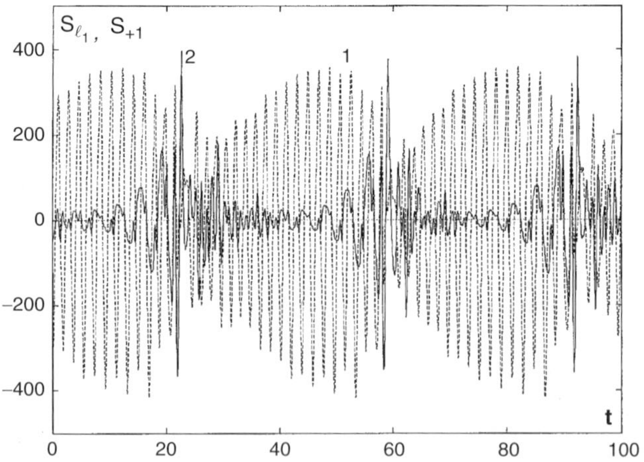

FIGURE 2.23. Dependences of $S_{l1}(t)$ and $S_{+1}(t)$; $M = 1.80 \times 10^4$.

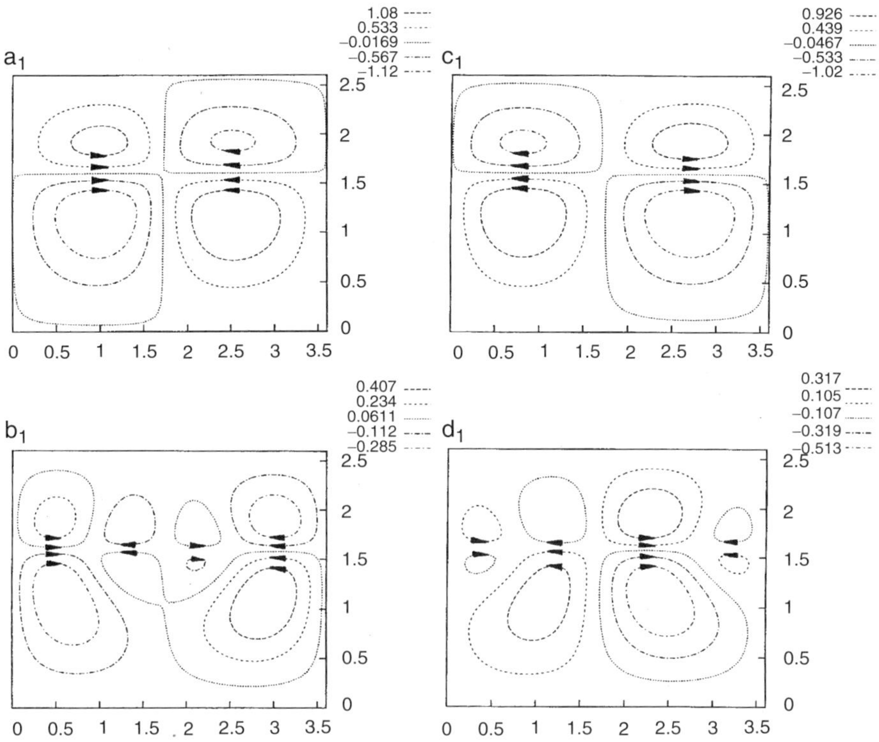

FIGURE 2.24. Streamlines $(a_1) - (d_1)$ for the motion with a weakly violated symmetry; $M = 1.80 \times 10^4$.

In the opposite case (as $a < a_0$), it was found that the first bifurcation generates a spatially periodic steady flow, and the second bifurcation produces a spatially quasiperiodic, time-periodic flow (Colinet et al., 1996).

For a specific choice of wave numbers $k_1 = 3$ and $k_2 = 2k_1 = 6$ of the interacting monotonic and oscillatory disturbances (resonant case; the monotonic and oscillatory instability boundaries coincide as $a_0 = 1.335$), a much wider variety of dynamical behaviors, including quasiperiodic relaxational oscillations and temporal chaos, have been found. The chaos appears with the growth of the Marangoni number after an infinite sequence of "loop-doubling" homoclinic gluing bifurcations (Lyubimov and Zaks, 1983). When increasing the Marangoni number again, a reverse cascade is observed.

The appearance of the Marangoni oscillations was studied also in the presence of a heat release at the interface (Nepomnyashchy and Simanovskii, 1990a, 1990c; Gilev *et al.*, 1991).

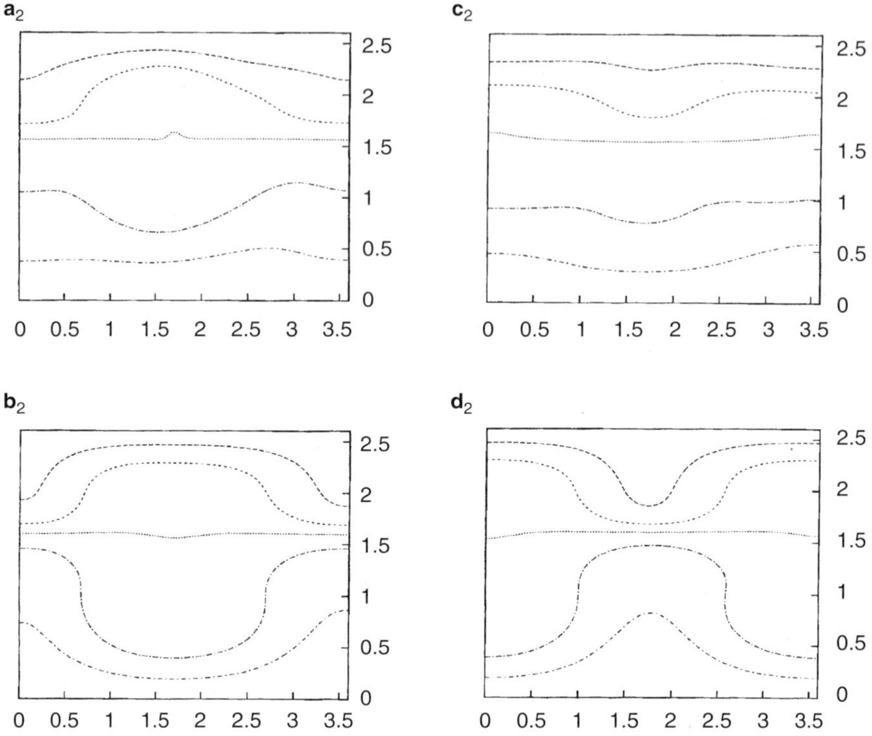

FIGURE 2.25. Isotherms $(a_2) - (d_2)$ for the motion with a weakly violated symmetry; $M = 1.80 \times 10^4$.

2.5.2 Competition between Marangoni and Rayleigh Instability Mechanisms

The pure Marangoni convection is observed only in systems that consist of very thin layers. In reality, the stability of the mechanical equilibrium state is determined by the combined action of the thermocapillary effect and the buoyancy effect. When the system is heated *from below*, the competition between both effects may lead to the appearance of the specific type of oscillations, which will be described in the present subsection. The cases of the normal thermocapillary effect ($d\sigma/dT < 0$) and the anomalous thermocapillary effect ($d\sigma/dT > 0$) will be considered separately.

The Normal Thermocapillary Effect.

For the normal thermocapillary effect, the Marangoni number is positive.

Recall that the buoyancy effect in each layer may be characterized by the "local" Rayleigh numbers $R_m = g\beta_m|A_m|a_m^4/\nu_m\chi_m$, $m = 1, 2$, which are not

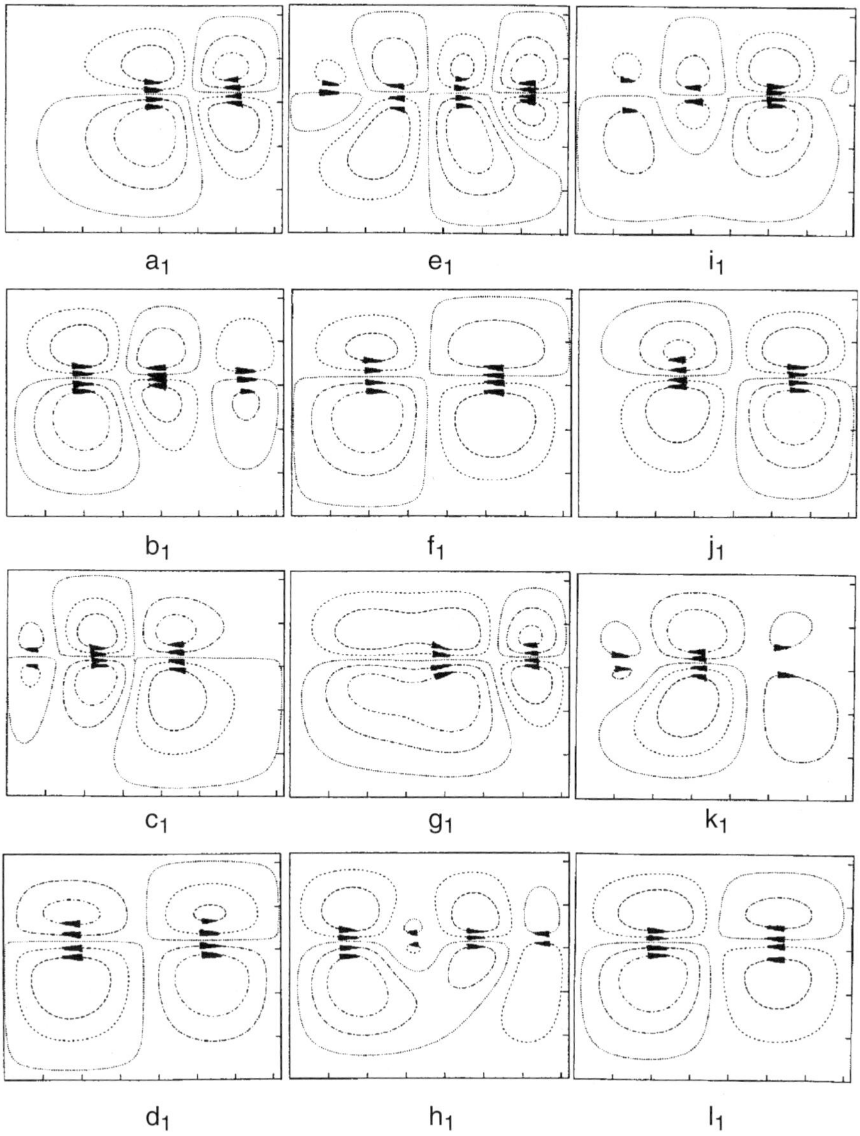

FIGURE 2.26. Streamlines $(a_1) - (l_1)$ for the asymmetric motion; $M = 1.80 \times 10^4$.

independent (see Eq. (2.16)). If the ratio $R_2(G)/R_1(G)$ is essentially different from 1, the buoyancy convection appears first in the layer with the larger value of the "local" Rayleigh number. In another layer, a weak induced flow arises (see Sec. 2.2).

The case $R_1(G) < R_2(G)$, where the buoyancy convection takes place mainly in the bottom layer, was considered in Sec. 2.4 (see Fig. 2.19(a)). In this situation,

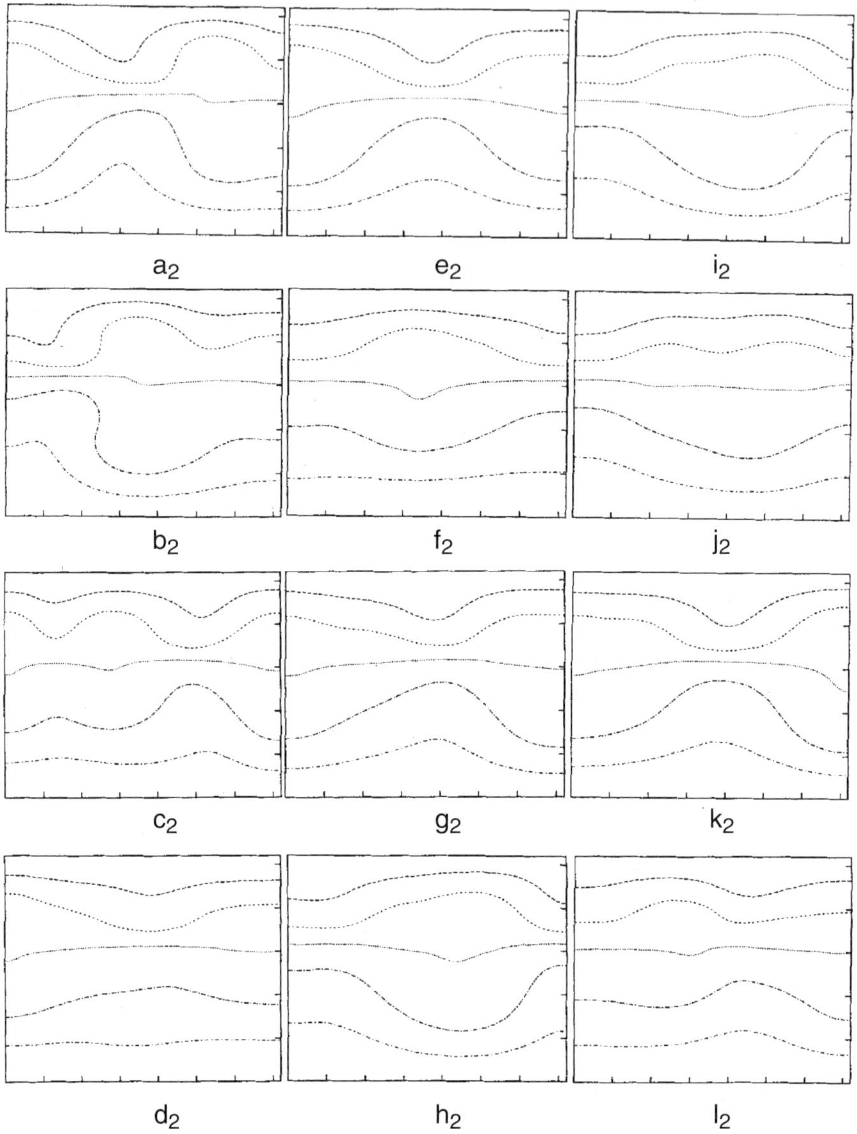

FIGURE 2.27. Isotherms $(a_2) - (l_2)$ for the asymmetric motion; $M = 1.80 \times 10^4$.

buoyancy volume forces and thermocapillary tangential stresses act *in the same direction*; therefore, the minimized critical Grashof number decreases (see Fig. 2.20).

In the case $R_1(G) > R_2(G)$, the buoyancy convection arises first in the upper layer (see Fig. 2.19(b)). Therefore, the buoyancy volume forces and thermocapillary tangential stresses act *in the opposite way*. Indeed, let us assume that there is a local negative temperature fluctuation ("cold spot") on the interface. The buoyancy generates a downstream flow above the cold spot and a divergent flow on the interface near the cold spot. At the same time, the normal thermocapillary effect produces thermocapillary stresses that tend to form a convergent flow on the interface near the cold spot and an upstream flow in the upper layer. Thus, thermocapillary stresses tend to suppress the buoyancy convection. The competition between the buoyancy and the thermocapillary effect leads to the stabilization of the monotonic instability. However, the asynchronic action of two factors working in opposite directions can produce an *overstability*.

Note that if $R_1(G) > R_2(G)$, but $R_1(G)$ and $R_2(G)$ are close, the buoyancy acting in the lower layer should be also taken into account. It is obvious (see Fig. 2.19(c)) that the buoyancy in the lower layer, which acts in the same direction as the thermocapillarity, can contribute to the appearance of oscillations.

The most important factors, which influence the appearance and disappearance of oscillations, are the ratio M/G and the ratio of the layers' thicknesses a. We will discuss these factors separately.

Influence of M/G for a fixed ratio of the layers' thicknesses. As an example, let us consider the appearance of convection in the system water / silicone oil DC200 (system 5 in Table 1) at $a = 1.6$ (in this case, $R_2/R_1 \approx 0.23$). If $M = 0$, the neutral curve consists of two separate monotonic fragments. The lower neutral curve has the minimum at $G = 270$; it corresponds to convection arising in the upper layer. The convection in the lower layer is excited at the upper neutral curve, which has the minimum at $G = 2860$.

As M increases, the neutral curve for the disturbances in the upper layer slowly goes upward (see lines 1 ($M = 1200$) and 3 ($M = 1380$) in Fig. 2.28), and that for the disturbances in the lower layer quickly goes downward (see lines 2 ($M = 1200$) and 4 ($M = 1380$)). For the larger value of M, the monotonic curves coincide in a certain point, and then they are separated into a "long-wave" (line 5) and a "short-wave" (line 6) part (both curves have been calculated as $M = 1380$). The dependence of the minimized value of the Grashof number G_c on the Marangoni number M for the monotonic neutral curve is shown in Fig. 2.29 (line 1).

Even earlier, at $M \approx 860$, a fragment of an oscillatory neutral curve appears. It is situated below the monotonic neutral curve for the disturbances in the upper layer in a finite interval of the wave numbers k. On the ends of this interval, the frequency of oscillations tends to zero. The location of these points in the plane (G, M) is shown in Fig. 2.29 (lines 2 and 3). After the separation of the monotonic neutral curves into a long-wave and a short-wave fragment, the oscillatory neutral curve connects both "monotonic" fragments (see line 7 in Fig. 2.28). The minimum of the oscillatory neutral curve (line 4 in Fig. 2.29) becomes lower than that of the short-wave monotonic neutral curve (line 1 in Fig. 2.29) as $M > 2750$.

In the case $G = 0$, the stationary Marangoni convection appears as $M > 22,000$. When G increases, the critical Marangoni number grows (see line 5 in Fig. 2.29). The lines 5 and 4 cross at $M \approx 36,400$. For $M > 36,400$, the equilibrium state is unstable for any G.

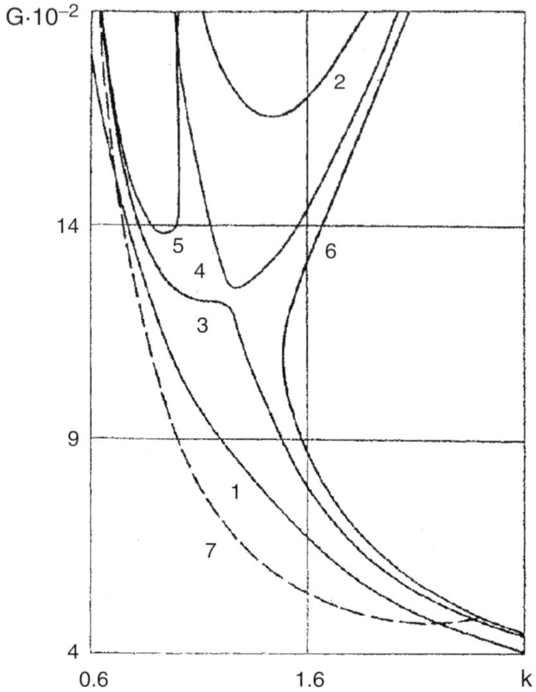

FIGURE 2.28. The neutral curves for the water / silicone oil DC200 system; $a = 1.6$.

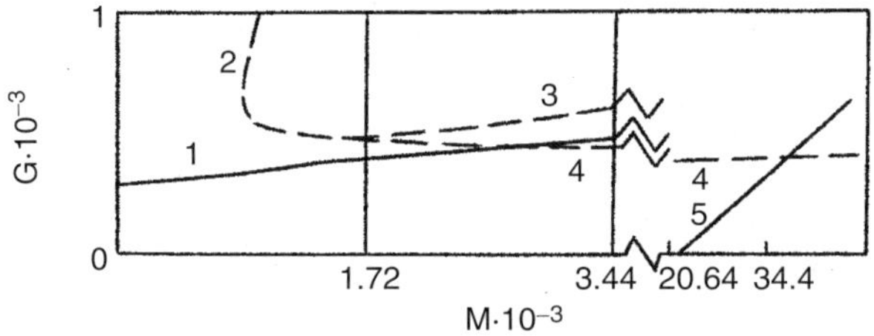

FIGURE 2.29. Dependence G_c on M for the water / silicone oil DC200 system; $a = 1.6$.

Thus, there is a large interval of the Marangoni number values, where the oscillatory instability is the most "dangerous" one. Let us emphasize that in the case of the normal thermocapillary effect, the oscillations appear when the buoyancy convection is excited mainly in the upper layer.

Obviously, the mechanism of oscillations described above is different from that of pure thermogravitational oscillations in two-layer systems described in Sec. 2.2. Let us note, however, that if the neutral curve has an oscillatory fragment as $M = 0$, the buoyancy/thermocapillary oscillations are developed according to another scenario (Gilev *et al.*, 1987a; Simanovskii and Nepomnyashchy, 1993) .

Let us consider the transformer oil / formic acid system (system 6 in Table 1) with $a = 0.54$ ($R_2/R_1 \sim 0.57$). In the absence of the thermocapillary effect ($M = 0$), a monotonic instability appears first in the relatively thick upper layer (see Fig. 2.33, line 1), but because of the "crossing" of neutral curves corresponding to the onset of convection in different layers (lines 1 and 2), there exists a small fragment of the oscillatory neutral curve (line 3). The thermocapillary effect, which acts against the buoyancy convection in the upper layer, suppresses the monotonic instability in the upper layer (line 4) but extends essentially the oscillatory neutral curve (lines 6 and 8). In a certain interval of M, the minimum value of G corresponds to the oscillatory instability. However, the simultaneous action of the buoyancy and the thermocapillary effect in the lower layer leads to its destabilization with respect to the monotonic instability (lines 5 and 7). For M sufficiently large, the minimum of the neutral curve corresponds to the monotonic instability in the lower layer.

Recently, the appearance of oscillations near the linear instability threshold was observed by Degen *et al.* (1998) in the system 47v2 silicone oil / water. This result can be explained by the combined action of the buoyancy and the thermocapillary effect (Nepomnyashchy and Simanovskii, 2004; Simanovskii, 2005). Under the conditions of the experiment, when the geometric configuration of the system is fixed while the temperature difference θ is changed, the Marangoni number M and the Grashof number G are proportional. It is convenient to define a parameter that does not change when θ is changed. In the literature, the dynamic Bond number GP/M is typically used. We define the inverse dynamic Bond number,

$$K = \frac{M}{GP} = \frac{\alpha}{g\beta_1\rho_1 a_1^2},$$

and investigate the evolution of neutral curves with the change of parameter K.

Let us consider the case $a = 1.6$ ($R_2/R_1 = 0.328$). For $K = 0$, the neutral curves are presented in Fig. 2.2. As explained above, with the growth of K, the instability boundary for the convection in the top layer (line 1) moves upward, whereas the instability boundary for the convection in the bottom layer (line 2a) moves downward. The oscillatory branch 5 expands, and the minimum value of the Grashof number for the oscillatory instability curve decreases (Fig. 2.30). For $K_- < K < K_+$, $K_- \approx 0.328$, and $K_+ \approx 0.411$, the minimum value of the Grashof number is achived at the oscillatory branch of the neutral curve; therefore, the oscillations become observable. A typical neutral curve in the region $K_- < K < K_+$ is shown in Fig. 2.31.

Thus, although the fluid system under consideration does not reveal any oscillations in the absence of the interfacial heat release and thermocapillary effect, it is subject to an oscillatory instability when the latter effects are taken into account.

In the case of the combined action of buoyancy and thermocapillary effect, the nonlinear simulations with periodic boundary conditions justify the results of the linear theory presented above. It was found by Nepomnyashchy and Simanovskii (2004) (see also Simanovskii and Nepomnyashchy, 2006) that the oscillatory

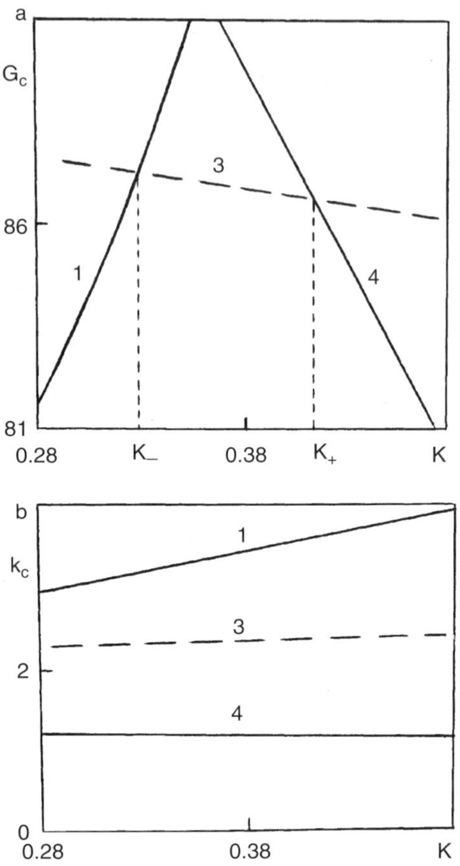

FIGURE 2.30. (a) The critical Grashof number and (b) the critical wave number for the monotonic instabilities (lines 1 and 4) and for the oscillatory instability (lines 3); the 47v2 silicone oil / water system (system 1 of Table 2.1);

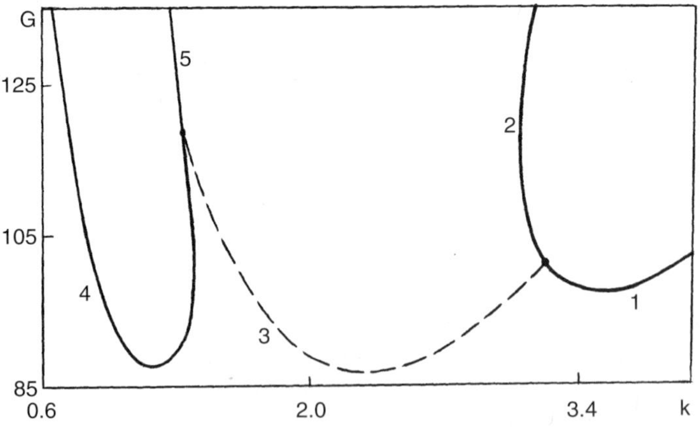

FIGURE 2.31. The neutral curves for the 47v2 silicone oil / water system (system 1 of Table 2.1); $a = 1.6$; $K = 0.4$.

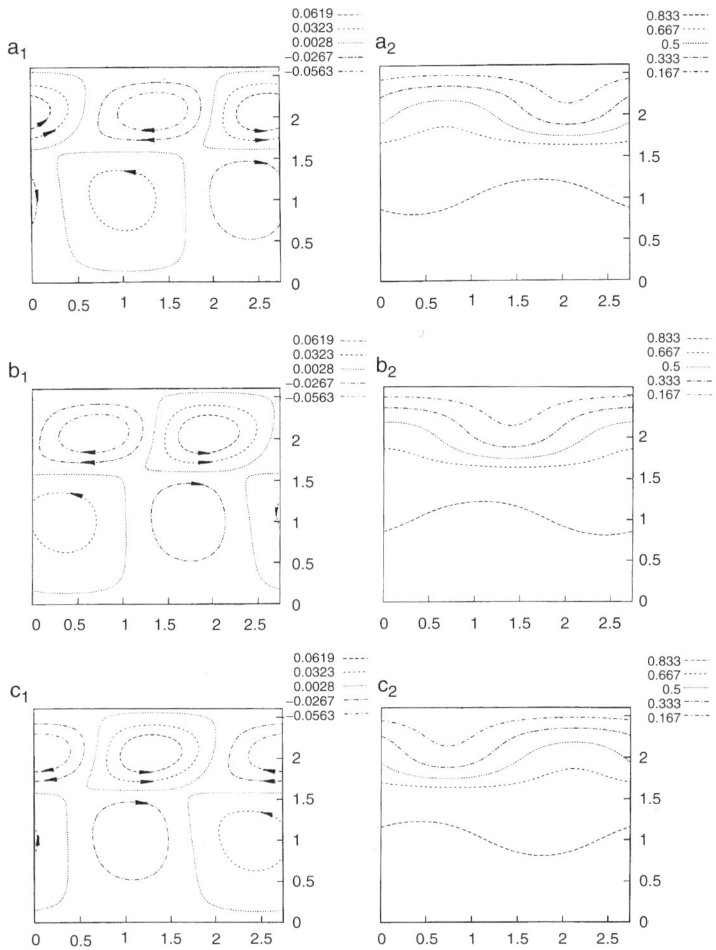

FIGURE 2.32. (a_1 - c_1) Snapshots of the stream lines and (a_2 - c_2) isotherms for the traveling wave; $a = 1.6$; $L = 2.74$; $G = 100$; $K = 0.4$. The wave moves from the right to the left.

instability generates the traveling waves (see Fig. 2.32). The direction of the wave propagation is casual.

Influence of a. The influence of the thicknesses ratio on the appearance and disappearance of oscillations was studied by Juel *et al.* (2000). The calculations have been performed for the system n-hexane / acetonitrile characterized by the ratio of thermal diffusivities $\chi = 0.743 < 1$. The total thickness of both fluid layers was kept constant, whereas the ratio of layers' thicknesses was changed.

The oscillations have been found by heating from below in the interval $0.5 < a < 1.2$ (the width of this interval slightly depends on the total thickness). The ratio of the "local" Rayleigh numbers $R_2/R_1 \sim 0.456a^4$. Thus, in the interval

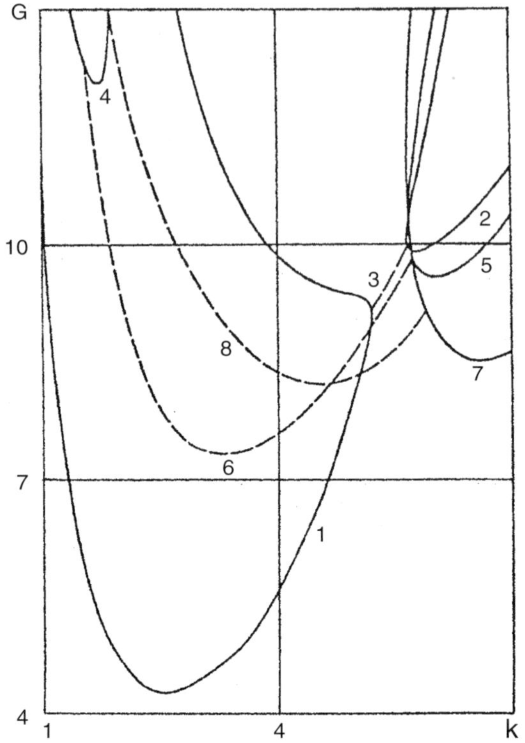

FIGURE 2.33. The neutral curves for the transformer oil / formic acid system; $a = 0.54$.

where the oscillations are observed, $0.03 < R_2/R_1 < 0.95$. One can expect the appearance of oscillations caused by the competition of the buoyancy in the upper layer, from one side, and the thermocapillarity and the buoyancy in the lower layer, from another side (see Fig. 2.19(c)).

As it was shown in Sec. 2.4.2, the stationary Marangoni convection is developed only by heating from above if $a < a_* = 1/\sqrt{\chi} \sim 1.16$. This circumstance is favorable for the appearance of the oscillatory convection by heating from below. In the case $a > a_*$, the monotonic Marangoni instability with small wave numbers becomes possible (see Fig. 2.17(d)), which can compete with the oscillatory instability. Indeed, when a increases, one observes a transition to relatively long-wave monotonic instability. The transition value of a is rather close to a_*, despite the influence of the buoyancy, which is not taken into account by the calculation of a_*. The oscillatory instability disappears also by diminishing a.

The oscillations in the system n-hexane / acetonitrile were studied also in experiments by Juel *et al.* (2000). The first observed convection pattern was a time-periodic wave, which was initially localized in a small region close to the side boundary. As the temperature difference increased, the system displayed a mixed state of wave fronts and steady cells. In Fig. 2.34, one can see a collection

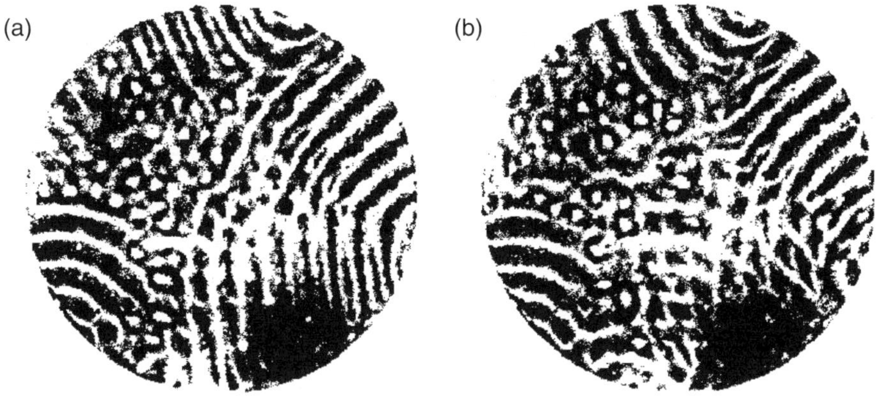

FIGURE 2.34. Snapshots of oscillatory convection patterns in the system n-hexane / acetonitrile. The time interval between (a) and (b) is 232 s. (Juel *et al.*, 2000).

of standing waves oriented in different directions, which overlap forming rolls, squares, and diamonds. The experimental values of wave number and frequency were in a satisfactory agreement with the linear stability calculations. The critical temperature difference for the onset of convection that was extrapolated from the power spectral density measurements turned out to be essentially higher than the theoretical critical temperature difference calculated for the system of laterally infinite layers.

The Anomalous Thermocapillary Effect.

In this subsection, we investigate the interaction between buoyancy and thermocapillary instability mechanisms in a two-layer system in the case of an *anomalous* thermocapillary effect (the interfacial tension increases with temperature). The possibility of the anomalous thermocapillary effect should be taken into account in realistic models of multilayer convection, because there are indications that the occurrence of an anomalous thermocapillary effect might be a typical property of various liquid/liquid systems (Eckert and Thess, 2001). Specifically, it was observed in aqueous alcohol solutions, nematic liquid crystals, binary metallic alloys, etc. (see, e.g., Legros (1986) and references therein).

In the case of the anomalous thermocapillary effect, the oscillatory instability caused by the competition between the buoyancy force and the thermocapillary tangential stresses will appear when the buoyancy convection is generated mainly in the lower layer.

As an example, we present results obtained for the 10cS silicone oil / ethylenglycol system with the thickness ratio of $a = 1.8$ with the local Rayleigh numbers' ratio $R_2/R_1 = 3.19$. Some typical neutral curves are shown in Fig. 2.35(a). Corresponding dependences of the frequency λ_i on the wave number k along the neutral curves are presented in Fig. 2.35(b). One can see that in the region $15 < G < 15.5$ the monotonic neutral curves (solid lines) change rather slowly with G. As

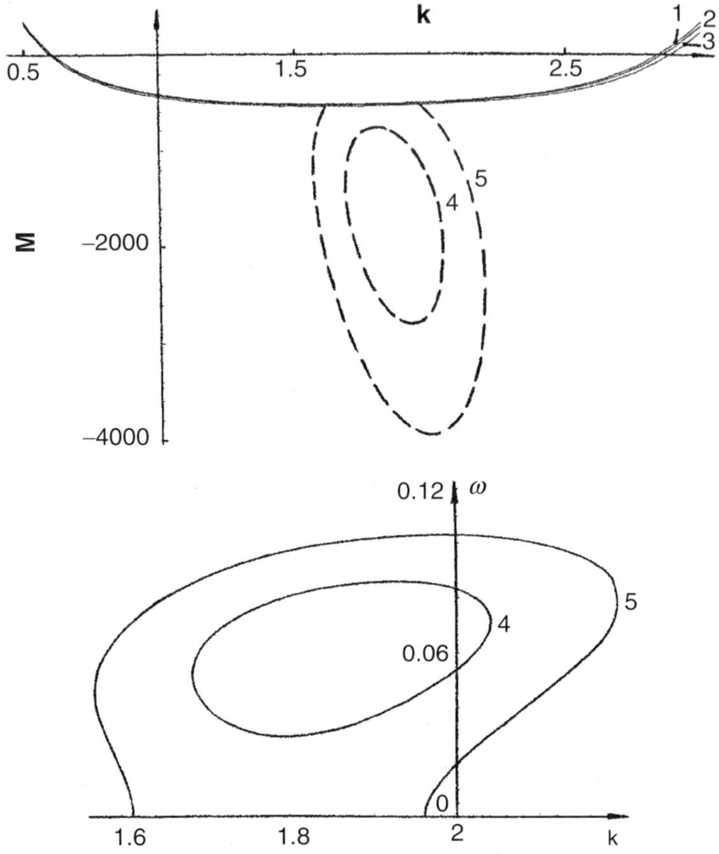

FIGURE 2.35. Neutral curves in the case $a = 1.8$: $G = 15$ (line 1), $G = 15.2$ (lines 2 and 4), $G = 15.5$ (lines 3 and 5); dependences $\omega(k)$: $G = 15.2$ (line 4), $G = 15.5$ (line 5).

$G < G_1 = 15.1$, the oscillatory neutral curve is absent. At $G > G_1$, a closed region of the oscillatory instability (dashed line) appears. It grows rapidly with G and at last touches the monotonic neutral curve as $G = G_2 = 15.3$. The stability boundaries in the $(M\text{-}G)$-plane for the monotonic and oscillatory instabilities are shown in Fig. 2.36. One can conclude that in the case $K = |M|/G < K_* = 31.7$, only the monotonic instability will be observed. Using the physical parameters of the system, we find that the latter case will take place if the total thickness of the two-layer system is larger than 5.9 mm. If the thickness of the two-layer system is smaller than the critical one, some slow oscillations appear near the threshold. The dependences of the frequency $\omega = \lambda_i$ and of the wave number on the ratio K for the critical oscillatory mode are shown in Fig. 2.37. Let us note that the *dimensional* frequency $\Omega = \omega\nu_1/a_1^2$. For instance, in the case of the total thickness $a_1 + a_2 = 4$ mm ($K = 70.3$), we find $\Omega = 0.21\ \text{s}^{-1}$.

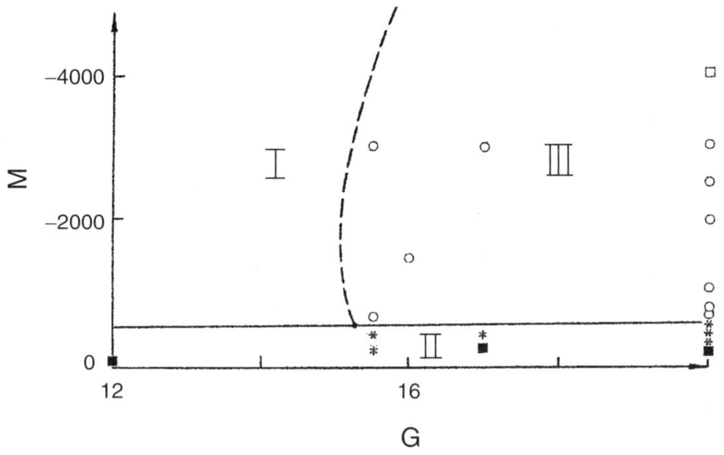

FIGURE 2.36. Diagram of stability regions: I : stability, II : monotonic instability, III : oscillatory instability. Filled squares correspond to stationary rolls, asterisks correspond to hexagons, circles correspond to oscillating rolls, and empty squares correspond to the spatio-temporal chaos.

In order to describe the main stages of oscillations, let us present some results of 2D simulations performed in the case $G = 18$, $M = -3933$ (for more details, see Braverman *et al.* (2000)).

In a certain moment of time, the structure of the convective motion is shown in Figs. 2.38(a_1) and 2.39(a_2) (let us call it structure A).

The upward motion in the lower layer generates the temperature field on the interface that has a maximum in the middle of the interface. Because of the anomalous thermocapillary effect, the tangential stresses appear that are directed toward this maximum. These stresses produce a four-vortex motion near the interface, so that a three-storied structure is produced (see Figs. 2.38(b_1) and 2.39(b_2)). Because the Prandtl numbers of both fluids are rather large, the temperature field is much more inertial than that of the stream function. That is why the temperature field generated by structure A exists during some time and supports both the buoyancy-induced motion in the lower part of the second layer and the thermocapillarity-induced motion around the interface. Finally the former motion is completely ousted by the latter one (see Figs. 2.38(c_1) and 2.39(c_2)). Consequently, the temperature maximum in the middle of the interface disappears. The thermocapillary motion near the interface decays, whereas in the lower layer, convection of another type is developed (see Figs. 2.38(d_1) and 2.39(d_2)); it will be called structure B. The transition between the structures A and B takes place during the first half of the period.

The subsequent evolution can be understood in a similar way: The temperature field generated by structure B produces a thermocapillary motion near the interface (see Figs. 2.38(e_1) and 2.39(e_2)), which replaces the buoyancy-induced motion in the lower layer (see Figs. 2.38(f_1) and 2.39(f_2)), but afterward, the

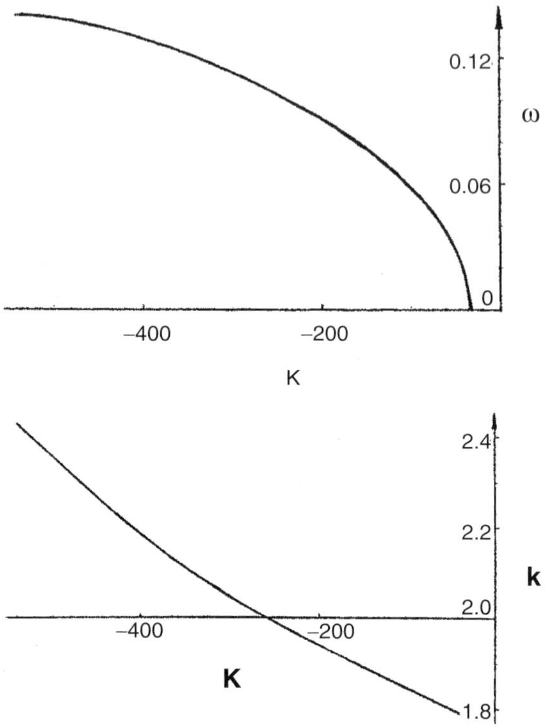

FIGURE 2.37. Dependences of critical frequency ω and critical wave number k on the ratio K in the case $a = 1.8$.

temperature field in the lower layer is rearranged and structure A (see Figs. 2.38(a_1) and 2.39(a_2)) is restored.

In reality, the patterns generated by the instability are typically three-dimensional. Let us describe results of three-dimensional nonlinear simulations fulfilled by means of a pseudospectral numerical method (for details, see Boeck and Thess (2001)).

First, we will describe the results for convective regimes generated by the monotonic instability (region II, Fig. 2.36). Generally, two types of stationary pattern have been found. For relatively large values of the Grashof number G, the roll pattern is observed, whereas for relatively large values of the Marangoni number M, the typical pattern is the hexagonal one (see Fig. 2.36). Note that this result corresponds to the weakly nonlinear theory for the Rayleigh-Benard-Marangoni convection in one-layer systems (Parmentier *et al.*, 1996).

Typical roll patterns are shown in Figs. 2.40 (a) and (b). One can see that rolls with different orientations can appear in the system. The selection of the orientation is connected with the wave number selection. In a finite periodic domain, the set of allowed wave vectors is discrete, and the possible values of wave numbers depend on the orientation of rolls. Specifically, the main wave

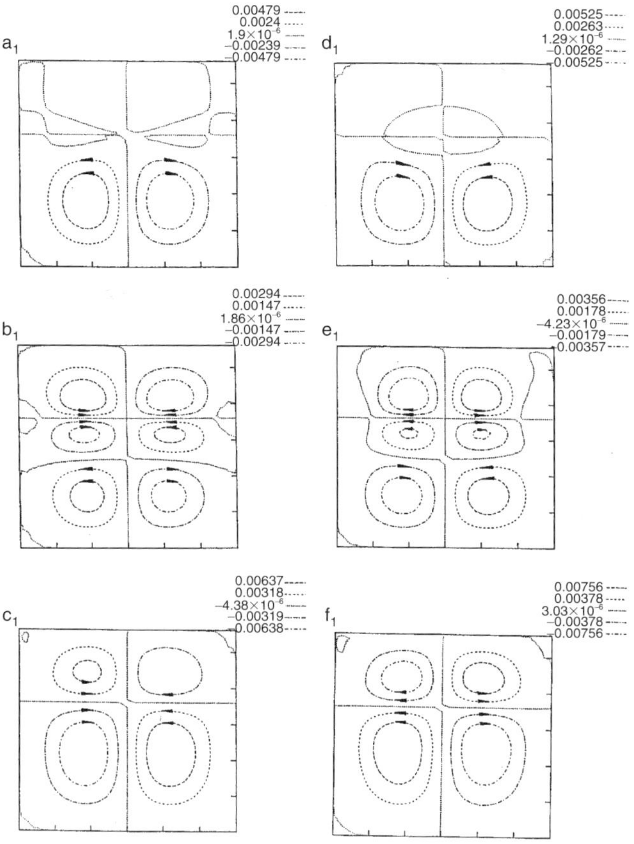

FIGURE 2.38. (a_1 - f_1) Streamlines for the oscillatory motions in the system with $a = 1.8$; $G = 18$, $M = -3933$.

vector of the pattern shown in Fig. 2.40(a) is ($k_x = 7k_b$, $k_y = 0$) (therefore, $k = |\mathbf{k}| = \left(k_x^2 + k_y^2\right)^{1/2} = 7$), whereas the main wave vector of the pattern in Fig. 2.40(b) is ($k_x = 5k_b$, $k_y = 3k_b$) ($k \approx 5.8k_b$), where $k_b = 2\pi/L_x \approx 0.251$ is the basic wave number of the computations.

Hexagonal patterns are presented in Fig. 2.40(c). In the middle of the convective cell, the interfacial temperature is lower than that at the boundary between different cells (the dashed lines correspond to negative values of the temperature perturbation θ). In contrast to the roll patterns, which were all free of defects, in some of the hexagonal patterns defects persisted even after long integration times. This was observed for $M = -400$ and $M = -500$ with $G = 20$, whereas for $M = -400$ and lower G, the pattern was purely hexagonal. Both the proximity of the II/III boundary and the stronger forcing at higher G may be responsible for the defects. An example of a pattern with persistent defects is shown in Fig. 2.40(d).

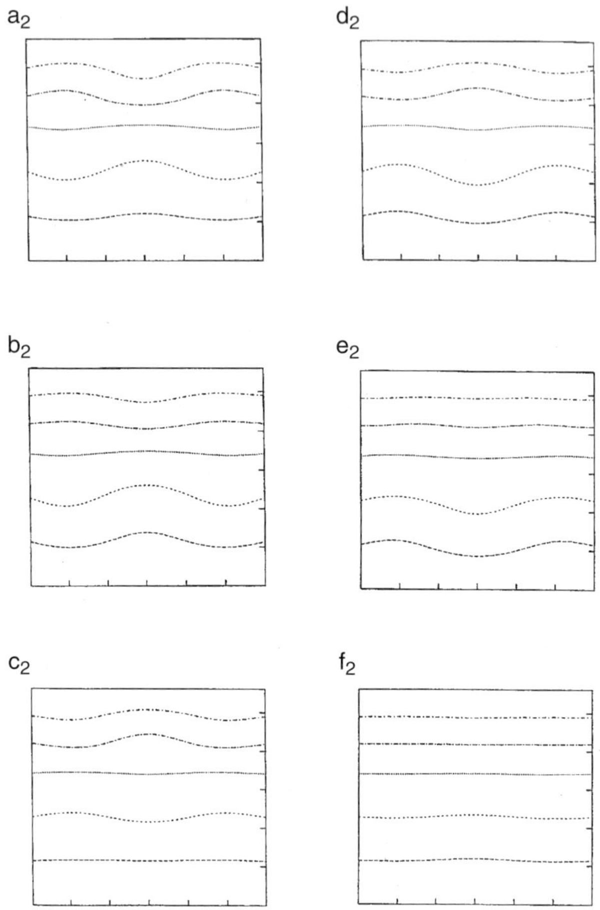

FIGURE 2.39. (a_2 - f_2) Isotherms for the oscillatory motions in the system with $a = 1.8$; $G = 18$, $M = -3933$.

The pattern consists of two coexisting stripes of hexagons and squares with some transition layers between them.

Further increase of the Marangoni number leads to the appearance of the alternating roll pattern (see Fig. 2.41), which turns out to be the typical kind of oscillatory pattern in region III. One can see that this kind of pattern is a nonlinear superposition of two systems of standing waves with orthogonal wave vectors. The temporal phase shift between standing waves of different spatial orientations is equal to $T/4$, where T is the full period of oscillations (see the corresponding time series shown in Fig. 2.42 and the phase diagrams in Figs. 2.43 and 2.44). That is why one observes some kind of roll patterns that change their orientation with the time interval $T/4$.

The orthogonal wave vectors of the alternating roll pattern shown in Fig. 2.41 are ($k_x = 7k_b$, $k_y = 0$) and ($k_x = 0$, $k_y = 7k_b$) since each of the roll patterns

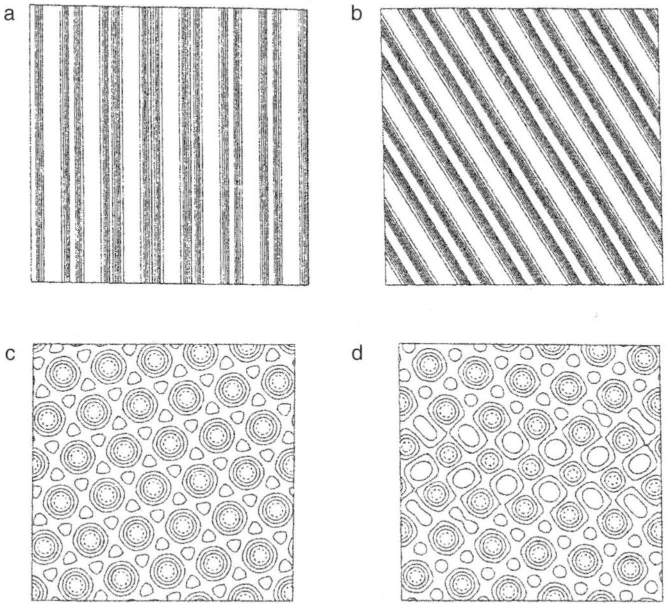

FIGURE 2.40. Contour plots of the interfacial temperature: (a) $G = 20$, $M = -200$;(b) $G = 17$, $M = -200$; (c) $G = 15.5$, $M = -400$; (d) $G = 20$, $M = -400$.

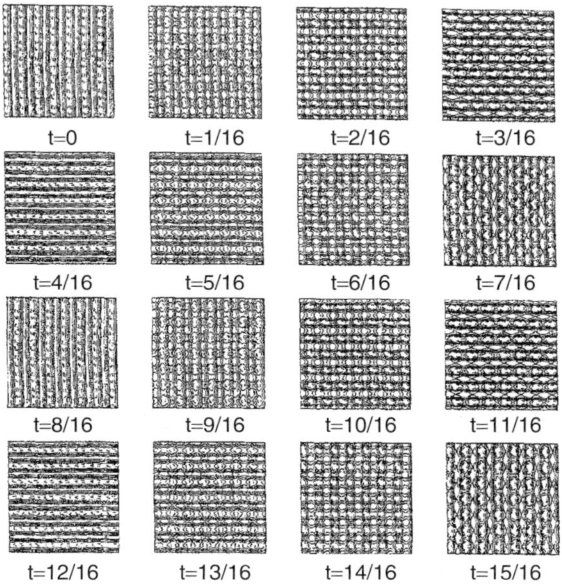

FIGURE 2.41. Snapshots of the interfacial temperature field during one period of oscillations of alternating rolls for $G = 20$ and $M = -1000$.

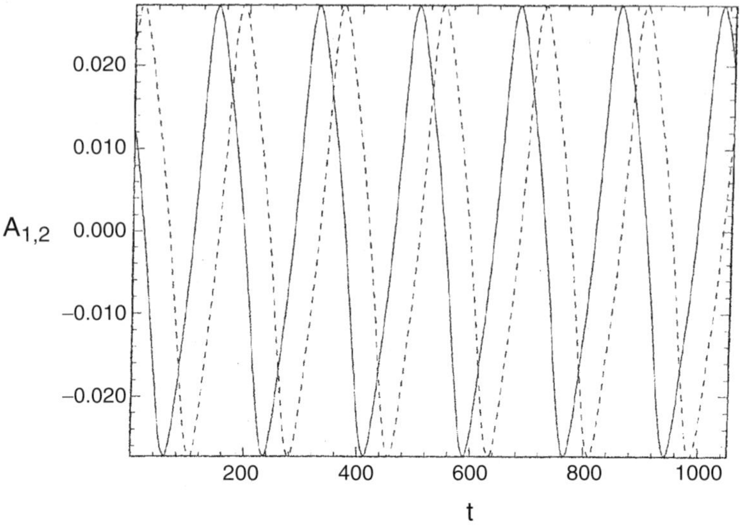

FIGURE 2.42. Time series of the two major Fourier amplitudes of the interfacial temperature field shown in Fig. 2.41; A_1 is the amplitude corresponding to $k_x = 7k_b$, $k_y = 0$; A_2 is the amplitude corresponding to $k_x = 0$, $k_y = 7k_b$ ($G = 20$, $M = -1000$).

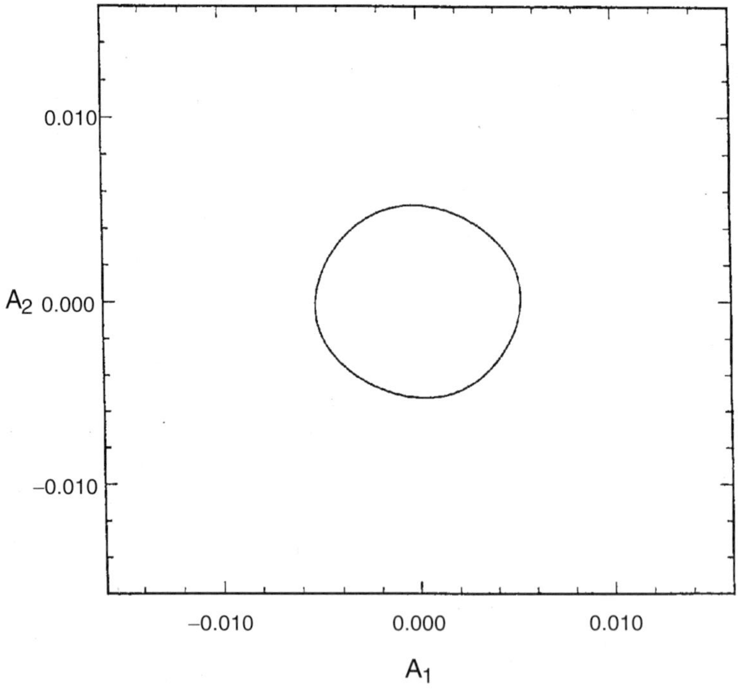

FIGURE 2.43. Phase trajectory on the plane (A_1, A_2) for $G = 15.5$, $M = -2000$. The amplitudes A_1 and A_2 have been defined in the caption of Fig. 2.42.

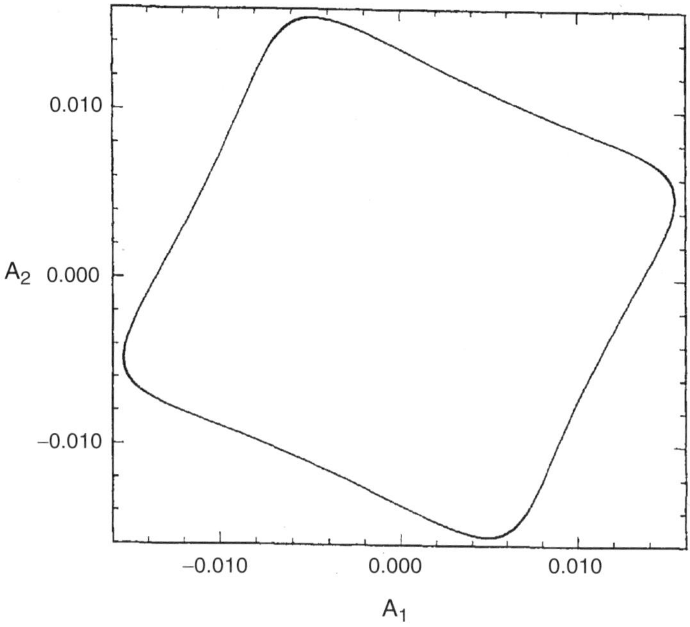

FIGURE 2.44. Phase trajectory on the plane (A_1, A_2) for $G = 20$, $M = -3000$.

appearing during the cycle consist of seven pairs of rolls. However, patterns with the wave vectors parallel to the coordinate axes are typically not attained when starting from random initial conditions.

The alternating roll patterns obtained for $G = 20$ are oblique. For example, for $M = -2000$, the wave vector of one of the standing waves has the components $(k_x = 5k_b,\ k_y = 5k_b)$; for $M = -2500$, these components are $(k_x = 4k_b,\ k_y = 6k_b)$; and for $M = -3000$, they are $(k_x = 7k_b,\ k_y = 2k_b)$. In each of these cases, the wave vector of the second standing wave is perpendicular to the first and of equal length. Strict orthogonality of the wave vectors is not always observed. For $G = 20$ and $M = -1000$, the basic wave vectors of the pattern turn out to be $(k_x = 7k_b,\ k_y = 0)$ and $(k_x = k_b,\ k_y = 7k_b)$.

As noted earlier for the rolls, the different orientations of the patterns indicate the preference of a certain wave number by the system. The patterns for $G = 20$ described above suggest that the wave number of the constitutive standing waves increases with $|M|$ because it takes values $k = \sqrt{50}\,k_b$ at $M = -2000$, $k = \sqrt{52}\,k_b$ at $M = -2500$, and $k = \sqrt{53}\,k_b$ at $M = -3000$. However, for $M = -700$, the same orientation is observed as for $M = -3000$; that is, k is initially decreasing with $|M|$ (for fixed G). This conclusion should be regarded with some caution since it is based merely on a single simulation at each point (G, M). In any case, the numerical value of k of about $7k_b \approx 1.76$ is in good agreement with linear stability results obtained by Braverman et al. (2000). Linear theory predicts not only a critical wave number $k_c \approx 1.8$ for $|M|/G \approx 50$ but also that k_c grows with $|M|/G$ (cf. Fig. 2.37).

As can be expected, the flow amplitude in each layer grows with G and $|M|$. The flow in the bottom layer is stronger since both the Nusselt and the Peclet numbers in the bottom layer are larger than those in the top layer. The oscillation period T decreases as $|M|$ grows (for fixed G). This observation again coincides with the predictions of the linear stability theory for the critical mode. The period of the critical mode is a decreasing function of $|M|/G$.

Naturally, a finite flow amplitude causes anharmonicity of the oscillations. This effect is obvious from the two phase diagrams of Figs. 2.43 and 2.44, which are constructed from the two major Fourier amplitudes A_1 and A_2 explained in the caption of Fig. 2.42. Near the I/III boundary ($G = 15.5$, $M = -2000$), the flow amplitude is weak and the phase diagram (Fig. 2.43) is almost circular. It is significantly deformed for $G = 20$ and $M = -3000$ (Fig. 2.44).

The transition from steady to oscillatory convection as M and G are varied has been investigated by means of additional simulations. They were again performed with the large aspect ratio $L_x = L_y = 25$ for $G = 20$ and Marangoni numbers in the vicinity of the boundary between regions II and III in Fig. 2.36.

Typically, spatial irregularity and irregular timedependence of the patterns persist, but there are exceptions. Fig. 2.45 shows the final state attained in a simulation for $M = -600$, which was started from random initial conditions. In this case, one observes a competition, rather than a coexistence, of rolls with different orientation, which leads to their spatial separation. The pattern consists of two domains of rolls with different orientations separated by domain walls that look like strips of squares. The square cells are "absorbed" on one end of the roll and "emitted" on another one; thus, the domain walls move as a whole and oscillate. Remarkably, integral quantities such as the kinetic energy of each layer and the Nusselt numbers are constant in this case.

The alternating roll pattern is also a stable solution for $M = -600$. We have verified the multistability of patterns by two additional simulations with hexagons and alternating rolls as initial conditions.

Conversely, for $M = -500$, hexagons are observed when a simulation is started with alternating rolls as the initial condition. They are also obtained from random initial conditions (with defects). For the intermediate value $M = -550$, more

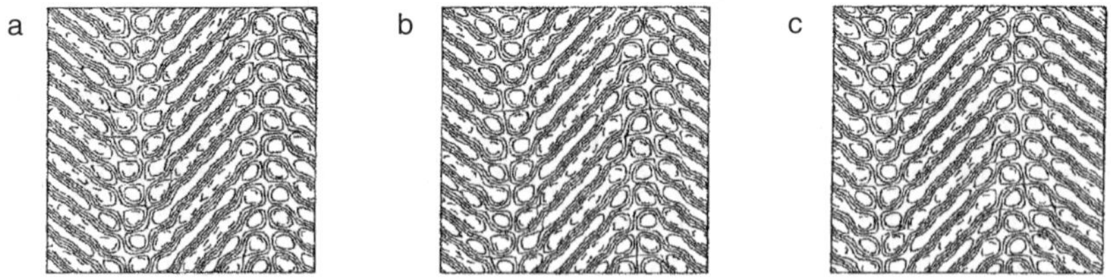

FIGURE 2.45. Snapshots of the interfacial temperature field for the modulated alternating rolls obtained for $G = 20$, $M = -600$. The period between the snapshots is $\Delta t = 12.6$.

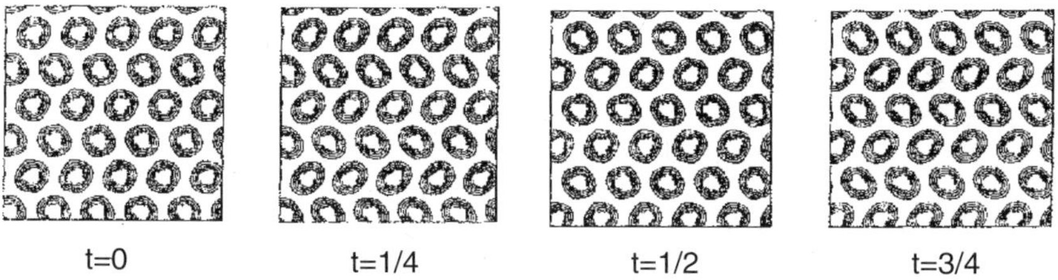

FIGURE 2.46. Snapshots of the interfacial temperature field during one period of oscillations for $G = 20$, $M = -550$ (oscillating hexagons).

complicated solutions are found. Starting from stationary hexagons, we observe periodic pulsations of the hexagonal pattern (see Fig. 2.46). The hexagonal structure is conserved, but some periodic changes of the isotherms' shapes are evident. They are caused by sinusoidal oscillations of the amplitudes of the wave vectors ($k_x = 5k_b, k_y = \pm 6k_b$).

In addition to the oscillation of these modes, the amplitudes of the basic modes of the hexagonal pattern also oscillate slightly about a nonzero average. This regular pattern was not observed when a simulation was started from random initial conditions. In this case, an irregular hexagonal pattern with chaotic time dependence persists. It is very similar to the case of $M = -570$, for which Fig. 2.47 shows a snapshot of the interfacial temperature and the chaotic evolution of a Fourier amplitude.

2.5.3 Mode Mixing of Interfacial and Internal Waves

In this subsection, we consider the stability of the mechanical equilibrium state under the combined action of the thermocapillary effect and the buoyancy effect by heating *from above*. It will be shown that in the case where both the buoyancy and the surface tension gradients are expected to play a stabilizing role, their combined action may lead to a specific oscillatory instability. This instability is caused by the resonant interaction and mode mixing for two physically different types of *decaying* wave: longitudinal Marangoni waves and internal waves.

Until now, the investigation of the above-mentioned instability mechanism was done in a one-layer approach (Rednikov *et al.*, 2000). However, there is no doubt that this type of instability can be important in two-layer systems. That is why it is quite reasonable to describe it here. The definitions of the Marangoni number M and of the Rayleigh number R used in this subsection are as follows: $M = \alpha A a^2/\eta\chi$ and $R = g\beta A a^4/\nu\chi$, where $\alpha = -d\sigma/dT$ is the temperature coefficient of the surface tension, η is the dynamic viscosity, ν is the kinematic viscosity, χ is the thermal diffusivity, g is the acceleration of gravity, β is the heat expansion coefficient, a is the thickness of the layer, and A is the temperature gradient. The parameters M and R are positive, which corresponds to the normal thermocapillary effect and the positive heat expansion coefficient. In this subsection, we consider the case of heating from above ($A > 0$).

FIGURE 2.47. Snapshot of the interfacial temperature for chaotic hexagons. The parameters are $G = 20$, $M = -570$.

Let us describe the physical meaning of each type of the mixed waves. The first type is the *longitudinal* waves, which are discussed in Sec. 2.5.1. Recall that in the limit of large M and k, the nondimensional frequency ω is determined by the relation

$$\omega^2 = \omega_{long}^2 = \frac{Ma}{P(\sqrt{P}+1)}k^2. \tag{2.54}$$

Another type of oscillations is produced by the buoyancy. Assume $M = 0$. By heating from above, there exists a density gradient which generates *internal waves* with the Brunt-Väisälä frequency

$$\omega^2 = \omega_{int}^2 = \frac{Rak^2}{P(k^2+k_z^2)}, \tag{2.55}$$

where k is the horizontal component of the wave vector and k_z is the vertical component of the wave vector (Landau and Lifshitz, 1987). The nondeformability of the surface and of the rigid bottom leads to a quantization of the vertical wave number: $k_z = n\pi$, $n = 1, 2, \ldots$. Because of the viscosity and thermal diffusivity, the internal waves decay.

Let us consider now the case of the combined action of the thermocapillary stresses and buoyancy, where both M and R are not equal to zero. Let us define the dynamic Bond number $Bd = R/M$. The time behavior of disturbances is governed by a general dispersion relation that contains M and R and describes both branches of waves: Marangoni longitudinal waves and internal waves. However, if the frequencies ω_{sf} and ω_{int} given by expressions (2.54) and (2.55) are essentially different from each other, then both kinds of waves are clearly distinguishable, and both of them decay. The situation is different near the points of *resonance* where $\omega_{dil} = \omega_{int}$. The resonance between both kinds of waves takes place for the values of the Bond number

$$Bd_n = \frac{k^2 + \pi^2 n^2}{P^{1/2} + 1}, \quad n = 1, 2, \ldots. \tag{2.56}$$

The asymptotic analysis performed by Rednikov *et al.* (2000) in the limit $\epsilon = (P/R)^{1/4} \ll 1$ have shown that near the resonance point, there is a continuous transition by changing $Bd - Bd_n$ from the *dilational* mode to the *internal* one, and vice versa; thus, there is a *mode mixing*. Moreover, the analysis of the dispersion relation near the point $Bd = Bd_n$ shows that one of the eigenvalues (corresponding to a complex frequency) has a positive imaginary part and another one has a negative imaginary part. Thus, near the resonance point, one of the modes is unstable while another one is stable.

The results of the asymptotic analysis have been verified by the direct numerical analysis of the full linear stability problem (Rednikov *et al.*, 2000). The stability diagram in the plane (M, R) obtained by minimization of neutral curves with respect to wave numbers is shown in Fig. 2.48. Fig. 2.48(a) corresponds to the case $n = 1$. For a given Bd, the intersection of the straight line $R = BdM$ with the line 1 of Fig. 2.48(a) yields the critical temperature gradient for the onset of instability. The lower branch has an oblique asymptote with slope 0.715; thus, there is no instability as $Bd < 0.715$. The critical wave number tends to zero along the lower branch and to infinity along the upper branch. Fig. 2.48(b) also contains the results for $n = 2$ and $n = 3$.

The weakly nonlinear analysis have been performed for several values of P and Bd by Rednikov *et al.* (2000). In all of the cases, a standing wave was found to be stable, and traveling waves turned out to be unstable.

It should be noted that a certain kind of mixing of interfacial Marangoni waves and internal waves was observed in experiments by Wierschem *et al.* (2000). In these experiments, the absorption of pentane in a liquid toluene out of the vapor phase was studied. The Marangoni waves appeared due to the dependence of the surface tension on the pentane concentration (the concentration/capillary effect), and the internal waves were caused by the dependence of the density on the pentane concentration. However, the theory described above is not directly applicable to the experimental situation. The theory was based on the assumption that the density gradient was constant, whereas in the experiment, there was a sharp interface inside the liquid layer, which separated the upper (pentanerich) and the lower (pentanepoor) sublayers. Indeed, it was found by measurements that the dispersion relation for the internal waves was quite similar to that of a gravity wave on the interface between two liquids with different densities. To our knowledge, the mixing between the interfacial and internal waves was never studied theoretically in that limit.

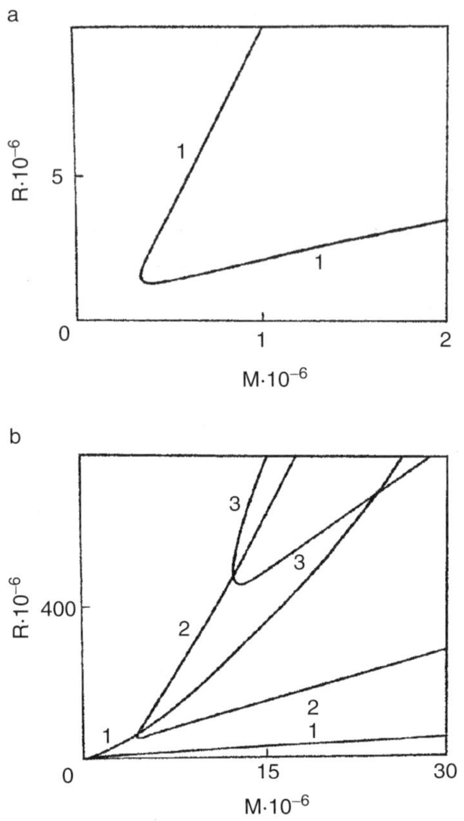

FIGURE 2.48. Stability boundary at $P = 6$ and $Bi = 0$. (a) The fundamental mode ($n = 1$); (b) modes $n = 1$ (line 1), $n = 2$ (line 2) and $n = 3$ (line 3).

2.5.4 *Oscillatory Instability in the Presence of a Thermal Gradient and a Surfactant*

In this subsection, we discuss the influence of surface-active agents on the thermal convection in systems with an interface. It is well known that if the amount of a surfactant absorbed at the interface is sufficiently large, it forms a "rigid" film and suppresses completely any motion at the interface (Levich, 1962). However, if the surface concentration of the surfactant Γ is small, its molecules form a "surface gas". The transfer of the surfactant along the interface is governed by the equation (Levich, 1962)

$$\frac{\partial \Gamma}{\partial t} + \nabla_{\perp} \cdot (\mathbf{v}_{\perp} \Gamma) = D_0 \Delta_{\perp} \Gamma + j_n. \tag{2.57}$$

Here, $\mathbf{v}_{\perp} = \mathbf{v} - v_n \mathbf{n}$ is the tangential component of the fluid velocity at the interface, $\nabla_{\perp} = \nabla - \mathbf{n}(\mathbf{n} \cdot \nabla)$, $\Delta_{\perp} = \nabla_{\perp}^2$, D_0 is the surface diffusion coefficient, and $j_n = \mathbf{j} \cdot \mathbf{n}$ is the mass flux of the surfactant from the bulk to the interface.

Typically, the presence of the surfactant diminishes the surface tension and that may lead to the suppression of stationary convective flows and to the generation of a specific kind of oscillations (Berg and Acrivos, 1965; Palmer and Berg, 1972). This effect can be explained in the following manner. Let us consider a flow in the bulk that generates at the interface a velocity field with a nonvanishing interfacial divergence $q = \nabla_{\perp} \cdot \mathbf{v}_{\perp}$. Assume that the surface diffusion coefficient is small, so that the diffusion of the surfactant is negligible compared to the advection of the surfactant by the flow, and the surfactant is insoluble ($j_n = 0$). Assume also that the initial distribution of the surfactant is homogeneous. Obviously, because of the advection of the surfactant, the surface concentration decreases in the region of a divergent interfacial flow where $q > 0$, and it increases in the region of a convergent flow where $q < 0$. The tangential stresses generated by the inhomogeneity of the surfactant concentration are directed opposite to the fluid motion. Thus, they will prevent the development of a monotonic instability and suppress stationary flows. At the same time, such a "negative feedback" can lead to an oscillatory instability ("overstability").

Note that the arguments given above do not depend on the physical nature of the flow. Thus, we can expect that for both buoyancy-driven convection and the surface-tension-driven convection, the monotonic instability will be suppressed and replaced by the oscillatory instability.

The quantitative analysis of the influence of surfactants on the convective stability in two-layer systems was given by Nepomnyashchy and Simanovskii (1986, 1988, 1989a, 1989b) and Gilev *et al.* (1986).

In the case of Marangoni instability (Nepomnyashchy and Simanovskii, 1986), the linear stability of the mechanical equilibrium state with the constant surface concentration Γ_0 of the insoluble surfactant is governed by the eigenvalue problem (2.3) – (2.6), (2.11), (2.12) with the modified boundary condition for tangential stresses at the interface:

$$z = 0: \quad \eta \psi_1'' - ik[(\eta M/P)T_1 + B\Gamma] - \psi_2'' = 0 \tag{2.58}$$

and with the linearized equation (2.57)

$$(\lambda + D_s k^2)\Gamma = -ik\psi_1'(0), \tag{2.59}$$

where

$$B = \frac{\alpha_s \Gamma_0 a_1}{\eta_1 \nu_1}, \quad \alpha_s = -\left(\frac{\partial \sigma}{\partial \Gamma}\right)_{\Gamma=\Gamma_0}; \quad D_s = \frac{D_0}{\nu_1}.$$

Eliminating Γ, we get the following boundary condition:

$$z = 0: \quad \eta \psi_1'' - ik\left[(\eta M/P)T_1 - \frac{ikB}{\lambda + D_s k^2}\psi_1'\right] = \psi_2''. \tag{2.60}$$

For the monotonic instability, one obtains the following analytical expression for the neutral curve $sM = sM_m(k)$:

$$sM = sM_m(k) = 8k^2 \frac{(1+\kappa a)(\kappa D_1 + D_2)(\eta B_1 + B_2 + B/2kD_s)}{\eta\kappa(\chi E_2 - E_1)}. \tag{2.61}$$

One can see that the presence of surfactant always leads to the increase of the monotonic instability threshold proportional to B/D_s, which is large, as D_s is small.

It can be shown that for arbitrary small B, an oscillatory neutral curve appears in the long-wave region ($0 < k < k_*(B)$). In that region, this curve is always situated lower than the monotonic one. It is interesting that the long-wave asymptotics of the oscillatory neutral curve does not depend on B in the leading order, and it coincides exactly with the long-wave asymptotics of expression (2.61) taken with $B = 0$:

$$sM = sM_o = -\frac{1}{1-\chi a^2}\frac{80(1+\kappa a)^2(1+\eta a)}{\eta a^2 \kappa}k^{-2}. \tag{2.62}$$

Thus, in the long-wave region, the oscillatory instability just replaces the monotonic one. The long-wave asymptotics of the frequency of oscillations ω is

$$\omega \sim \pm\omega^{(1)}k,$$

$$\begin{aligned}\omega^{(1)} &= \pm B^{1/2} a \left\{ \frac{2}{15}(\eta a^2 + \nu^3) + \frac{1}{315}\frac{1+a\eta}{1-\chi a^2} \right.\\ &\times \left[10P \frac{11(1-\kappa\chi^2 a^5) + 53a(\kappa - \chi^2 a^3) + 42\chi a^2(1-\kappa a)}{1+\kappa a} \right.\\ &\left.\left. +19(1-\nu\chi a^4) \right]\right\}. \end{aligned} \tag{2.63}$$

The interval of the oscillatory instability grows with B, and eventually the oscillatory instability becomes the most "dangerous" one.

Let us discuss now the case of Rayleigh instability (Gilev *et al.*, 1986). For small values of B, the behavior of the oscillatory neutral curve is quite similar to that in the case of the Marangoni instability. The oscillations appear in the long-wave region $0 < k < k_*(B)$, $k(B)$ grows with B, and the oscillatory instability becomes the most "dangerous" one when B exceeds a certain value B_1. However, for large values of B, when the surfactant film becomes "rigid" (the boundary condition (2.60) is $\psi_1' = 0$ to the leading order), there is an essential difference between the Marangoni and the Rayleigh instabilities. In contradistinction to the monotonic Marangoni instability, the monotonic Rayleigh instability does not disappear when the boundary between two layers is rigid. The oscillatory instability, which is essentially connected with the fluid motion at the interface, cannot exist in the limit of large B. That is why as B increases, $k(B)$ eventually begins to

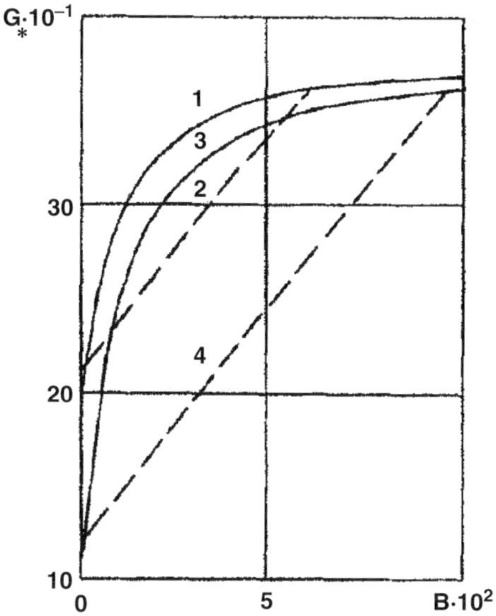

FIGURE 2.49. Dependences of the minimized critical Grashof number on parameter B for $M = 0$ (lines 1 and 2) and $M = 420$ (lines 3 and 4).

decrease, and for some $B > B_2$, the monotonic mode again becomes the most dangerous (see Fig. 2.49).

The transition from the monotonic to the oscillatory instability in the presence of a surface-active agent takes place for a certain $B = B_1$ also in the case of the anticonvective instability by heating from above (Simanovskii and Nepomnyashchy, 1993). Because the monotonic anticonvective instability is essentially connected with the hydrodynamic interaction of two fluid layers and cannot exist when the boundary between fluid layers is rigid, it is not restored at large B, similar to the case of the Marangoni convection.

The influence of the solubility of the surfactant was investigated by Nepomnyashchy and Simanovskii (1988, 1989a).

2.6 Marangoni Waves in Systems with a Deformable Interface

In this section, we consider oscillatory instabilities in two-layer systems in the case where the deformation of the interface is essential. This kind of instabilities can be called *transverse* instability. The problem is governed by the system of equations and boundary conditions (2.3) – (2.10). As it was discussed in Sec. 1.2.3, the Grashof number G should be taken equal to zero, while the Galileo number Ga can be non-zero.

2.6.1 The Transverse Marangoni Instability in One-Layer Systems

Infinitely Deep Layer

In order to understand the nature of the transverse Marangoni instability, we consider first an infinitely deep layer of the fluid heated from above (Levchenko and Chernyakov, 1981; Garcia-Ybarra and Velarde, 1987); the temperature gradient is equal to A. In this case, we cannot use the thickness of the layer as the length scale. Instead, we will use the *capillary length* $l = (\sigma_0/\rho g)^{1/2}$ and define

$$M = \frac{\alpha A l^2}{\eta \chi} \quad \text{(Marangoni number)}, \tag{2.64}$$

and

$$Ga = \frac{g l^3}{\nu^2} \quad \text{(Galileo number)}. \tag{2.65}$$

As the unit of time, we take l^2/ν.

Two kinds of wavy motion are possible in such a system. The first kind is the *longitudinal*, or dilational, wave, which was discussed in subsection 2.5.1. This kind of waves is caused by the thermocapillary effect. The corresponding dispersion relation is

$$\omega^2 = \omega_{long}^2 = \frac{M}{P(\sqrt{P}+1)} k^2. \tag{2.66}$$

Another type of wave that is possible in a system with a deformable surface is the *transverse*, or capillary-gravity wave. It is caused by the joint action of the gravity and the surface tension. The corresponding dispersion relation is

$$\omega^2 = \omega_{tr}^2 = Ga(k + k^3). \tag{2.67}$$

In the high-frequency limit, when the decay of waves is neglected, the full dispersion relation of the system is just

$$(\omega^2 - \omega_{long}^2)(\omega^2 - \omega_{tr}^2) = 0. \tag{2.68}$$

However, the next order corrections to the dispersion relation (2.68) describe a certain *mixing* between two kinds of wave (cf. subsection 2.5.3), which is especially strong in the resonant case where the frequencies of both waves are close. Also, if the next order corrections are taken into account, any solution $\omega(k)$ has a certain *imaginary part* that corresponds to decay or growth of waves in time. The neutral curve is

$$M = 2^{5/2}(GaP^2)^{3/4} \frac{(1+k^2)^{3/4}}{k^{1/4}}. \tag{2.69}$$

It has the minimum as

$$k = k_c = \sqrt{5}/5. \tag{2.70}$$

Stress-Free Lower Boundary

Another analytically tractable case is the case of the transverse Marangoni instability in a fluid layer of a finite thickness a with a *stress-free* lower boundary (see

Garazo and Velarde, 1991, 1992). Now, we will use a and a^2/ν as the units of length and time, respectively, and define the Marangoni number and the modified Galileo number as $M = \alpha A a^2/\eta\chi$ and $Ga = ga^3/\nu^2$, respectively.

In this case, the minimum of the neutral stability curve is situated at $k = 0$. The problem under consideration belongs to the class of the problems where the growth of a spatially homogeneous disturbances is precluded by a *conservation law*. Indeed, the homogeneous change of the layer's thickness is impossible because of the conservation of the fluid volume. In this case, there exists the *Goldstone mode* with the growth rate $\lambda(k, M) = \lambda_r(k, M) + i\lambda_i(k, M)$, which satisfies the condition

$$\lambda(0, M) = 0. \tag{2.71}$$

This mode generates an instability when

$$\frac{\partial^2 \lambda_r}{\partial k^2}(0, M)$$

becomes positive, which happens as $M > M_c = 12$. Near the threshold point $k = 0$, $M = M_c$, the growth rate can be expanded into the Taylor series:

$$\lambda_r(k, M) = \lambda_{21} k^2 (M - M_c) + \lambda_{40} k^4 + \cdots, \tag{2.72}$$

$$\lambda_i(k, M) = \lambda_{10} k + \lambda_{11} k (M - M_c) + \lambda_{30} k^3 + \cdots, \tag{2.73}$$

where

$$\lambda_{21} = \frac{1}{6}, \quad \lambda_{40} = -\frac{2}{105}(17GP^2 + 204P + 134 + 22P^{-1}),$$

$$\lambda_{10} = \sqrt{\frac{GP + 12}{P}}, \quad \lambda_{11} = \frac{1}{2\sqrt{(GP + 12)P}},$$

$$\lambda_{30} = \frac{C^{-1} - GP(8P/5 + 1/3) - (96P/5 + 56/5)}{2\sqrt{(GP + 12)P}}.$$

The instability interval is $0 < k < k_m$, where

$$k_m = \sqrt{\frac{\lambda_{21}(M - M_c)}{-\lambda_{40}}}.$$

Solid Lower Boundary

In the case of a layer with a finite thickness a and the solid lower boundary, the existence of transverse Marangoni waves by heating from the gas side was justified by Takashima (1981). A remarkable phenomenon found by the numerical solution of the eigenvalue problem is the nonmonotonic behavior of the critical Marangoni number M_c, and the nonsmooth behavior of the critical wave number k_c and the critical frequency ω_c as functions of the crispation number $C = \eta\chi/\sigma_0 a$. Note that the appearance of the oscillatory instability was investigated for both positive and negative values of the static Bond number $Bo = \rho_0 g a^2/\sigma_0$ (i.e., for a liquid resting on solid support or hanging from it). Some additional data concerning the transverse Marangoni waves can be found in the paper of Zimmerman (2001).

2.6.2 The Limit of Large *Ga* and *M*

We will start now the description of oscillatory instabilities in two-layer systems with a deformable interface. The present subsection is devoted to the analytically tractable limit of large Ga, W_0 and M (Rednikov *et al.*, 1998). It will be assumed that $m = M/Ga = O(1)$ and $Bo = Ga/W_0 = O(1)$. In this limit, the frequencies of both modes, the longitudinal one and the capillary-gravity one, are large (except the case of small wave numbers), because $\omega_{long} \sim M^{1/2}$, and $\omega_{cg} \sim Ga^{1/2}$. Because of that, there appear boundary layers ("skinlayers") at the interface and near the rigid boundaries, whereas the bulk of each layer can be well approximated by inviscid flow. The thickness of boundary layers is proportional to $\omega^{-1/2} \sim Ga^{-1/4}$. Thus,

$$\epsilon \equiv Ga^{-1/4} \tag{2.74}$$

is the natural small parameter of the problem. We will expand all of the variables that appear in the boundary value problem (2.3) – (2.10) as a power series in ϵ and take $\omega = Ga^{1/2}\Omega$. For the description of the fields of variables in boundary layers, we will define stretched coordinates $Z_0 = z/\epsilon$ in the interfacial boundary layer, $Z_1 = (z-1)/\epsilon$ in the top boundary layer, and $Z_2 = (z+a)/\epsilon$ in the bottom boundary layer.

In the leading order, the longitudinal and transverse interfacial modes are nondissipative and nonmixed. The dispersion relation has the following form:

$$(\Omega^{(0)2} - \Omega^2_{long})(\Omega^{(0)2} - \Omega^2_{tr}) = 0, \tag{2.75}$$

where

$$\Omega^2_{long} = \frac{sm(\sqrt{\nu} - \sqrt{\chi})\kappa\eta k^2}{P(1+\sqrt{P})(\sqrt{\nu}+\sqrt{\chi P})(\eta+\sqrt{\nu})(\kappa+\sqrt{\chi})(1+\kappa a)} \tag{2.76}$$

(see (2.51)) and

$$\Omega^2_{tr} = \frac{\left(\rho^{-1}-1\right)k + B^{-1}k^3}{\coth k + \rho^{-1}\coth ka}. \tag{2.77}$$

Later, we assume that the condition of the existence of longitudinal waves for arbitrary large M,

$$s(\sqrt{\nu} - \sqrt{\chi}) > 0, \tag{2.78}$$

is satisfied. In this case, $\Omega^2_{long} = \Omega^2_{tr}$ when $m = m_{res}$,

$$m_{res} = s\Omega_{tr}^2 \frac{P(1+\sqrt{P})(\sqrt{\nu}+\sqrt{\chi P})(\eta+\sqrt{\nu})(\kappa+\sqrt{\chi})(1+\kappa a)}{(\sqrt{\nu}-\sqrt{\chi})\kappa\eta k^2}. \tag{2.79}$$

In the next order, one obtains complex corrections to the modes' frequencies, which correspond to the waves' decay or growth as well as to the shift of frequencies. If m is not too close to $m = m_{res}$, the growth rate $\lambda^{(1)}$ (the imaginary part of the above-mentioned correction) satisfies the following equations:

$$\begin{aligned} &2\lambda^{(1)}\left(1-\frac{m_{res}}{m}\right)(\coth k+\rho^{-1}\coth ka) \\ &+\ k\sqrt{\frac{\Omega_{long}^{(0)}}{2}}\left[\frac{1}{1+\rho\sqrt{\nu}}+\frac{\sqrt{\nu}+\sqrt{\chi P}}{\rho\sqrt{\nu P}(\sqrt{\nu}-\sqrt{\chi})}\right] \\ &\times[(\rho-1)(1+\nu^{1/2})\coth k\,\coth ka \\ &+\frac{m_{res}}{m}(\coth ka-\sqrt{\nu}\coth k)(\coth ka+\rho\coth k)]=0 \end{aligned} \tag{2.80}$$

for the longitudinal mode and

$$\begin{aligned} &2\lambda^{(1)}\left(1-\frac{m}{m_{res}}\right)(\coth k+\rho^{-1}\coth ka)+k\sqrt{\frac{\Omega_{tr}^{(0)}}{2}}\left(1-\frac{m}{m_{res}}\right) \\ &\times[\coth k(\coth k+\coth ka)+\csc^2 k+1/(\rho\sqrt{\nu})\csc^2 ka] \\ &+k\sqrt{\frac{\Omega_{tr}^{(0)}}{2}}\left[1+\frac{m}{m_{res}}\cdot\frac{(1+\rho\sqrt{\nu})(\sqrt{\nu}+\sqrt{\chi P})}{\sqrt{P}(\sqrt{\nu}-\sqrt{\chi})}\right] \\ &\times\frac{(\coth ka-\rho\sqrt{\nu}\coth k)(\coth k+\coth ka)}{1+\rho\sqrt{\nu}}=0 \end{aligned} \tag{2.81}$$

for the transverse mode. Formulas (2.80) and (2.81) determine the regions of the excitation and the decay for both modes of waves. Note that formula (2.80) is correct only if $\Omega_{long}^2 > 0$ (i.e., in the region where the signs of s and $\sqrt{\nu}-\sqrt{\chi}$ coincide). Also, the derived expressions for $\lambda^{(1)}$ are correct if $k \gg G^{-1/2} \sim M^{-1/2}$, because by their derivation it was assumed that the thicknesses of the boundary layers are small with respect to the thicknesses of the fluid layers.

Expressions (2.80) and (2.81) diverge as $m \to m_{res}$. In that limit, the *mixing* of both types of mode should be taken into account. The details can be found in the paper of Rednikov *et al.* (1998).

2.6.3 Linear Theory of Transverse Instability: Numerical Results

The asymptotic theory described in the previous subsection presents the conditions for the existence of the oscillatory instability in the limit of large Ga and M.

However, it does not provide any criteria for the appearance of instability at finite values of Ga and M. In order to obtain neutral stability curves, it is necessary to solve numerically the boundary value problem (2.3) – (2.10). In this subsection, we will present some examples that show that the asymptotic theory is not always sufficient for the prediction of the oscillatory instability. Specifically, it is quite natural to expect that the influence of the deformation will be especially strong in the long-wave region $k \sim Ga^{-1/2}$ (see section 2.4.1), where the asymptotic theory described above is not valid.

Air/water system by heating from above First, let us consider the appearance of the oscillatory instability in the air/water system (the system of parameters is presented in Table 1) with $a = 1$. For this system on the Earth, the parameters W and Ga are not independent and are connected with each other by the relation $Ga = (W/\gamma)^3$, where the constant $\gamma = (\sigma_0/\rho_1)(g\nu_1^4)^{-1/3}$ depends on the physical properties of the fluids only. However, we will consider Ga and W as independent parameters, with the gravity acceleration g reduced due to the microgravity conditions. We choose $W = 10^6$ (which corresponds to the thickness of each layer about 3 mm) and vary the parameter Ga.

Because the Prandtl number of air is smaller than that of water, the asymptotic theory predicts the possibility of longitudinal oscillations by heating from above. Moreover, expression (2.80) predicts the appearance of the longitudinal oscillations for arbitrary small $m = M/Ga$ (i.e., in the limit $Ga \to \infty$) (see criterion (2.53)). However, this prediction is physically meaningless because the dimensional temperature difference is too large. It is necessary to take into account that the variation of the surface tension due to the temperature variation $\alpha|\theta|$ must be smaller than the mean value of the surface tension σ_0 (i.e., $M < W_0 P$). It turns out that this relation is satisfied only if $Ga < 20$, which corresponds to microgravity conditions. Oscillatory neutral curves and the dependences of frequency on the wave number are shown in Fig. 2.50 and Fig. 2.51, respectively.

Model system by heating from below Let us consider now the case $\eta = \nu = 0.5$, $\chi = \kappa = P = a = s = 1$, where an oscillatory longitudinal instability was found in the limit $Ga \to \infty$ despite the violation of the criterion $s(\sqrt{\nu} - \sqrt{\chi} > 0)$. Because the influence of the interface deformation is expected to be especially strong in the case of fluids with nearly equal densities, let us choose $\rho = 0.999$. Also, we take $W = 1$.

At finite values of Ga, the monotonic instability mode appears, according to equation (2.41). The minimum of the monotonic neutral curve is situated at $k = 0$, $M(0) = 4 \times 10^{-3}$. Both monotonic and oscillatory neutral curves are shown in Fig. 2.52. The dependence of the oscillation frequency on the wave number is presented in Fig. 2.53. One can see that the oscillatory instability disappears in the long-wave region. As $Ga \leq 5 \times 10^6$, the monotonic disturbances become more dangerous than the oscillatory ones.

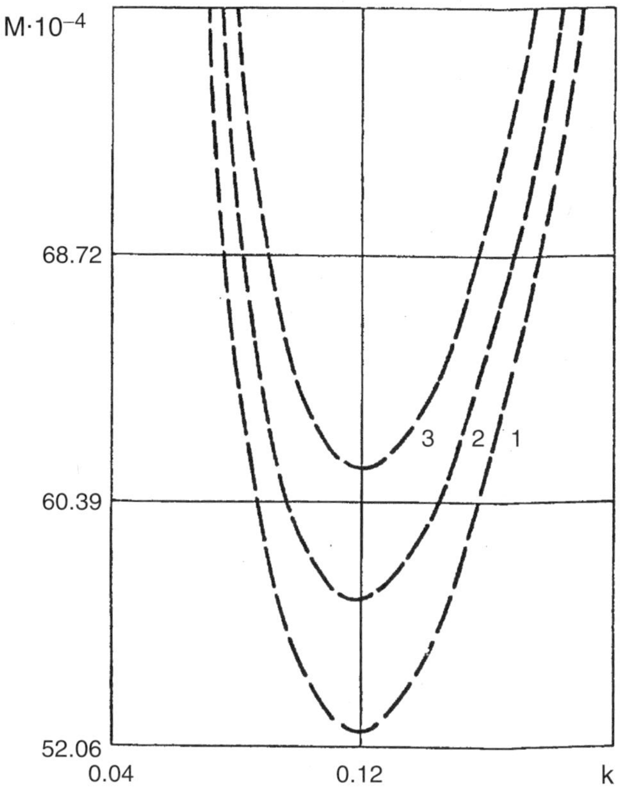

FIGURE 2.50. The neutral curves for the air/water system heated from above. $Ga = 0$ (line 1); 3 (line 2); 6 (line 3).

Model system by heating from above If the same model system is heated from above, the asymptotic theory predicts the appearance of transverse oscillatory instability characterized by the relation $\omega^2 \sim Ga$. However, the onset of instability is caused by another type of disturbances (Nepomnyashchy and Simanovskii, 1991; 1995). The neutral curve looks like a "sack" (Fig. 2.54(a)); (that is, the instability exists in the long-wave region $k < k_m(Ga)$ in a certain interval $M_-(k) < M < M_+(k)$. The dependence of the frequency of oscillations ω on the wave number k for both branches of the neutral curve is shown in Fig. 2.54(b). As $k \to 0$, the upper and the lower branches of the neutral curve have asymptotic limits $M \to C_\pm k^{-2}$ and $\omega \to \omega_\pm$. Let us emphasize that the values $C_\pm$ and $\omega_\pm$ *do not depend* on Ga. The width of the instability interval k_m decreases monotonically with increasing Ga, and the threshold value M tends to infinity as $Ga \to \infty$. This result proves that the instability is caused by the interface deformation.

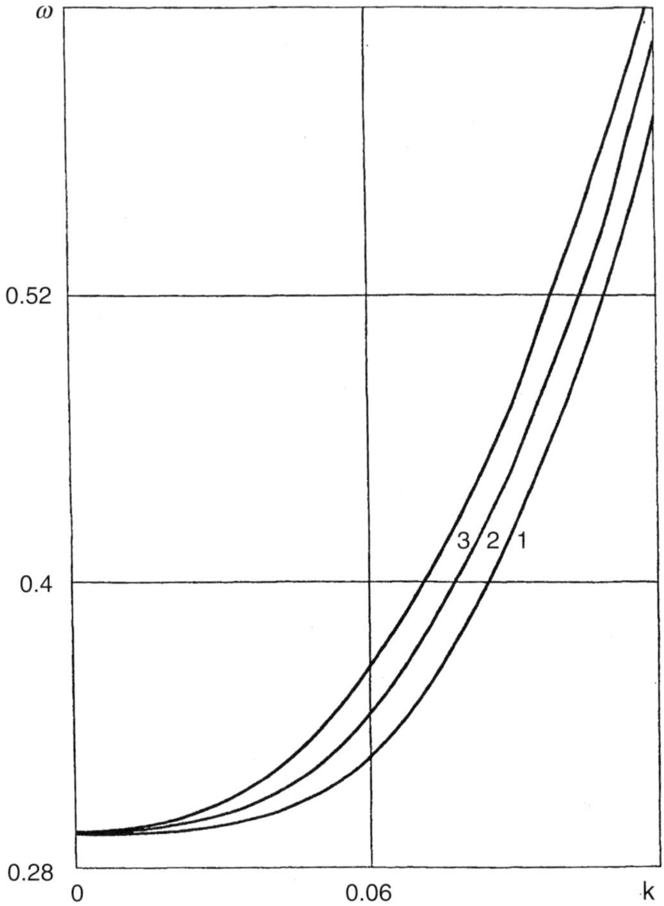

FIGURE 2.51. Dependence of the frequency on the wave number (the notation is the same as in Fig. 2.50).

2.6.4 Nonlinear Theory of Transverse Instability

The transverse Marangoni instability described in the previous subsections is the origin of the spontaneous generation of capillary-gravity waves. The problem of theoretical description of nonlinear wavy patterns produced by the instability is very difficult, and it is not yet completed. In the present subsection, we will present some attempts of a simplified description of the problem based on the long-wave approximation.

First, let us discuss the model problem of the transverse wave generation in a layer with a stress-free lower boundary. As was noted in Sec. 2.6.1, in this case the minimum of the neutral stability curve is situated at $k = 0$, which provides

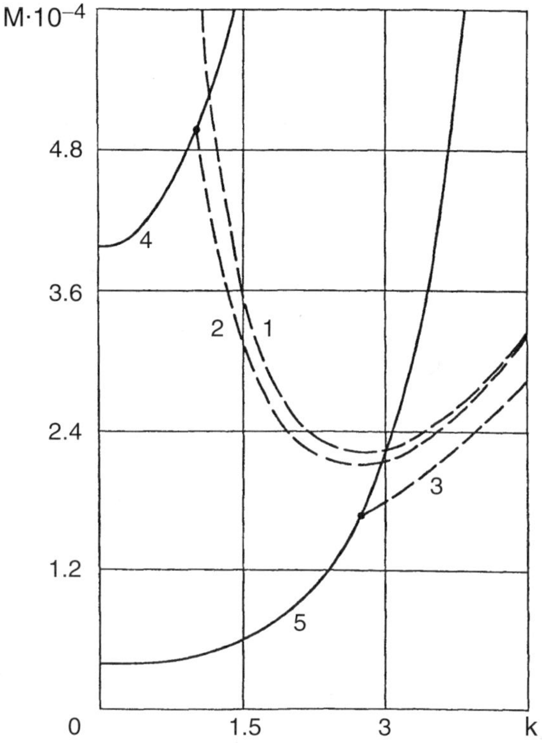

FIGURE 2.52. Oscillatory ($Ga = 10^9$ (line 1); 10^7 (line 2); 10^6 (line 3)) and monotonic ($Ga = 10^7$ (line 4); 10^6 (line 5)) neutral curves for the model system.

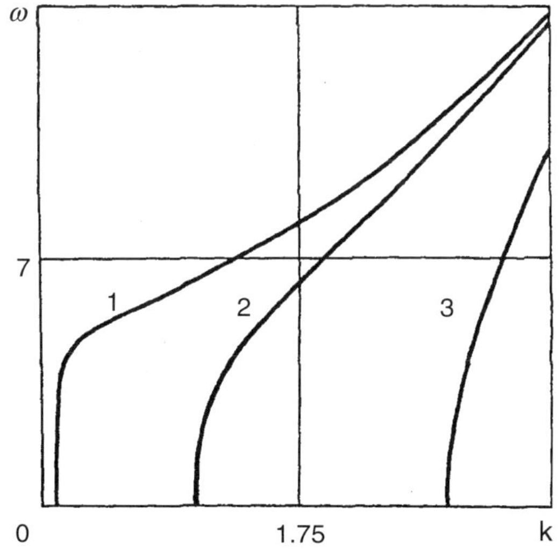

FIGURE 2.53. Dependence of the frequency on the wave number ($Ga = 10^9$ (line 1); 10^7 (line 2); 10^6 (line 3)).

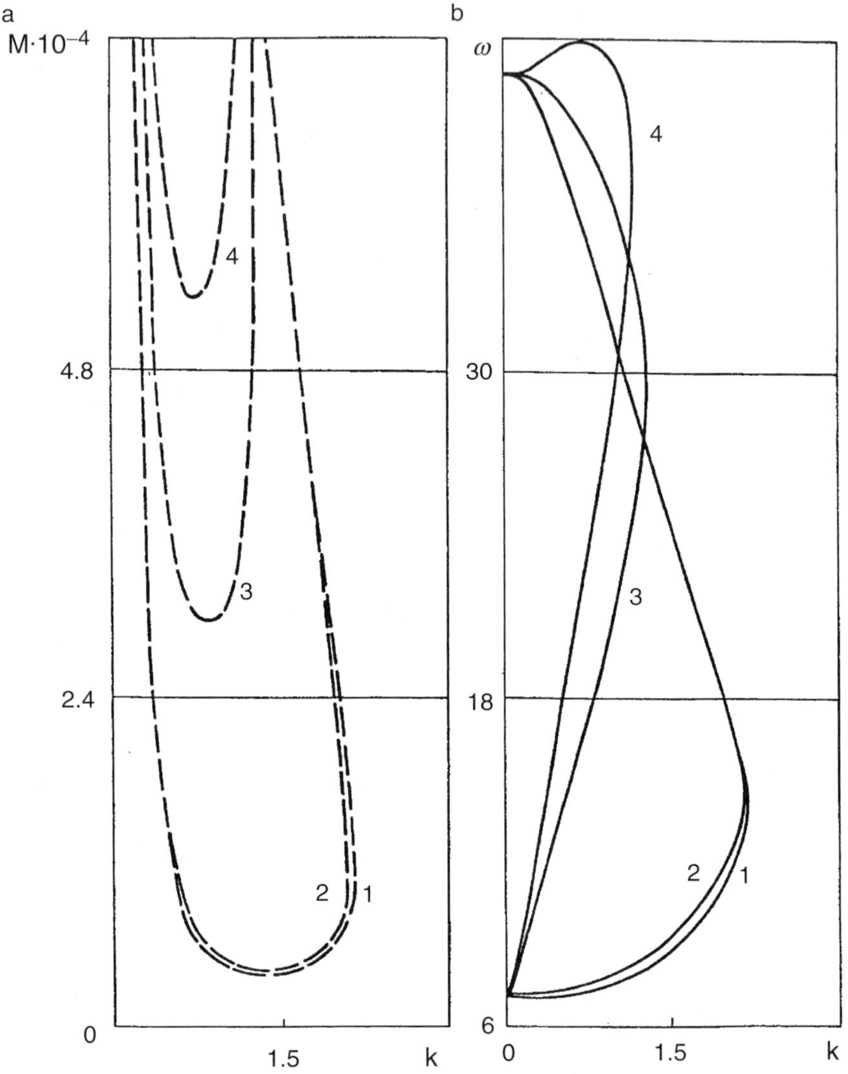

FIGURE 2.54. Dependences (a) of the Marangoni number and (b) of the frequency on the wave number ($Ga = 0$, line 1; 10^4, line 2; 10^6, line 3; 2×10^6, line 4).

the possibility to apply longwave expansions (Chu and Velarde, 1991; Garazo and Velarde, 1991, 1992; Nepomnyashchy and Velarde, 1994).

In the simplest case of a one-dimensional wave, the long-wave expansions give rise to an evolution equation for the surface deformation $h(x, t)$ ("order parameter equation"), which can be written in the form

$$\frac{\partial h}{\partial t} = Lh + N(h). \tag{2.82}$$

Here,

$$L = [\lambda_{10}+\lambda_{11}(M-M_c)]\frac{\partial}{\partial x} - \lambda_{21}(M-M_c)\frac{\partial^2}{\partial x^2} - \lambda_{30}\frac{\partial^3}{\partial x^3} + \lambda_{40}\frac{\partial^4}{\partial x^4} \tag{2.83}$$

is the linear operator corresponding to expressions (2.72) and (2.73). The nonlinear terms compatible with the conservation law have the structure

$$N(h) = \frac{\partial}{\partial x}Q(h), \;\; Q(h) = \delta_1 a^2 + \delta_2 \frac{\partial}{\partial x}(a^2) + \cdots, \tag{2.84}$$

where

$$\delta_1 = -\frac{3(GP+8)}{4\sqrt{(GP+12)P}}, \;\; \delta_2 = -2.$$

Equation (2.82) is called *dissipation-modified Korteweg-de Vries* (KdV) equation. By means of a scaling transformation of variables, in a moving reference frame, it can be reduced to the standard form (see Bar and Nepomnyashchy, 1995)

$$H_T + H_{XXX} + 3(H^2)_X + \delta[H_{XX} + H_{XXXX} + D(H^2)_{XX}] = 0, \tag{2.85}$$

where

$$\delta = \frac{\sqrt{-\lambda_{21}\lambda_{40}(M-M_c)}}{|\lambda_{30}|}, \;\; D = -\frac{3\lambda_{30}\delta_2}{\lambda_{40}\delta_1}. \tag{2.86}$$

When δ vanishes, the ideal KdV equation is recovered (Whitham, 1974). It is known that this equation has an infinite number of conservation laws. Recall that this equation possesses a family of traveling wave solutions

$$H(X,T) = H(\xi), \quad \xi = X - cT \tag{2.87}$$

with arbitrary spatial period $L = 2\pi/k$:

$$H(\xi + 2\pi/k) = H(\xi). \tag{2.88}$$

If we take

$$\langle H \rangle = \int_0^{2\pi/k} H(\xi)d\xi = 0 \tag{2.89}$$

(the mean surface deformation is equal to zero by the definition), the corresponding solutions (cnoidal waves) are determined by the formulas (Nepomnyashchy, 1976; Bar and Nepomnyashchy, 1995)

$$u(\xi) = \frac{2k^2K^2}{\pi^2}\left(dn^2\left(\frac{(\xi-\xi_0)kK}{\pi}\right) - \frac{E}{K}\right), \tag{2.90}$$

$$c = \frac{4k^2K^2}{\pi^2}\left(2 - \kappa^2 - \frac{3E}{K}\right), \tag{2.91}$$

where dn is Jacobi's delta amplitude function with the modulus κ and $E = E(\kappa)$ and $K = K(\kappa)$ are complete elliptic integrals. The limit $\kappa \to 1$ corresponds to soliton solutions. Thus, in the case $\delta = 0$, the wavelength $L = 2\pi/k$ and the wave amplitude

$$A = H_{max} - H_{min} = \frac{2k^2 K^2 \kappa^2}{\pi^2} \tag{2.92}$$

are independent parameters that can take arbitrary values.

When $\delta \neq 0$, the governing equation (2.85) contains three additional terms that describe correspondingly the long-wave instability ("negative viscosity"), the short-wave dissipation, and the nonlinear dissipation. The conservation laws characteristic for the pure KdV equation, are violated (except that of the conservation of the fluid volume). For instance, in the case where $H(X,T)$ is spatially periodic with the period $L = 2\pi/k$, the time evolution of the squared deflection of the interface ("momentum") is governed by the equation

$$\frac{d}{dT}\int_0^L H^2 dX = \delta\left(\int_0^L H_X^2 dX - \int_0^L H_{XX}^2 dX + 2D\int_0^L H H_X^2 dX\right). \tag{2.93}$$

The right-hand side of equation (2.93) vanishes only for some definite values of the wave amplitude $A(k)$.

In the limit of small δ, the stationary values of the wave amplitude are calculated analytically (see Nepomnyashchy, 1976; Kawahara and Toh, 1985; Bar and Nepomnyashchy, 1995). There are three different types of behavior of $A(k)$, depending on D:

(i) If $D \leq 5/4$, the function $A(k)$ is uniquely defined in the whole region $0 < k < 1$. However, the spatially periodic solutions with the stationary value of the amplitude $A(k)$ are stable only inside a certain subinterval $k_- \leq k \leq k_+$.

(ii) If $5/4 < D < 2$, there are two solutions for $A(k)$ in a certain region $k_{min}(D) < k < 1$ (the lower branch and the upper branch): one solution for $k > 1$ and no solutions for $0 < k < k_{min}(D)$. Only the solutions on the lower branch in the interval $k_{min}(D) < k < 1$ are stable with respect to strictly periodic disturbances with the same spatial period $2\pi/k$. However, even these solutions are unstable with respect to disturbances violating the periodicity of the solution.

(iii) If $D \geq 2$, there is a unique solution for $k > 1$, which is unstable.

For finite values of δ, the traveling wave solutions of equation (2.85) were studied analytically by Nekorkin and Velarde (1994) and Velarde *et al.* (1995). Some numerical simulations of equation (2.85) were fulfilled by Christov and Velarde (1995) and Oron and Rosenau (1997).

In the nearly one-dimensional case, when some transverse modulations of one-dimensional waves are taken into account, the problem is governed by the *dissipation – modified Kadomtsev–Petviashvili* equation (Garazo and Velarde, 1992). By means of a scaling transformation, this equation is reduced to the following form (Bar and Nepomnyashchy, 1999):

$$\{H_T + H_{XXX} + 3(H^2)_X + \delta[H_{XX} + H_{XXXX} + D(H^2)_{XX}]\}_X - 3sH_{YY} = 0, \tag{2.94}$$

where δ and D are determined by formulas (2.86),

$$s = \operatorname{sign}\left(\frac{\lambda_{30}}{\lambda_{10}}\right). \tag{2.95}$$

In the framework of equation (2.95), it can be shown (see Bar and Nepomnyashchy, 1999) that any one-dimensional waves are unstable with respect to transverse modulations.

In the general case, the waves generated by the instability can propagate in arbitrary directions simultaneously. The full system of long-wave equations that describes such waves is rather complicated (see Nepomnyashchy and Velarde, 1994). This system was used for the investigation of collisions of onedimensional solitary waves at arbitrary incident angles.

The problem considered above is unrealistic in two aspects. First, the friction of the fluid at the bottom is completely neglected. As it was shown by Zimmerman (2001), the friction at the solid lower boundary suppresses the instability in the limit $k \to 0$, so that the critical wave number k_c is nonzero. Also, the phenomena in the gas phase adjacent to liquid are ignored.

The development of the nonlinear theory in the general case is a formidable task. However, the problem can be essentially simplified in the quite realistic limit $G \gg 1$, because, in this case, the critical wave number k_c is small and the long-wave approach can be applied. Also, one can take into account the fact that the kinematic viscosity and thermal diffusivity of a gas are typically much larger than corresponding parameters of a fluid (i.e., $\nu \gg 1$, $\chi \gg 1$ for a two-layer gas/liquid system).

An attempt to use the above-mentioned circumstances for the construction of the nonlinear theory of long transverse Marangoni waves was made by Velarde *et al.* (1999). The authors considered the limit where the characteristic wave number $k \sim \epsilon \ll 1$ but $\nu \sim \chi \sim G^{1/2}\epsilon \gg 1$ (actually, the relation $\epsilon \sim G^{-1/10}$ was selected). In this limit, another kind of the dissipation-modified KdV equation was obtained, which after rescaling of variables can be written as

$$H_T + H_{XXX} + 3(H^2)_X + \int_{-\infty}^{+\infty} Q(X - X')dX' = 0. \tag{2.96}$$

The Fourier transform $\hat{Q}(k)$ of the kernel $Q(X - X')$ is a complicated function of k, $m = M/G$, P, a, and ν_*, where $\nu_* = \nu/(\epsilon G^{1/2}) = O(1)$.

Zimmerman (2001) performed a direct simulation of the nonlinear system of equations that describes the finite-amplitude Marangoni convection in a layer with a deformable free surface and a solid bottom surface. The flow was assumed to be two-dimensional. The Galerkin finite-element method with the grid adaptation procedure developed formerly by Goodwin and Schowalter (1995) was used. The fields of velocity and temperature generated by the surface elevation have been studied.

2.6.5 Oscillations Generated by a Surfactant

In Sec. 2.5.4, we discussed the influence of surface-active agents on the thermal convection in systems with a nondeformable interface. It was shown that in the presence of a "surface gas" of the surfactant molecules, a specific kind of oscillations replaces any kind of monotonic instability. This phenomenon was observed for the Marangoni instability, for the Rayleigh instability, and for the anticonvection.

In the present subsection, we consider the influence of surface-active agents on the thermal convection in systems with a deformable interface (see Nepomnyashchy and Simanovskii, 1992). The system of equations and boundary conditions (2.3) – (2.7), (2.9), (2.10) is not changed; equation (2.8) is replaced by

$$z = 0: \ \eta\left(\psi_1'' + k^2\psi_1\right) - \left(\psi_2'' + k^2\psi_2\right)$$

$$-ik\left[(\eta M/P)\left(T_1 - \frac{s}{1+\kappa a}h\right) - \frac{ikB}{\lambda + D_s k^2}\psi_1'\right] = 0 \qquad (2.97)$$

(see Sec. 2.5.4).

Analytical results

Monotonic instability. The monotonic instability boundary can be determined analytically:

$$sM = sM_m(k) =$$

$$8k^2 \frac{(1+\kappa a)(\kappa D_1 + D_2)(\eta B_1 + B_2 + B/2kD_s)}{\eta\kappa\left[(\chi E_2 - E_1) - 8k^5(D_1 + D_2)(F_1 - \eta^{-1}F_2)P^{-1}[Ga(\rho^{-1} - 1) + Wk^2]^{-1}\right]}. \qquad (2.98)$$

As in the case of the nondeformable interface (see equation (2.61)), the presence of a surfactant always leads to the increase of the monotonic instability threshold proportional to B/D_s, which is large, as D_s is small. Note that the location of the discontinuities of the neutral curve determined by zeros of the denominator of equation (2.98) is not changed by the appearance of a surfactant. In the limit of large k, expression (2.98) tends to equation (2.61).

Let us consider the limit $W \gg 1$ (strong interfacial tension) in more detail. The boundary condition (2.7) for the normal stresses shows that in the latter case, the deformation of the boundary is essential only in the long-wave region, $k \ll 1$. For $k = O(W^{-1/2})$, the monotonic neutral curve is determined by the expression

$$sM = sM_m(k) = \frac{2P[Ga(\rho^{-1} - 1) + Wk^2](1 + \eta a + Ba/4D_s)(1+\kappa a)^2 a}{(1+a)(1-\eta a^2)}. \qquad (2.99)$$

Oscillatory instability. It turns out that in the presence of the surfactant on the interface, this monotonic mode is replaced by an oscillatory one, like in the cases studied in Sec. 2.5.4. Analytical results can be obtained in the limit $W \gg 1$, $k \ll 1$. The cases $k = O(W^{-1/2})$ and $k = O(W^{-1/4})$ should be considered separately.

In the region $k = O(W^{-1/2})$, we obtain the following expressions for the oscillatory neutral curve $sM = sM_o(k)$ and for the frequency ω of neutral disturbances:

$$sM = sM_o(k) = \frac{P(1+\kappa a)^2}{\eta a^2 \kappa (1+a)(1-\eta a^2)} \left\{ \frac{2}{3} \eta a^3 (1+\eta a)[Ga(\rho^{-1}-1) + Wk^2] \right.$$
$$\left. + 2aB(1+\eta a^3) + 2D_s(1 + 4\eta a + 6\eta a^2 + 4\eta a^3 + \eta^2 a^4) \right\}; \qquad (2.100)$$

$$\omega^2 = B \frac{a^2 k^4}{1 + 4\eta a + 6\eta a^2 + 4\eta a^3 + \eta^2 a^4} \left\{ \frac{\eta a^2}{6} [Ga(\rho^{-1}-1) + Wk^2] \right.$$
$$\left. - 2D_s(1+\eta a^3) \right\} - D_s^2. \qquad (2.101)$$

The comparison of expressions (2.99) and (2.100) shows that $M_o < M_m$ in the whole region of the existence of the oscillatory instability, $\omega^2 > 0$.

In order to match expressions (2.100) and (2.101) with expressions (2.62) and (2.63), it is necessary to investigate the intermediate region $k = O(W^{-1/4})$, where we find

$$sM = \frac{80(1+\kappa a)^2 (1+\eta a}{\eta a^2 \kappa (\chi a^2 - 1)} \frac{k^2}{k^4 + k_*^4 S}; \qquad (2.102)$$

$$\omega^2 = B a^2 k^2 [L_1 + (L_2 + L_3 k^4)(k^4 + k_*^4 S)^{-1}]^{-1}. \qquad (2.103)$$

Here,

$$k_*^4 = \frac{120(1+a)}{WP\eta a^3} \left| \frac{1-\eta a^2}{\chi a^2 - 1} \right|; \quad S = \mathrm{sign}\left(\frac{1-\eta a^2}{\chi a^2 - 1} \right);$$

$$L_1 = \frac{a^2}{15} \left[(\eta + a\nu) + \eta(1+a) \frac{1-\nu a^2}{1-\eta a^2} \right]; \quad L_2 = \frac{160(1+\eta)(1-\kappa)(1-\eta a^2)}{\eta a(1+\kappa a)W};$$

$$L_3 = -\frac{2aP(1+\eta a^2)}{63(\chi a^2-1)(1+\kappa a)} \left\{ P[11(1-\kappa\chi a^5) + 53a(\kappa - \chi a^3) \right.$$
$$+ 42\chi a^2(1-\kappa a^2)]$$
$$\left. + (1+\kappa a) \left[4(1-\chi\nu a^4) - \frac{21(1-\nu a^2)(1-\chi\eta a^4)}{10(1-\eta a^2)} \right] \right\}.$$

If $S > 0$, the oscillatory neutral curve is continuous and the function $M_o(k)$ has a maximum as $k = k_*$. If $S < 0$, the oscillatory neutral curve $sM = sM_o(k)$ has a discontinuity, as $k = k_*$.

Expressions (2.102) and (2.103) tend to expressions (2.100) and (2.101), as $k \ll W^{-1/4}$, and to expressions (2.62) and (2.63), as $W^{-1/4} \ll k \ll 1$.

Numerical results. Let us now describe some results of the numerical solution of the eigenvalue problem (2.3) – (2.7), (2.9), (2.10), (2.97). The computation has been fulfilled for the same systems as in Sec. 2.6.3.

Model system by heating from below. First, let us consider the model system with $\eta = \nu = 0.5$, $\chi = \kappa = P = a = W = 1$, $\rho = 0.999$. In Sec. 2.6.3, it was shown that by heating from below ($s = 1$), this system is characterized by the coexistence of two instability modes: a short-wave oscillatory longitudinal instability mode and a long-wave monotonic deformational instability mode (see Fig. 2.52). In the presence of the surfactant, the neutral curve is split into two fragments: the monotonic one and the oscillatory one. The threshold value of M for the monotonic instability is determined by formula (2.99), and it grows rapidly with B. The dependence of the oscillatory instability threshold on B is much weaker. In Fig. 2.55, the monotonic and oscillatory neutral curves are shown for

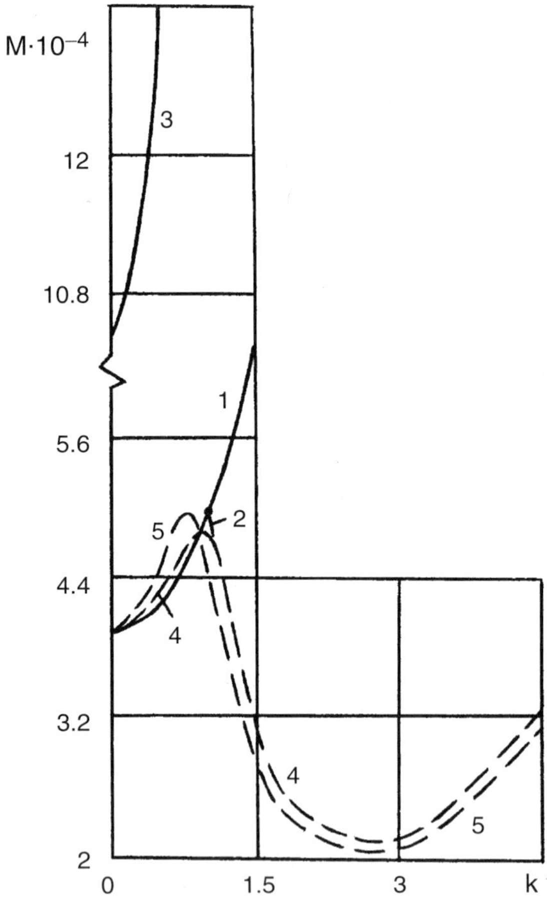

FIGURE 2.55. The monotonic and oscillatory neutral curves by heating from below.

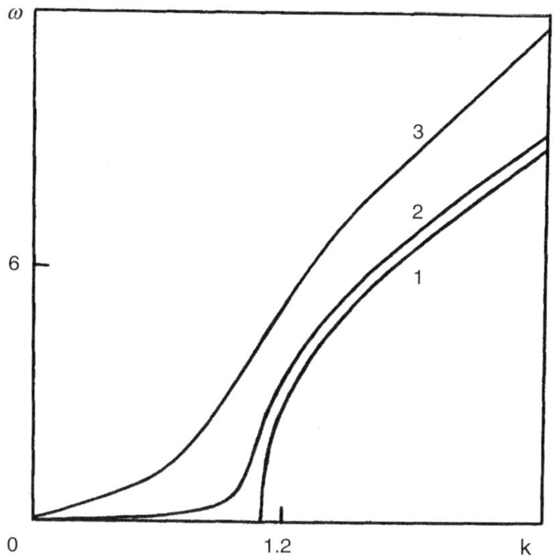

FIGURE 2.56. The dependence of the frequency on the wave number for $B = 0$ (line 1), $B = 0.1$ (line 2), and $B = 5$ (line 3). $Ga = 10^7$.

$B = 0$ (lines 1 and 2) and $B = 0.1$ (lines 3 and 4). Also, the oscillatory neutral curve is shown for $B = 5$ (line 5). The dependence of the frequency ω on the wave number k is shown in Fig. 2.56.

Model System by Heating from Above

When heating is from above ($s = -1$), another kind of oscillatory instability, the transverse one, takes place (see Sec. 2.6.3). The neutral curve looks like a "sack" (see Fig. 2.57(a)); that is, there are two different values of sM for each value of the wave number in a certain interval $0 < k < k_1(B)$. The corresponding dispersion curves $\omega(k)$ are shown in Fig. 2.57(b). The width of the instability region decreases, while the threshold value of M increases with B.

Air/water system by heating from below

Let us now consider the influence of a surfactant on the instability in the air/water system by heating from below (the set of parameters is presented in Table 1). Fix $a = 1$. As in Sec. 2.6.3, we choose $W = 10^6$ (which corresponds to the thickness of each layer about 3 mm) and vary the parameter Ga.

The neutral curves obtained in the case of heating from below are shown in Fig. 2.58 ($G = 10$; lines 1 – 4 correspond to $B = 0, 1, 5, 10$). In the absence of the surfactant ($B = 0$), the instability is monotonic. The neutral curve has two minima: the long-wave one and the short-wave one. When $B \neq 0$, the neutral curve is split into two neutral curves: a monotonic one and an oscillatory one.

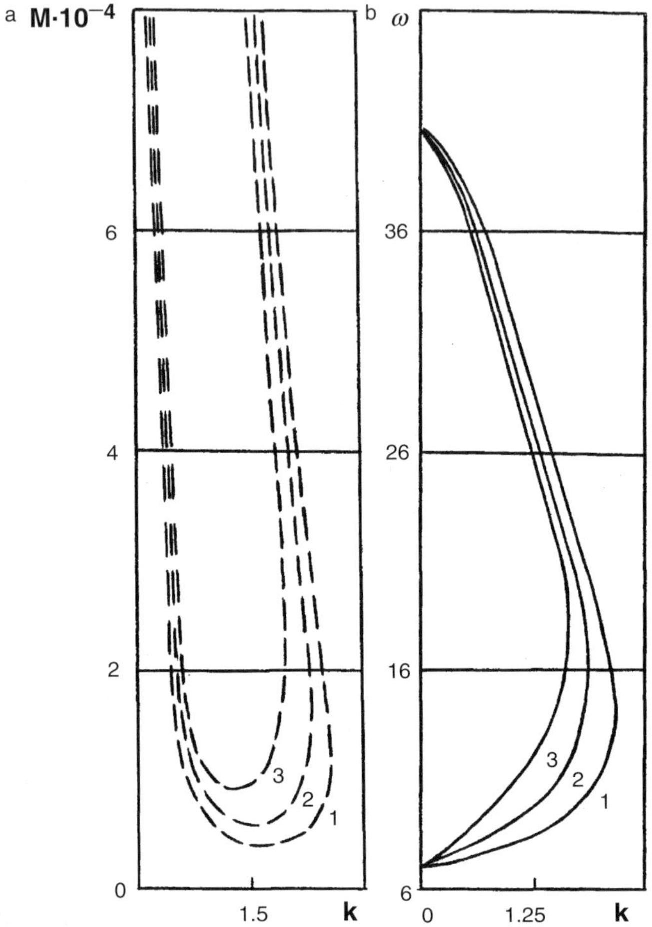

FIGURE 2.57. (a) The oscillatory neutral curves by heating from above and (b) the dependence of the frequency on the wave number for $B = 0$ (lines 1), $B = 5$ (lines 2), and $B = 15$ (lines 3). $Ga = 10^4$.

The monotonic neutral curve determined by formula (2.98) is located in the region of large M, and it is not shown in Fig. 2.58. The oscillatory neutral curve, which replaces the monotonic one, has two minima. The short-wave oscillations are mainly longitudinal, whereas the long-wave oscillations are essentially influenced by the deformation of the interface.

When B grows, the critical M grows for both minima. However, the stabilization of the short-wave oscillations is stronger than that of long-wave oscillations. Because of that, the long-wave mode provides a lower maximum, as B is larger than a certain value B_* (see Fig. 2.58). The frequency of oscillations grows monotonically with B (see Fig. 2.59; lines 1 - 3 for $B = 1, 5, 10$).

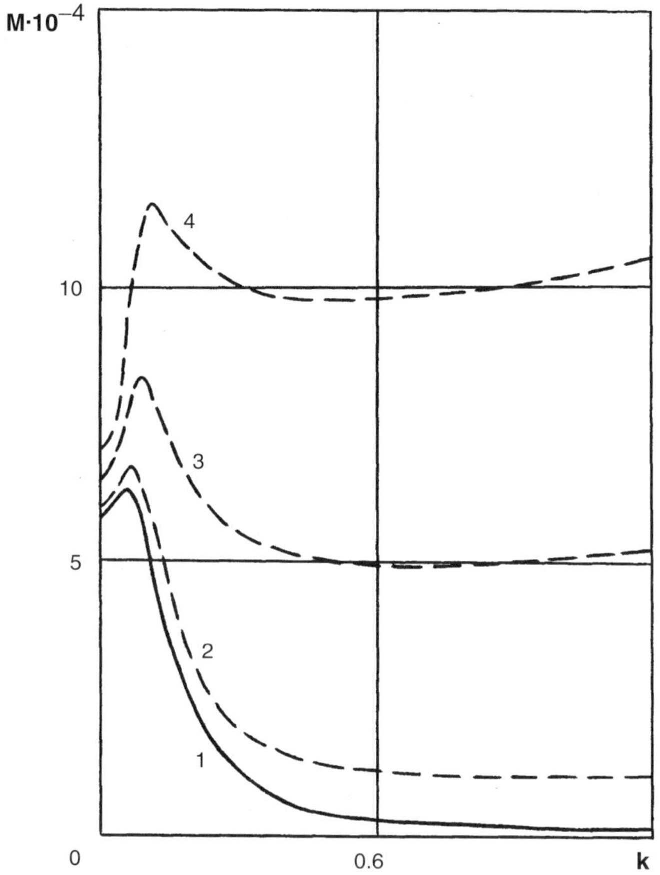

FIGURE 2.58. The monotonic and oscillatory neutral curves for the air/water system by heating from below; $B = 0$ (line 1), $B = 1$ (line 2), $B = 5$ (line 3), and $B = 10$ (line 4). $Ga = 10$.

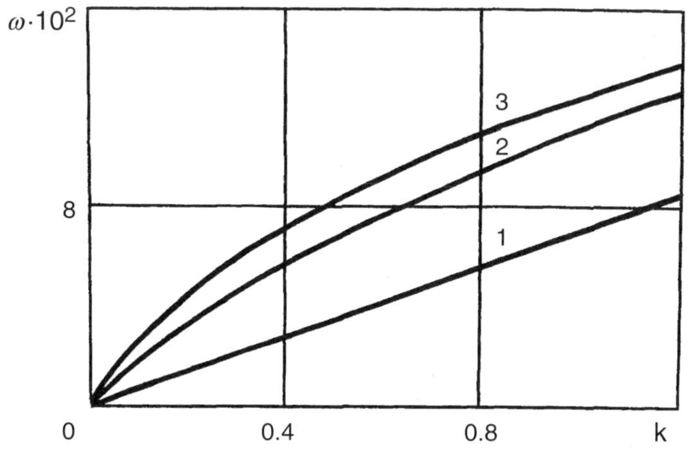

FIGURE 2.59. The dependence of the frequency on the wave number; $B = 0$ (line 1), $B = 1$ (line 2) and $B = 5$ (line 3). $Ga = 10$.

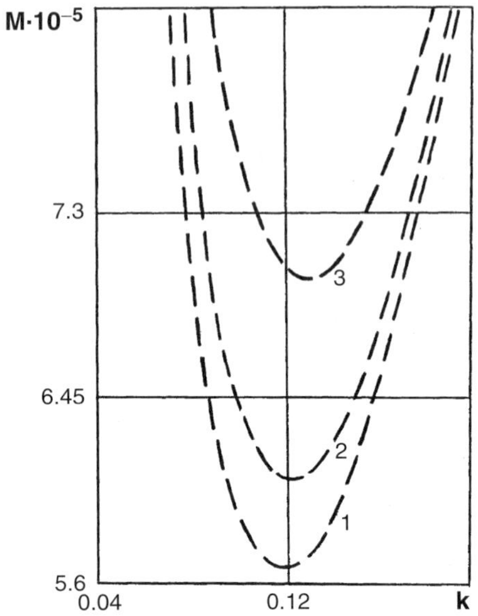

FIGURE 2.60. The oscillatory neutral curves for the air/water system by heating from above; $B = 0$ (line 1), $B = 5$ (line 2) and $B = 15$ (line 3).

Air/water system by heating from above

When the air/water system is heated from above, a long-wave transverse oscillatory instability appears even in the absence of a surfactant (see Sec. 2.6.3). The influence of a surfactant leads to the growth of the threshold value of M (see Fig. 2.60) and to the growth of the oscillation frequency (see Fig. 2.61). Note that the neutral curve does not have a shape of a "sack" for this system.

2.6.6 Transverse and Longitudinal Marangoni Instabilities in the Case of the Mass Transfer

In the previous sections, we considered in detail the case where the Marangoni instability was caused by the heat transfer through the interface. Recall that the influence of a surfactant on the heat-transfer-generated instability was studied in Secs. 2.5.4 and 2.6.5. Let us emphasize however that in the cases considered in those sections, no gradient of the surfactant concentration was applied, and there was no transfer of the surfactant through the interface. Now, we will study the case when the Marangoni instability is *caused* by the mass transfer of a surfactant through the interface. The temperature disturbances will be neglected.

In Sec. 1.2, we discussed the similarities and the differences between the heat transfer and the mass transfer of a surfactant. First, the field of the temperature is continuous at the interface, whereas the concentration of a surfactant has a

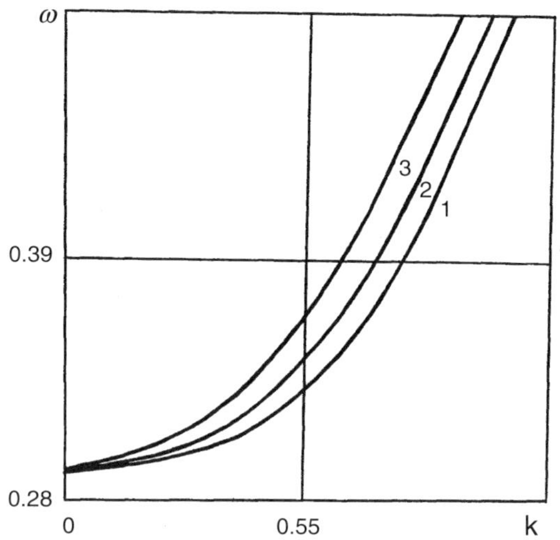

FIGURE 2.61. The dependence of the frequency on the wave number for the air/water system by heating from above; $B = 0$ (line 1), $B = 5$ (line 2), and $B = 15$ (line 3).

jump across the interface. However, this difference is easily eliminated by a certain transformation of variables (see Sec. 1.2). That is why the basic results of Sternling and Scriven (1959) are equally valid for the longitudinal Marangoni instabilities generated by heat transfer and by mass transfer.

The most important difference between the mass transfer of a surfactant and the heat transfer is the possibility of the accumulation of the surfactant at the interface, which has no analog in the case of the heat transfer. The distribution of a surfactant on the interface is governed by an *additional* evolution equation (2.57), which essentially influences the development of instabilities.

The investigation of instabilities in the case of solute accumulation was started by Hennenberg *et al.* (1979, 1980). An extensive analysis of the instability problem in a system of two semi-infinite fluid layers was fulfilled by Chu and Velarde (1989) (see also the paper by Chu and Velarde (1988), where the problem was studied by means of the one-layer approach, and that by Hennenberg *et al.* (1992), where the finite-time ratio between the adsorption and diffusion processes was taken into account). The crucial parameters that determine the type of instability are the ratio of kinematic viscosities ν and the ratio of diffusion coefficients $D = D_1/D_2$.

In the case $\nu < 1$, both longitudinal and transverse instabilities appear in the system. Typical neutral curves are shown in Fig. 2.62. The low-frequency branch is associated with longitudinal waves, whereas the high-frequency branch corresponds to transverse waves. In the case $\nu > 1$, $D > 1$, only one oscillatory instability mode is typically observed (Fig. 2.63). If $\nu > 1$ and $D < 1$, the system can be monotonically unstable, but no oscillatory instability was found.

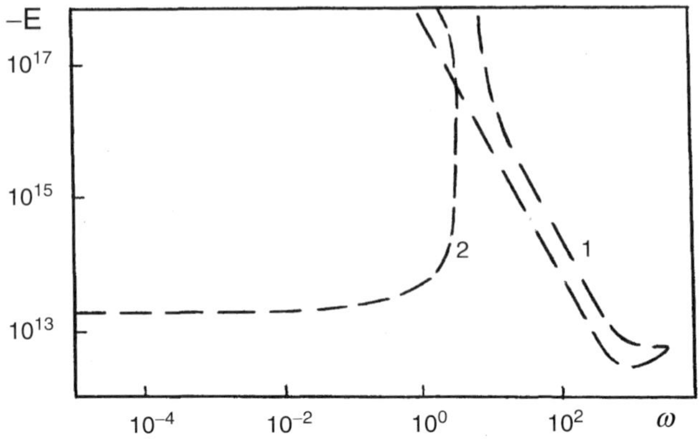

FIGURE 2.62. Typical neutral curves for the transverse waves (line 1) and longitudinal waves (line 2) in the case $\nu < 1$, $D < 1$.

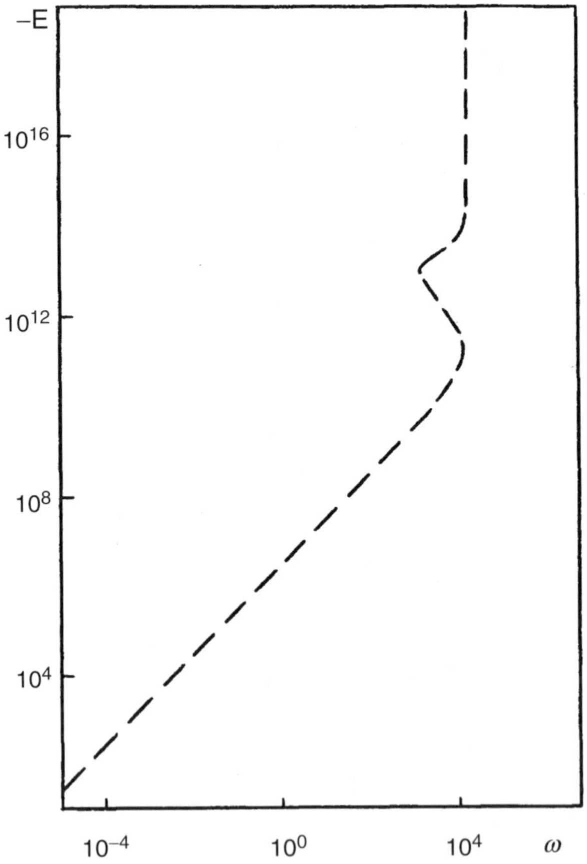

FIGURE 2.63. Typical neutral curve for the longitudinal wave in the case $\nu > 1$, $D > 1$.

Kovalchuk and Vollhardt (2000, 2002, 2004) considered a specific kind of oscillations produced by the dissolution of a surfactant droplet situated at the tip of a capillary under the surface. Experiments and numerical simulations revealed auto-oscillations of the surface tension caused by the Marangoni convection together with adsorption/desorption processes at the liquid-gas interface.

The onset of stationary solutal Marangoni convection by the mass transfer across the interface in a two-layer system *of finite width* has been studied by Slavtchev *et al.* (1998). Let us mention also a semi-empirical criterion for transfer of surfactants suggested by Nakache *et al.* (1983), Nakache and Raharimalala (1988).

An important problem is the development of the Marangoni convection in a system of two mutually solvable liquids. In the isothermic case, the interfacial concentration is always equal to the equilibrium one, and the Marangoni convection is impossible. However, the interfacial convection can be induced by the temperature disturbances caused by the heat of solution. The stability criteria in this case have been obtained theoretically (Perez de Ortiz and Sawistowski, 1973a) and compared with the experimental observation (Perez de Ortiz and Sawistowski, 1973b). The problem has been recently revisited by Slavtchev and Mendes (2004). They confirmed the results by Perez de Ortiz and Sawistowski for the stationary instability and revised them in the oscillatory case. The effect of surfactants on Marangoni convection in binary liquid-liquid systems was studied by Mendes-Tatsis and Agble (2000), Agble and Mendes-Tatsis (2000, 2001).

Let us discuss now the spontaneous motion of a droplet located at the interface of a liquid solution. This phenomenon can be caused by a solutal Marangoni effect (Bekki *et al.*, 1990, 1992; Nagai *et al.*, 2005) or by a chemically-driven Marangoni effect (Yoshikawa and Magome, 1993). A similar phenomenon is observed for solid grains (e.g., those of camphor) which form a layer reducing the surface tension of the liquid (Nakata *et al.*, 1997; Nagayama *et al.*, 2004 and references therein). A spontaneous motion of droplets in the presence of a chemical reaction takes place also on solid substrates (Dos Santos and Ondarçuhu, 1995). Recently, Sumino *et al.* (2005a, 2005b) reported the appearance of a spontaneous motion of an oil droplet on a glass substrate in an aqueous phase, which can be generated by a chemically-induced wettability gradient (Thiele *et al.*, 2004).

3

Bénard Problem in Multilayer Systems with Undeformable Interfaces

In the previous chapter, we studied the convection in systems with a *single* interface. Now, we start with the investigation of the case where the fluid system has *two* interfaces. We will see that the interaction between different interfaces leads to the appearance of qualitatively new types of instability. In the present chapter, we will consider the case where the deformations of interfaces are negligible. The influence of the interfaces' deformations will be analyzed in Chapter 4.

In Section 3.1, we recall the mathematical formulation of the Bénard problem in three-layer systems. Section 3.2 contains the results of the linear stability analysis. The nonlinear simulations of convective regimes are described in Section 3.3. Section 3.4 is devoted to the description of space experiments on convection in three-layer systems.

3.1 General Equations and Boundary Conditions

Let the space between two rigid horizontal plates be filled by three immiscible fluids with different physical properties (see Fig. 1.2(b)). Even in the absence of gravity, we will call layer 1 "the top layer" and layer 3 "the bottom layer". The equilibrium thicknesses of the layers are a_m, $m = 1, 2, 3$. The deformable interfaces are described by the equations $z = h(x, y, t)$ and $z = -a_2 + h_*(x, y, t)$. The mth fluid has density ρ_m, kinematic viscosity ν_m, dynamic viscosity $\eta_m = \rho_m \nu_m$, thermal diffusivity χ_m, heat conductivity κ_m, and heat expansion coefficient β_m. The surface tension coefficients on the upper and lower interfaces, σ and σ_*, are linear functions of temperature T: $\sigma = \sigma_0 - \alpha T$, $\sigma_* = \sigma_{*0} - \alpha_* T$. The acceleration due to gravity is g. We do not take into account buoyancy effects, which are negligible in the case of thin layers or under microgravity conditions. The horizontal plates are kept at different constant temperatures. The temperature difference can be positive or negative and the overall temperature drop is θ.

The mathematical formulation of the problem is given in Sec. 1.2.4. Recall that the complete nonlinear equations governing the convection are written in the following dimensionless form:

$$\frac{\partial \mathbf{v}_m}{\partial t} + (\mathbf{v}_m \cdot \nabla)\mathbf{v}_m = -e_m \nabla p_m + c_m \Delta \mathbf{v}_m + b_m G T_m \boldsymbol{\gamma}, \tag{3.1}$$

$$\frac{\partial T_m}{\partial t} + \mathbf{v}_m \cdot \nabla T_m = \frac{d_m}{P} \Delta T_m, \tag{3.2}$$

$$\nabla \mathbf{v}_m = 0, \ m = 1, 2, 3, \tag{3.3}$$

where $e_1 = c_1 = b_1 = d_1 = 1$, $e_2 = \rho$, $c_2 = 1/\nu$, $b_2 = 1/\beta$, $d_2 = 1/\chi$, $e_3 = \rho_*$, $c_3 = 1/\nu_*$, $b_3 = 1/\beta_*$, $d_3 = 1/\chi_*$; $\Delta = \nabla^2$, $G = g\beta_1\theta a_1^3/\nu_1^2$ is the Grashof number, and $P = \nu_1/\chi_1$ is the Prandtl number.

The boundary conditions on the rigid walls are

$$\mathbf{v}_1 = 0, \ T_1 = 0 \ \text{ at } \ z = 1, \tag{3.4}$$

$$\mathbf{v}_3 = 0, \ T_3 = s \ \text{ at } \ z = -a - a_*, \tag{3.5}$$

with $s = \pm 1$. We will call the case $s = 1$ “heating from below” and $s = -1$ “heating from above”.

In the present chapter, we disregard the deformations of interfaces. The following system of boundary conditions is used: at $z = 0$,

$$\frac{\partial v_{1x}}{\partial z} - \eta^{-1}\frac{\partial v_{2x}}{\partial z} - \frac{M}{P}\frac{\partial T_1}{\partial x} = 0, \quad \frac{\partial v_{1y}}{\partial z} - \eta^{-1}\frac{\partial v_{2y}}{\partial z} - \frac{M}{P}\frac{\partial T_1}{\partial y} = 0, \tag{3.6}$$

$$v_{1x} = v_{2x}, \ v_{1y} = v_{2y}, \ v_{1z} = v_{2z} = 0, \tag{3.7}$$

$$T_1 = T_2, \tag{3.8}$$

$$\frac{\partial T_1}{\partial z} = \kappa^{-1}\frac{\partial T_2}{\partial z}, \tag{3.9}$$

and at $z = -a$,

$$\eta^{-1}\frac{\partial v_{2x}}{\partial z} - \eta_*^{-1}\frac{\partial v_{3x}}{\partial z} - \frac{M}{P}\frac{\partial T_3}{\partial x} = 0, \ \eta^{-1}\frac{\partial v_{2y}}{\partial z} - \eta_*^{-1}\frac{\partial v_{3y}}{\partial z} - \frac{M}{P}\frac{\partial T_3}{\partial y} = 0, \tag{3.10}$$

$$v_{2x} = v_{3x}, \ v_{2y} = v_{3y}, \ v_{2z} = v_{3z} = 0, \tag{3.11}$$

$$T_2 = T_3, \tag{3.12}$$

$$\kappa^{-1}\frac{\partial T_2}{\partial z} = \kappa_*^{-1}\frac{\partial T_3}{\partial z}. \tag{3.13}$$

The boundary value problem given above has the solution

$$\mathbf{v}_m = 0, \ p_m = 0, \ m = 1, 2, 3; \ h = 0; \ h_* = 0, \tag{3.14}$$

$$T_1 = T_1^0 = -\frac{s(z - 1)}{1 + \kappa a + \kappa_* a_*}, \tag{3.15}$$

$$T_2 = T_2^0 = -\frac{s(\kappa z - 1)}{1 + \kappa a + \kappa_* a_*}, \tag{3.16}$$

$$T_3 = T_3^0 = -s\frac{\kappa_* z - 1 + (\kappa_* - \kappa)a}{1 + \kappa a + \kappa_* a_*} \tag{3.17}$$

corresponding to the mechanical equilibrium state.

Depending on physical parameters of fluids, the mechanical equilibrium state may become unstable with respect to different instability modes, which are studied below.

3.2 Linear Stability Theory

In order to investigate the linear stability of the solution (3.14) – (3.17), we consider the infinitesimal disturbances

$$(\mathbf{v}_m,\ T_m,\ p_m)\exp(\lambda t + ikx),\ \ m = 1, 2, 3.$$

It is convenient to introduce the stream function ψ_m, $m = 1, 2, 3$:

$$v_{mx} = \psi_m',\ v_{mz} = -ik\psi_m.$$

Eliminating p_m, $m = 1, 2, 3$, we obtain the following eigenvalue problem:

$$\lambda(\psi_m'' - k^2\psi_m) = c_m(\psi_m'''' - 2k^2\psi_m'' + k^4\psi_m) + ikGb_mT_m, \tag{3.18}$$

$$\lambda T_m = \frac{d_m}{P}(T_m'' - k^2T_m) + ikA_m\psi_m; \tag{3.19}$$

$$\psi_1 = \psi_1' = T_1 = 0 \text{ at } z = 1, \tag{3.20}$$

$$\psi_3 = \psi_3' = T_3 = 0 \text{ at } z = -a - a_*, \tag{3.21}$$

$$\psi_1'' - \eta^{-1}\psi_2'' - ikMP^{-1}T_1 = 0, \tag{3.22}$$

$$\psi_1 = \psi_2 = 0, \tag{3.23}$$

$$\psi_1' = \psi_2', \tag{3.24}$$

$$T_1 - T_2 = 0, \tag{3.25}$$

$$T_1' - \kappa^{-1}T_2' = 0 \text{ at } z = 0, \tag{3.26}$$

and

$$\eta^{-1}\psi_2'' - \eta_*^{-1}\psi_3'' - ik\bar{a}MP^{-1}T_3 = 0, \tag{3.27}$$

$$\psi_2 = \psi_3 = 0, \tag{3.28}$$

$$\psi_2' = \psi_3', \tag{3.29}$$

$$T_2 - T_3 = 0, \tag{3.30}$$

$$\kappa^{-1}T_2' - \kappa_*^{-1}T_3' = 0 \text{ at } z = -a. \tag{3.31}$$

Here, primes denote differentiation with respect to z and A_m is the equilibrium temperature gradient in the mth fluid:

$$A_1 = -\frac{s}{1+\kappa a+\kappa_* a_*},\quad A_2 = -\frac{s\kappa}{1+\kappa a+\kappa_* a_*},\quad A_3 = -\frac{s\kappa_*}{1+\kappa a+\kappa_* a_*}. \tag{3.32}$$

The mechanical equilibrium state is unstable if the real part of an eigenvalue $\lambda(M, G, k)$ is positive for some k.

Note that the eigenvalue problem (3.18) – (3.31) is not self-adjoint, and one can expect the appearance of both monotonic instability ($\mathrm{Re}\lambda(M, G, k) = 0$) and oscillatory instability ($\mathrm{Re}\lambda(M, G, k) \neq 0$).

3.2.1 Marangoni Convection. The Case of a Symmetric System and Equal Layer Thicknesses

In the case of a pure Marangoni convection ($G = 0$), the monotonic stability problem can be solved analytically using the symbolic algebra package MATHEMATICA, but the expressions that describe the neutral curves are very cumbersome (Kats-Demyanets, 1996). It is more instructive to explain the peculiarities of the instabilities in three-layer systems through the consideration of a particular but typical case. In the present subsection, we will discuss the case where fluids 1 and 3 are identical ($\nu_* = \chi_* = \kappa_* = \alpha_* = 1$) and the thicknesses of all the layers are equal ($a = a_* = 1$).

Note that the geometric symmetry of the system with respect to a reflection in the plane $z = -1/2$ does not lead to a corresponding symmetry of eigenfunctions, because layers 1 and 3 are heated in a different way. Indeed, in one of those layers, the gradient of temperature is directed toward the rigid boundary, whereas in another one, it is directed toward the interface. The eigenfunctions $\psi_1(z)$ and $\psi_3(z)$ are similar:

$$\psi_1(z) = D_1\left[\sinh kz - \frac{S_1^2}{k} z\cosh kz + \left(\frac{S_1C_1}{k} - 1\right) z\sinh kz\right],\quad 0 \le z \le 1, \tag{3.33}$$

$$\psi_3(z) = D_3\left[\sinh k(z+1) - \frac{S_1^2}{k}(z+1)\cosh k(z+1) - \left(\frac{S_1C_1}{k} - 1\right)(z+1)\sinh k(z+1)\right],\quad -2 \le z \le -1, \tag{3.34}$$

where $S = \sinh k$ and $C = \cosh k$, but the coefficients D_1 and D_3 *are not equal* to each other (see below).

The symmetry of the problem manifests itself as the symmetry of the neutral curve; that is, the stability boundary does not depend on the way of heating (the sign of parameter s). The monotonic stability boundary $M = M(k)$ is determined by the analytical formula

$$M^2 = \frac{64k^4(2+\kappa)^2[(1+\kappa)^2C^2-1][(1+\eta)^2(SC-k)^2-(S-kC)^2]}{\eta^2\kappa^2[(1-\chi)^2(S^3-k^3C)^2-\chi^2k^2(kSC+k^2-2S^2)^2]}. \tag{3.35}$$

Note that in the case $\chi = 1$, expression (3.35) is *negative* for any values of other parameters; thus the monotonic instability *does not exist* (see Simanovskii *et al.*, 2003, 2004). If $3/5 < \chi < 3$, the monotonic neutral curve is located in the region $k > k_*(\chi)$ (see Fig. 3.1, line 1). The dependence $k_*(\chi)$ is determined by the equation

$$\frac{\chi}{|1-\chi|} = f(k_*), \tag{3.36}$$

where

$$f(k) = \frac{S^3-k^3C}{k(kSC+k^2-2S^2)}. \tag{3.37}$$

The graph of the function $f(k)$ is plotted in Fig. 3.2. If $\chi \le 3/5$ or $\chi \ge 3$, the monotonic instability exists for any k (Fig. 3.1, lines 2 and 3). In the latter case, the asymptotic behavior of the monotonic neutral curves in the longwave limit is given by the formula

$$M^2 = \frac{43200(2+\kappa)^3[2(1+\eta)^2-1]}{k^4\eta^2\kappa(3-\chi)(3-5\chi)}. \tag{3.38}$$

Let us present also the asymptotic behavior of the monotonic neutral curves in the short-wave limit:

$$M = \pm\frac{8k^2(1+\kappa)(1+\eta)(2+\kappa)}{\eta\kappa|1-\chi|}. \tag{3.39}$$

Recall that in the case of a *two-layer* system with equal thicknesses of the layers ($a = 1$) and equal thermal diffusivities ($\chi = 1$), the monotonic instability disappeared as well (see Sec. 2.4.2). In the latter case however, if $\chi \neq 1$, the

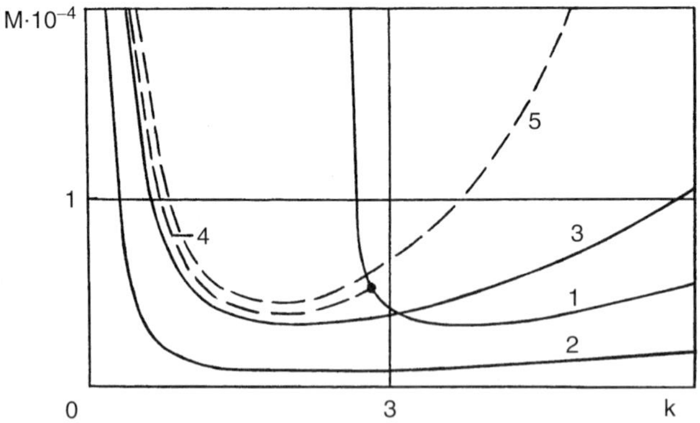

FIGURE 3.1. Neutral curves for the Marangoni instability in symmetric three-layer systems: $\chi = 2$ (lines 1 and 4), $\chi = 4$ (line 2), $\chi = 0.5$ (line 3), $\chi = 1$ (line 5).

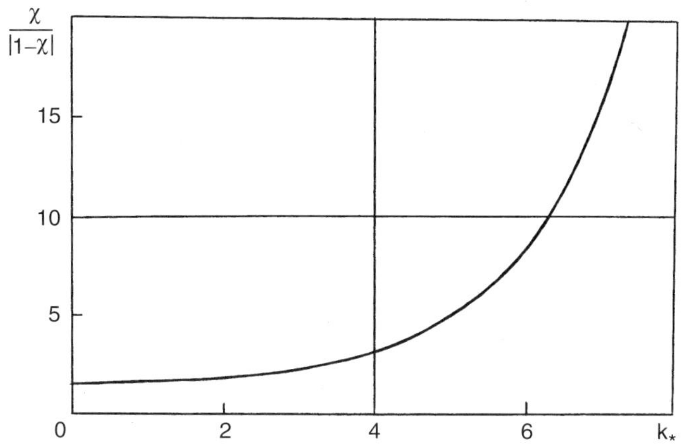

FIGURE 3.2. The dependence of $\chi/|1-\chi|$ on k_*.

monotonic instability exists at any k for a certain way of heating (either from below, or from above). In the case of a *three-layer* system, the monotonic instability is completely suppressed in a finite interval of wave number values if χ belongs to a certain interval around $\chi = 1$. This suppression is caused by the interaction of disturbances produced by the thermocapillary stresses at both interfaces.

The dependence of the monotonic critical Marangoni number M_m on the parameter χ calculated by Géoris *et al.* (1993) for $\eta = 0.5$, $\kappa = 1$, is shown in Fig. 3.3(a) (lines 1 and 2). The corresponding values of the critical wave number k_m are presented in Fig. 3.3(b) (lines 1 and 2). It turns out that on the left branch, $\chi < 1$, the ratio $D = D_3/D_1$ is negative and $|D| > 1$ (see Fig. 3.4(a)). Using expressions (3.33) and (3.34), one can find that the functions $\psi_1(z)$ and $\psi_3(z)$ have the same sign, but the maximum value of $\psi_3(z)$ is larger than that of $\psi_1(z)$. The flow generated by the monotonic instability in the case $\chi < 1$ is shown in the left inset in Fig. 3.3(a). One can see that the motion is produced by the thermocapillary effect on the lower interface (interface 2-3), whereas the upper interface is passive. On the right branch, $\chi > 1$, the ratio $D = D_3/D_1$ is positive and it is less than 1 (Fig. 3.4(b)). Thus, the maximum value of $\psi_1(z)$ is larger than that of $\psi_3(z)$. The signs of the functions $\psi_1(z)$ and $\psi_3(z)$ are different; that is why the flow in the middle layer has a two-storied structure (see the right inset in Fig. 3.3(a)). The upper interface (interface 1–2) plays an active role in the generation of the Marangoni convection.

The system Fluorinert FC70 / silicone oil 10 cSt / Fluorinert FC70 (see system 1 of Table 3.1), which is characterized by the value $\chi = 0.362$, provides an example of a real three-layer system where a monotonic instability takes place. In the minimum of the neutral curve, $k_c = 1.965$ and $M_c = 1263$. According to the theoretical prediction given above, the intensity of the convection in the bottom layer is much larger than that in the top layer (see Fig. 3.5).

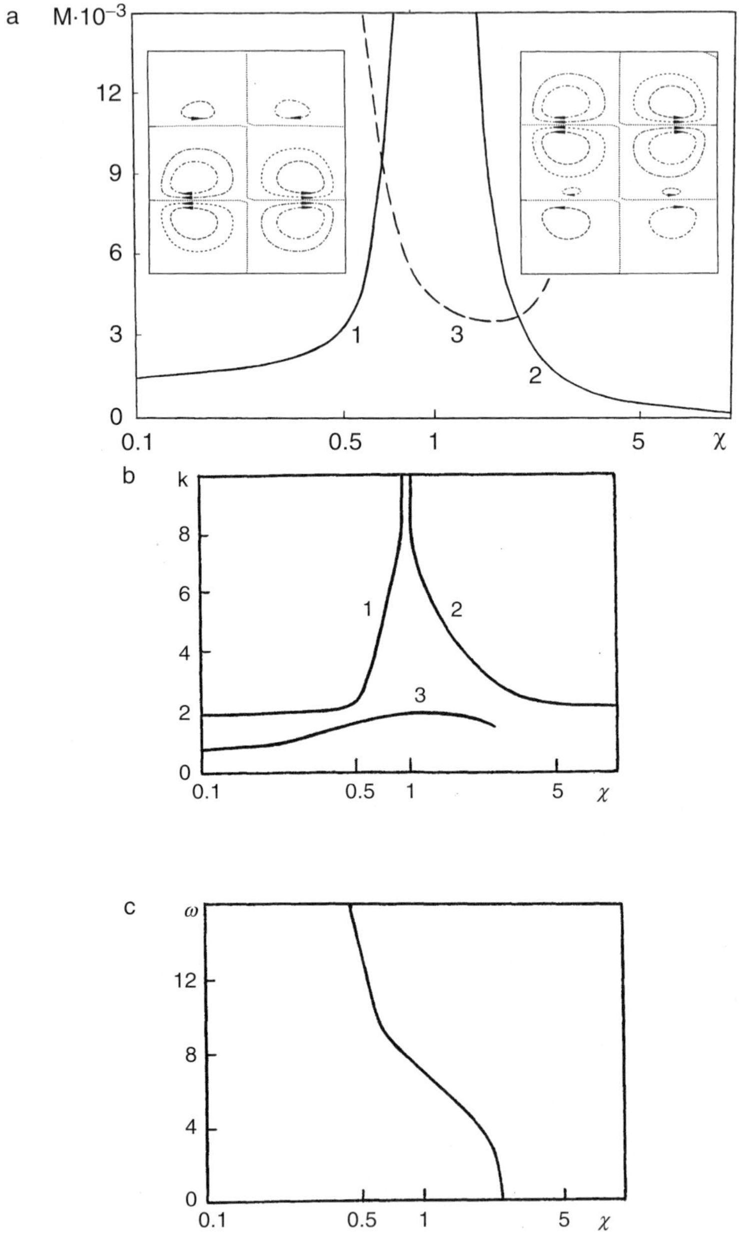

FIGURE 3.3. (a) The dependence of the monotonic (lines 1 and 2) and oscillatory (line 3) critical Marangoni numbers on χ. The left (right) insertion shows the stream lines in the case $\chi < 1$ ($\chi > 1$). (b) The dependence of the critical wave number k on χ.The numeration of the lines corresponds to that in (a). (c) The dependence of the critical frequency ω on χ.

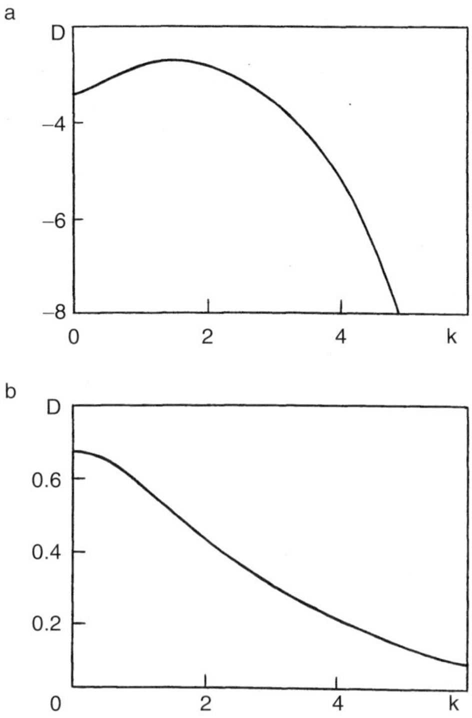

FIGURE 3.4. The dependence of the ratio D on k: (a) $\chi = 0.5$; (b) $\chi = 4$.

TABLE 3.1. Parameters of three-layer systems.

System	ν	ν_*	η	η_*	κ	κ_*
1	1.48	1	3.04	1	0.522	1
2	1.14	1	1.02	1	0.698	1
3	0.974	18.767	0.001	0.013	0.098	0.401
4	1.57	1	0.002	1	0.196	1
5	15.1	133	0.0182	0.0117	0.0396	0.00275
System	χ	χ_*	β	β_*	P	$\bar{\alpha}$
1	0.362	1		1	390	1
2	0.936	1	0.963	1	7.84	1
3	215.098	606.414	5.9	2.62	0.72	0.080
4	228.4	1		1	0.707	1
5	138	4.333	17.7	20.18	0.758	

As in the case of a *two-layer* system, the disappearance of the monotonic instability provides favorable conditions for the appearance of an *oscillatory* instability (see lines 4 and 5 in Fig. 3.1). The critical Marangoni number M_o for the oscillatory instability is given in Fig. 3.3(a) (line 3). The oscillatory instability is the only possible type of instability when $\chi = 1$ (line 5 in Fig. 3.1) and it is the most "dangerous" type of instability ($M_o < M_m$) in a certain interval of χ around 1

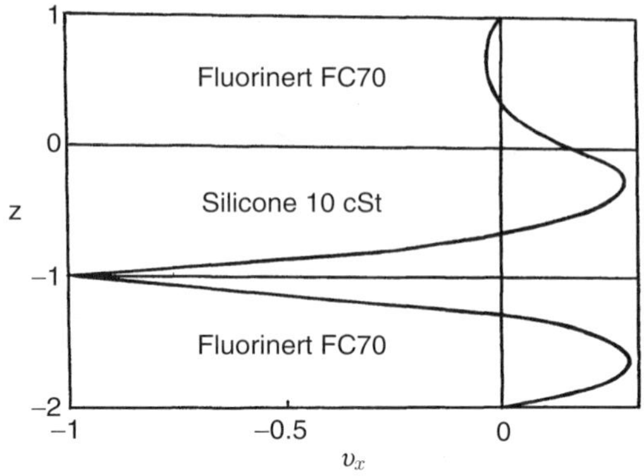

FIGURE 3.5. The critical eigenfunction $v_x(z) = \psi'(z)$ for the system Fluorinert F70 / silicone oil 10 cSt / Fluorinert F70.

($0.66 < \chi < 1.92$ for the model system $\nu = \eta = 0.5$, $\kappa = 1$). The oscillations can be considered as some transitions between two flow structures shown in the insets in Fig. 3.3(a) (for more details, see below). The critical wave number k_o of the oscillatory instability is always smaller than k_m (see Fig. 3.3(b), line 3). Note that the oscillatory neutral curve exists for arbitrary small χ, but the corresponding critical Marangoni number M_o is very high. The dependence of the critical frequency on χ is shown in Fig. 3.3(c). In the region $\chi > 2.55$, the minimum of the oscillatory neutral curve coincides with the codimension-2 "double-zero" point, where the frequency of oscillations tends to zero.

The oscillatory instability described above can be obtained for the real system of fluids if the corresponding parameter χ is sufficiently close to 1. For instance, it has been found for the system n-octane / methanol / n-octane (see Simanovskii *et al.*, 2003). The parameters of this system are presented in Table 3.1 (system 2). The oscillatory neutral curve $M(k)$ and the dependence of the oscillation frequency ω on the wave number k are presented in Fig. 3.6. In the minimum of the neutral curve, $k_c = 1.97$, $M_c = 4334.4$, and $\omega_c = 1.52$.

3.2.2 Onset of Marangoni Convection in Nonsymmetric Three-Layer Systems

As was mentioned above, the eigenvalue problem for the monotonic instability can be solved analytically. For any value of k, the quantity $\tilde{M}(k) = sM(k)$ satisfies the following quadratic equation:

$$A\tilde{M}^2 + B\tilde{M} + C = 0, \tag{3.40}$$

where A, B, and C are real functions of k and other parameters of the problem (the explicit expressions have been obtained by Kats-Demyanets (1996)). Function C is always different from zero, for $\tilde{M} = 0$ cannot be a root of equation (3.40) due to

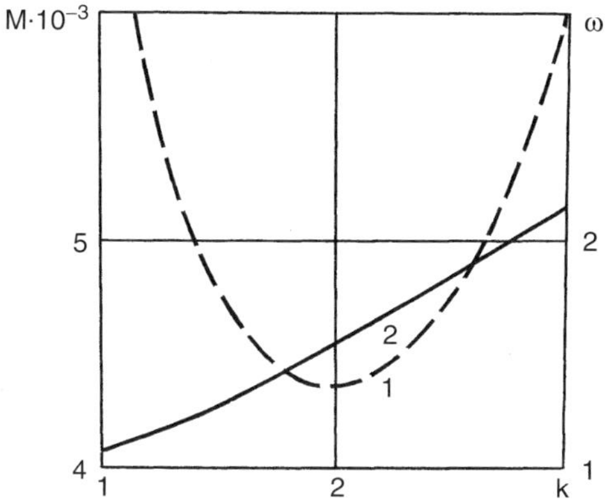

FIGURE 3.6. The oscillatory neutral curve (line 1) and the dependence of the oscillation frequency ω on the wave number k (line 2) for the system n-octane / methanol / n-octane.

physical reasons. Depending on the sign of the discriminant $D \equiv B^2 - 4AC$, equation (3.40) has typically either two real solutions corresponding to the boundary of monotonic instability or two complex conjugate solutions. In the latter case, there is no monotonic instability for a given set of parameters. Two monotonic instability curves merge in the point where $D(k) = 0$. The neutral curves are discontinuous in the points where $A(k) = 0$. Some typical configurations of the location of monotonic neutral curves are shown in Fig. 3.7.

For the calculation of an oscillatory instability boundaries, a numerical method must be used.

A nonsymmetric three-layer system is characterized by 12 thermophysical and geometrical parameters, and, therefore, a full parametric investigation of the linear stability problem seems to be impossible. We will concentrate on several typical examples. Some more details on the instabilities of three-layer systems can be found in the theses of Géoris (1994) and Kats-Demyanets (1996) and in the paper of Kats-Demyanets *et al.* (1997).

We consider the real system air / ethylene glycol (EG) / Fluorinert FC75. The relevant parameters for the chosen system, which are calculated on the basis of the physical data given in the paper by Prakash and Koster (1993), are presented in Table 3.1 (system 3). For the presentation of results, we use the Marangoni number $\bar{M}$ and the wave number $\bar{k}$, both normalized with respect to the total width of the system:

$$\bar{M} = M(1 + a + a_*), \; \bar{k} = k(1 + a + a_*).$$

For any value of k, only the lowest instability curve is shown.

As the first example, let us consider system 3 (Table 3.1) by heating from below ($s = 1$) with the fixed value of $a_* = 1$. Note that the thermocapillary effect acts much stronger on the interface between layers 1 and 2 (interface 1–2) than on the interface between layers 2 and 3 (interface 2–3), because $\bar{a}$ is small. If

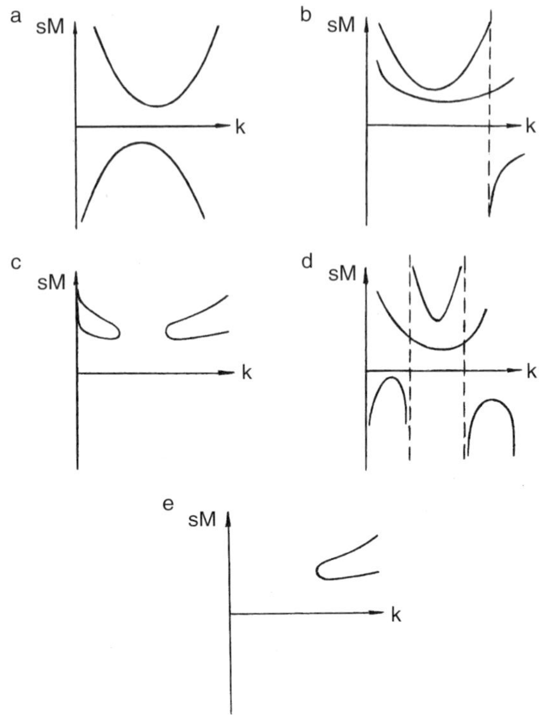

FIGURE 3.7. Typical configurations of monotonic neutral curves.

a is relatively large, the motions produced by different interfaces do not interact strongly, and only a monotonic instability is observed (Fig. 3.8, lines 1 – 3). The calculation of eigenfunctions shows that in this case, the lowest instability boundary corresponds to the onset of the Marangoni convection mainly in layers 1 and 2. This motion is caused by the thermocapillary effect on the interface 1–2. In layer 3, a weak motion is observed. For smaller values of a, the interaction between interfaces becomes significant, and an oscillatory neutral curve appears in the long-wave region, but still with $\bar{M}_o > \bar{M}_m$ (Fig. 3.8, line 4). For even smaller values of a, an oscillatory instability becomes critical ($\bar{M}_o < \bar{M}_m$) (Fig. 3.8, line 5) and replaces the monotonic instability (Fig. 3.8, line 6). The transition from the monotonic instability to the oscillatory instability leads to a jump of the critical wave number (Fig. 3.9). The critical wave number of the oscillatory instability is *smaller* than that of the monotonic instability.

Another scenario of transitions between the monotonic instability and the oscillatory instability is observed in the same system by heating from above ($s = -1$). For relatively small values of a, the neutral curve contains two fragments: a monotonic one and an oscillatory one (Fig. 3.10(a), lines 1 - 3). The monotonic instability is found to be driven at the interface 1–2, whereas the oscillatory one is at the interface 2–3. The critical mode is monotonic. When a increases, the monotonic neutral curve shrinks in the long-wave region (Fig. 3.10(a), line 4), and

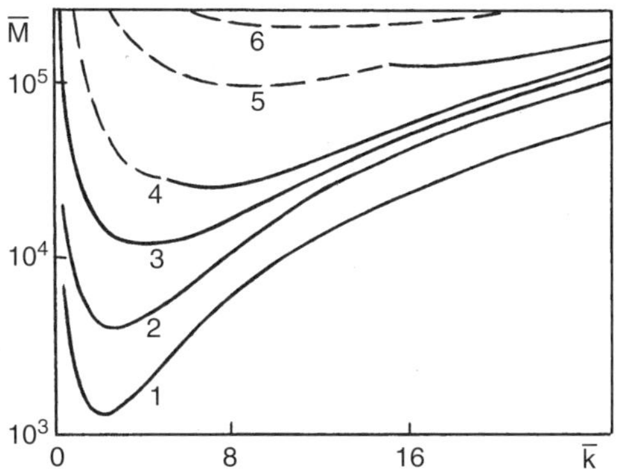

FIGURE 3.8. Neutral stability curves in the case $a_* = 1$ by heating from below; $a = 7$ (line 1); $a = 2$ (line 2); $a = 1$ (line 3); $a = 0.7$ (line 4); $a = 0.3$ (line 5); $a = 0.2$ (line 6).

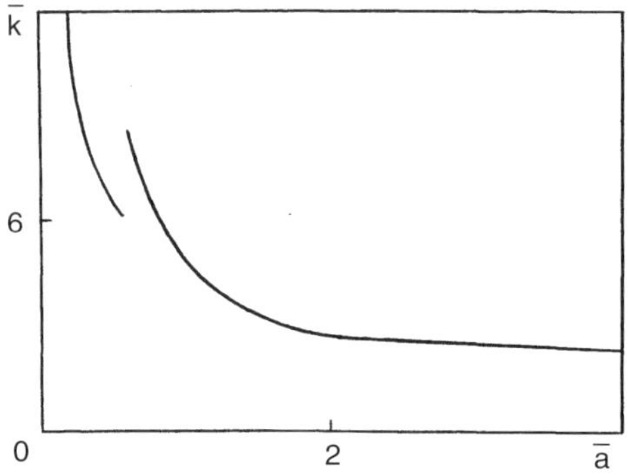

FIGURE 3.9. The critical wave number in the case $a_* = 1$. Heating from below.

the oscillatory instability becomes critical (Fig. 3.10(b), lines 1 - 3). For larger values of a, a new kind of the monotonic instability appears in the long-wave region (Fig. 3.10(b), lines 4 - 7). The motion contains two superimposed counterrotating convective cells emerging in the wide middle layer (Fig. 3.11). Their emergence is due to two in-phase disturbances of the velocity field that develop at the interfaces. The dependence of the critical wave number on a has two jumps (Fig. 3.12). Note that the critical wave number of the oscillatory instability is *larger* than that of the monotonic instability.

3.2.3 Combined Action of Marangoni and Rayleigh Instability Mechanisms

The results of this subsection have been obtained by Nepomnyashchy *et al.* (2005).

Let us consider the system air / ethylene glycol / Fluorinert FC75 (system 3 of Table 3.1) under a simultaneous action of the thermocapillary effect and the buoyancy. Fix the ratios of the layers thicknesses $a = a_* = 1$.

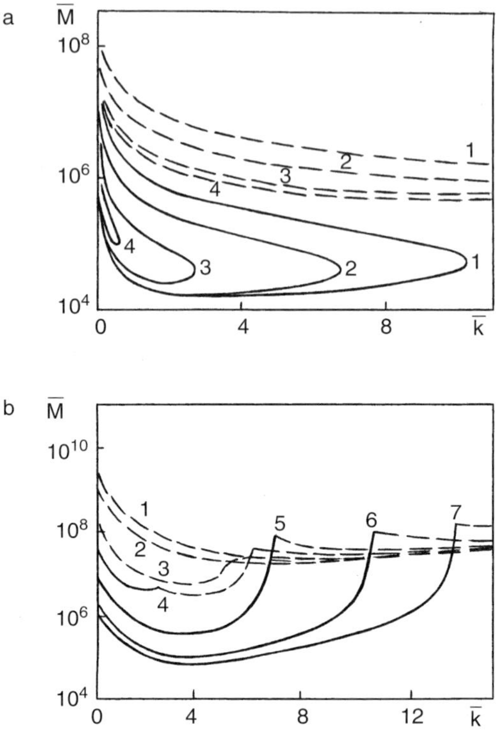

FIGURE 3.10. Neutral stability curves in the case of $a_* = 1$; (a) $a = 0.2$ (lines 1); $a = 0.3$ (lines 2); $a = 0.5$ (lines 3); $a = 0.6$ (lines 4); (b) $a = 0.7$ (line 1); $a = 1.0$ (line 2); $a = 2.0$ (line 3); $a = 2.4$ (line 4); $a = 3.0$ (line 5); $a = 4.0$ (line 6); $a = 7.0$ (line 7). Heating from above.

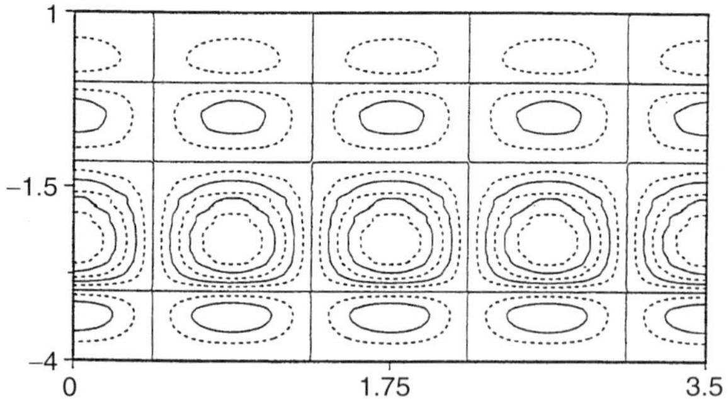

FIGURE 3.11. Stream lines of the flow field in the case of $a = 3$, $a_* = 1$. Heating from above.

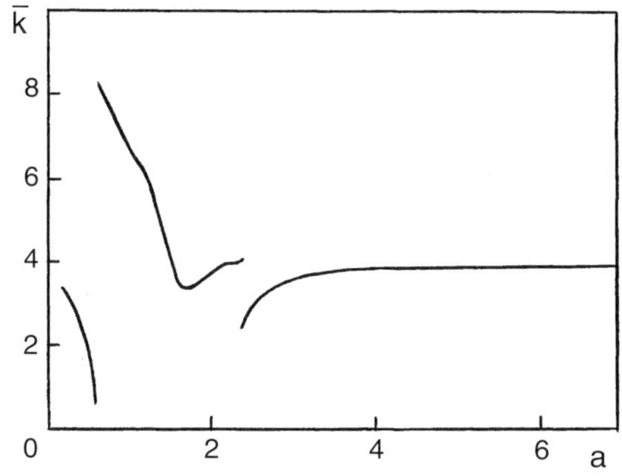

FIGURE 3.12. The critical wave number in the case of $a_* = 1$. Heating from above.

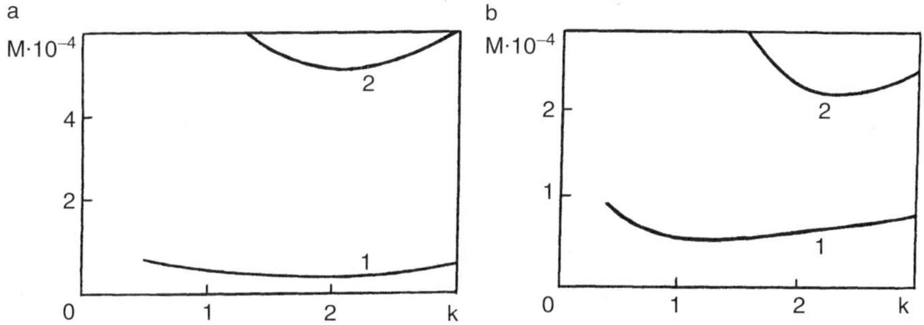

FIGURE 3.13. Monotonic neutral curves: (a) $G = 0$; (b) $G = 1.2$.

It was shown in Sec. 3.1.2 that a pure Marangoni convection is monotonic in this case. As the matter of fact, there are two monotonic neutral curves corresponding to the development of convection near the upper interface (line 1 in Fig. 3.13(a)) and near the lower interface (line 2 in Fig. 3.13(a)).

Recall also (Sec. 3.1.3) that the local Rayleigh numbers for this system differ considerably:

$$R_2/R_1 = 3.5, \quad R_3/R_1 = 1.7 \times 10^3.$$

Because of this, the buoyancy effect is essential only in the bottom layer.

First, let us discuss qualitatively the influence of the buoyancy on the Marangoni convection, which is developed near the upper interface (see Fig. 3.14(a)). Assume that the temperature is higher in the middle of the cell. One can see that an ascending flow is produced by the thermocapillary effect in the middle layer and by the buoyancy in the bottom layer. The viscous stresses generated by the bottom-layer buoyancy convection form a pair of vortices in the middle layer near the lower interface. These vortices prevent the thermocapillary flow. Thus, one can

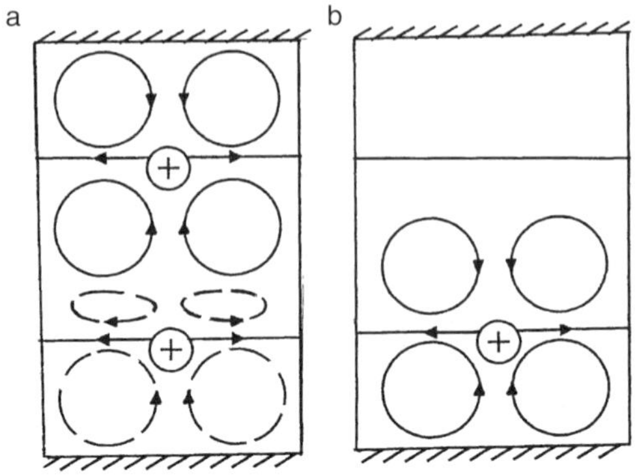

FIGURE 3.14. Influence of buoyancy on the Marangoni convection, which is developed (a) near the upper interface and (b) near the lower interface.

expect that the critical Marangoni number corresponding to the excitation of the Marangoni convection near the upper interface will increase when G increases.

If the Marangoni convection is developed near the lower interface, it is supported by the buoyancy (see Fig. 3.14(b)). Therefore, the upper neutral curve will decrease when G increases. Also, the critical Grashof number for the appearance of the buoyancy convection will decrease when the Marangoni number M increases.

The results of the calculations of neutral stability curves justify the argumentation given above. The comparison of neutral curves obtained for $G = 0$ (Fig. 3.13(a)) and for $G = 1.2$ (Fig. 3.13(b)) allows one to conclude that the lower neutral curve (line 1) goes up and the upper neutral curve (line 2) goes down with the increase of G.

When the Grashof number becomes larger than $G = 1.3$, a fragment of an *oscillatory* neutral curve appears as a small bump near $k \approx 2.2$ (see line 3 in Fig. 3.15(a_1)). The frequency of oscillations vanishes in two codimension-two points on the monotonic neutral curve (see Fig. 3.15(a_2)). The minimum of the oscillatory neutral curve becomes lower than that of the monotonic neutral curve as $G > 1.5$ (see Fig. 3.15(b)).

The unstable manifold of the fixed point corresponding to the mechanical equilibrium is twodimensional in the region between lines 1 and 3 and above line 2 (see Fig. 3.15(b)), whereas in the region between lines 1 and 2, it is onedimensional. Thus, the fragment of the monotonic neutral curve between the codimension-2 points corresponds to the *stabilization* of an unstable mode rather than to a destabilization. At $G \approx 1.552$, lines 1 and 2 approach each other, merge, and reconnect into two new lines: "long-wave" line 4 and "short-wave" line 5, as is shown in Fig. 3.15(c).

As $G \approx 1.77$, a closed monotonic neutral curve (line 6 of Fig. 3.16(a)) appears at $k \approx 3.1$. The instability region inside this curve grows rapidly. As $G > 1.776$,

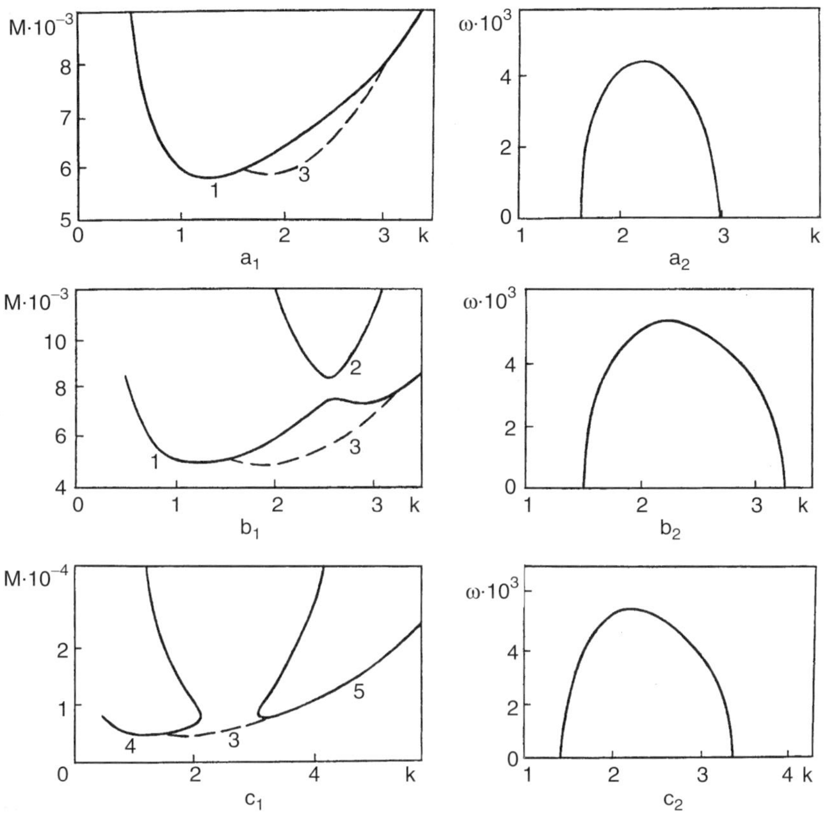

FIGURE 3.15. Neutral curves and dependences of the frequency ω on the wave number k: (a) $G = 1.43$; (b) $G = 1.55$; (c) $G = 1.6$. See the text for explanations.

the minimum of the closed monotonic neutral curve (line 6 of Fig. 3.16(b)) becomes lower than that of the oscillatory neutral curve (line 3 of Fig. 3.16(b)). Lines 6 and 5 couple at $G > 1.85$ (see Fig. 3.16(c)). The minimum of new line 5 decreases and touches the axis $M = 0$, as $G = G_* \approx 1.866$ in the point $k \approx 2.94$. That means that for $G = G_*$ a pure Rayleigh instability is developed as $M = 0$.

In Fig. 3.17, the dependence of the minimized Marangoni number M_c on G is presented. The boundary of the stability region consists of three fragments. Fragment 1 ($0 < G < 1.5$) corresponds to the Marangoni instability, which is generated around the upper interface (i.e., mainly in the top layer and in the middle layer) and is *stabilized* by the buoyancy. Fragment 2 ($1.776 < G < 1.866$) corresponds to the Rayleigh instability, which appears in the bottom layer and is *destabilized* by the thermocapillary effect. The oscillations caused by the competition of both instability mechanisms appear in fragment 3 ($1.5 < G < 1.776$).

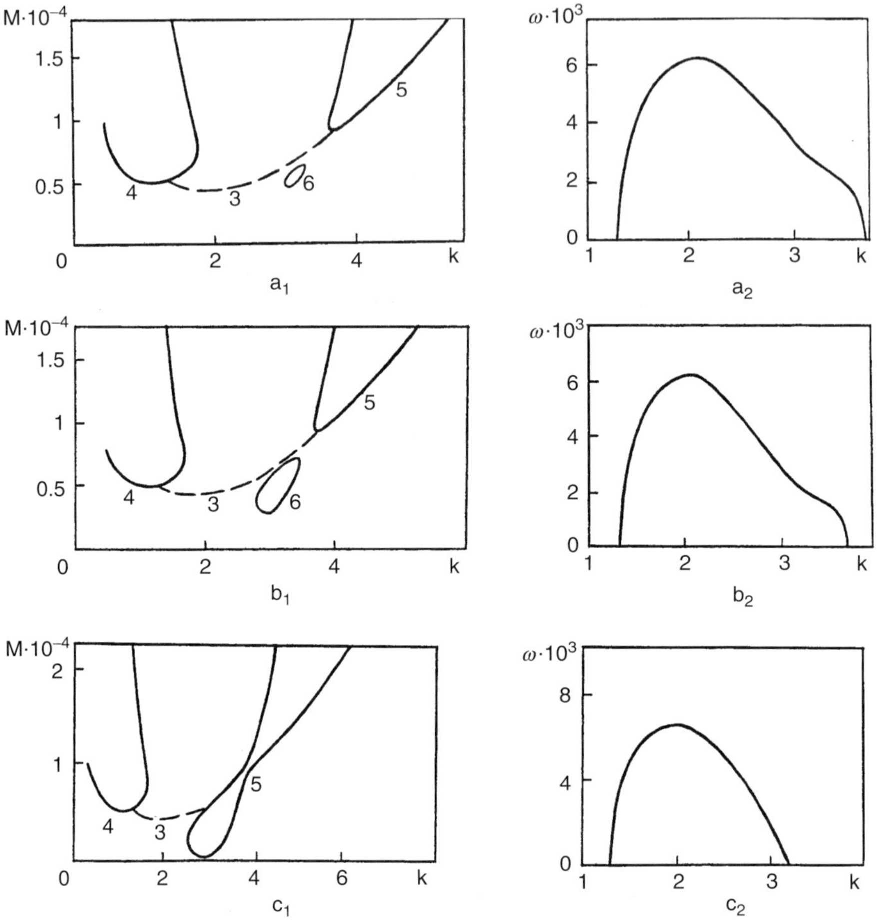

FIGURE 3.16. Neutral curves and dependences of the frequency ω on the wave number k: (a) $G = 1.775$; (b) $G = 1.8$; (c) $G = 1.85$.

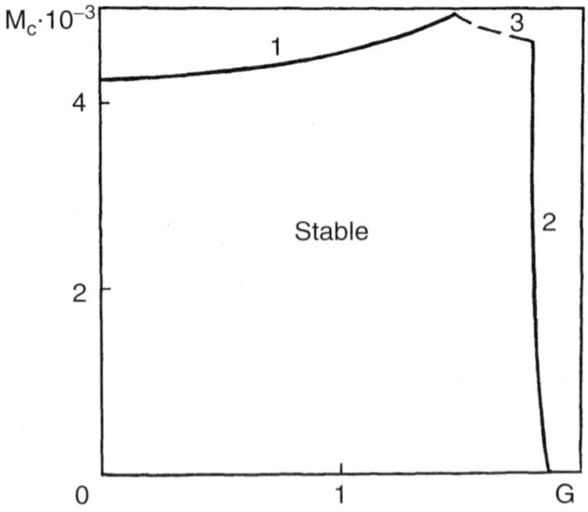

FIGURE 3.17. Stability diagram.

3.3 Nonlinear Simulations

Simulations of two-dimensional finite-amplitude flows are performed in a finite region $-l/2 \le x \le l/2$, $-a_2 - a_3 \le z \le a_1$ (see Fig. 3.18). We introduce the stream function ψ_m and the vorticity ϕ_m,

$$v_{m,x} = \frac{\partial \psi_m}{\partial z}, \quad v_{m,z} = -\frac{\partial \psi_m}{\partial x},$$

$$\phi_m = \frac{\partial v_{m,z}}{\partial x} - \frac{\partial v_{m,x}}{\partial z} \quad (m = 1, 2, 3),$$

and rewrite equations (3.1) and (3.2) in the following form:

$$\frac{\partial \phi_m}{\partial t} + \frac{\partial \psi_m}{\partial z}\frac{\partial \phi_m}{\partial x} - \frac{\partial \psi_m}{\partial x}\frac{\partial \phi_m}{\partial z} = d_m \Delta \phi_m + b_m G \frac{\partial T_m}{\partial x}, \tag{3.41}$$

$$\Delta \psi_m = -\phi_m, \tag{3.42}$$

$$\frac{\partial T_m}{\partial t} + \frac{\partial \psi_m}{\partial z}\frac{\partial T_m}{\partial x} - \frac{\partial \psi_m}{\partial x}\frac{\partial T_m}{\partial z} = \frac{c_m}{P}\Delta T_m \quad (m = 1, 2, 3). \tag{3.43}$$

At the interfaces, normal components of velocity vanish and the continuity conditions for tangential components of velocity, viscous stresses, temperatures, and heat fluxes also apply:

$$z = 0: \quad \psi_1 = \psi_2 = 0, \quad \frac{\partial \psi_1}{\partial z} = \frac{\partial \psi_2}{\partial z}, \quad T_1 = T_2, \tag{3.44}$$

$$\frac{\partial T_1}{\partial z} = \frac{1}{\kappa}\frac{\partial T_2}{\partial z}, \quad \frac{\partial^2 \psi_1}{\partial z^2} = \frac{1}{\eta}\frac{\partial^2 \psi_2}{\partial z^2} + \frac{M}{P}\frac{\partial T_1}{\partial x}, \tag{3.45}$$

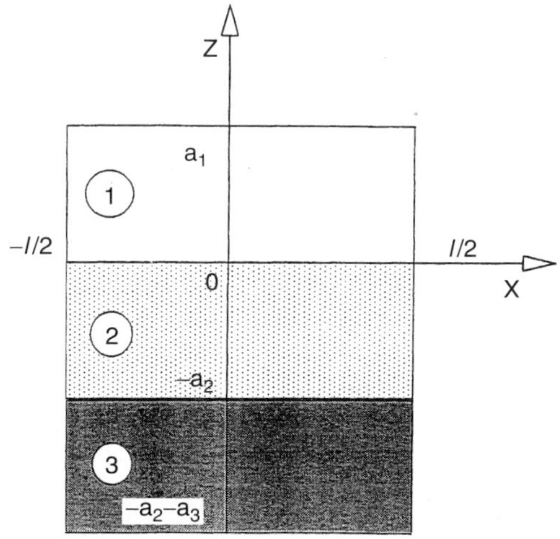

FIGURE 3.18. Computational domain for a three-layer system.

$$z = -a: \quad \psi_2 = \psi_3 = 0, \; \frac{\partial \psi_2}{\partial z} = \frac{\partial \psi_3}{\partial z}, \; T_2 = T_3, \tag{3.46}$$

$$\frac{1}{\kappa}\frac{\partial T_2}{\partial z} = \frac{1}{\kappa_*}\frac{\partial T_3}{\partial z}, \; \frac{1}{\eta}\frac{\partial^2 \psi_2}{\partial z^2} = \frac{1}{\eta_*}\frac{\partial^2 \psi_3}{\partial z^2} + \frac{\bar{a}M}{P}\frac{\partial T_2}{\partial x}. \tag{3.47}$$

On the horizontal solid plates, the boundary conditions read

$$z = 1: \quad \psi_1 = \frac{\partial \psi_1}{\partial z} = 0, \; T_1 = 0, \tag{3.48}$$

$$z = -a - a_*: \quad \psi_3 = \frac{\partial \psi_3}{\partial z} = 0, \; T_3 = s, \tag{3.49}$$

where $s = 1$ for heating from below, and $s = -1$ for heating from above.

Different kinds of boundary conditions are used on lateral walls: $x = \pm L/2$, where $L = l/a_1$.

For simulations of cellular motions in an infinite layer, one can use "free" boundary conditions:

$$x = \pm L/2: \quad \psi_m = \frac{\partial^2 \psi_m}{\partial x^2} = 0 \quad (m = 1, 2, 3). \tag{3.50}$$

In the case of a cavity with rigid walls, the boundary conditions for the stream function on the vertical boundaries are

$$x = \pm L/2: \quad \psi_m = \frac{\partial \psi_m}{\partial x} = 0. \tag{3.51}$$

There are two typical kinds of boundary condition for the temperature on the vertical walls:

(A) Well-conducting boundaries:

$$x = \pm L/2: \quad T_m = A_m z + B_m, \quad m = 1, 2, 3. \tag{3.52}$$

Here, A_m are determined by equation (3.32),

$$B_1 = B_2 = \frac{s}{1 + \kappa a + \kappa_* a_*}, \quad B_3 = s\frac{1 + (\kappa - \kappa_*)a}{1 + \kappa a + \kappa_* a_*}.$$

(B) Heat-insulated boundaries:

$$x = \pm L/2: \quad \frac{\partial T_m}{\partial x} = 0 \quad (m = 1, 2, 3). \tag{3.53}$$

The boundary value problem formulated above was solved by the finite-difference method (see Nepomnyashchy and Simanovskii, 1999). Equations (3.41) – (3.43) were approximated on a uniform mesh using a second-order approximation for the spatial coordinates. To solve the nonlinear equations, explicit scheme on a rectangular uniform mesh $I \times J$ was used. The time step was calculated by the formula

$$\Delta t = \frac{[\min(\Delta x, \Delta z)]^2[\min(1, \nu, \chi, \nu_*, \chi_*)]}{2[2 + \max|\psi_m(x, z)|]},$$

where Δx and Δz are the mesh sizes for the corresponding coordinates. For the solution of the Poisson equations, we used the iterative Liebman successive overrelaxation method on each time step. The Kuskova and Chudov formulas (Kuskova and Chudov, 1968) were used for the approximation of the vorticity on the solid boundaries. For example, on the boundary $x = 0$,

$$\phi_m(0, z) = \frac{\psi_m(2\Delta x, z) - 8\psi_m(\Delta x, z)}{2(\Delta x)^2}.$$

At the interfaces, the expressions for the vorticities at the exterior layers are approximated with the second-order accuracy for the spatial coordinates and have the form

$$\phi_1(x, 0) = -\frac{2[\psi_2(x, -\Delta z) + \psi_1(x, \Delta z)]}{(\Delta z)^2(1 + \eta)} - \frac{M}{P}\frac{\eta}{1+\eta}\frac{\partial T_1}{\partial x}(x, 0), \tag{3.54}$$

$$\phi_2(x, 0) = \eta\phi_1(x, 0) + \frac{\eta M}{P}\frac{\partial T_1}{\partial x}(x, 0), \tag{3.55}$$

$$\phi_2(x, -a) = -\frac{2[\psi_3(x, -a - \Delta z) + \psi_2(x, -a + \Delta z)]}{(\Delta z)^2(1 + \eta_*\eta^{-1})} - \frac{\bar{a}M}{P}\frac{\eta_*}{1 + \eta_*\eta^{-1}}\frac{\partial T_2}{\partial x}(x, -a), \tag{3.56}$$

$$\phi_3(x, -a) = \eta_*\eta^{-1}\phi_2(x, -a) + \frac{\eta_*\bar{a}M}{P}\frac{\partial T_2}{\partial x}(x, -a). \tag{3.57}$$

The temperatures on the interfaces were calculated by the second-order approximation formulas:

$$T_2(x, 0) = T_1(x, 0) = \frac{[4T_2(x, -\Delta z) - T_2(x, -2\Delta z)] + \kappa[4T_1(x, \Delta z) - T_1(x, 2\Delta z)]}{3(1 + \kappa)}, \tag{3.58}$$

$$T_2(x, -a) = T_3(x, -a) = \frac{\kappa_*[4T_2(x, -a + \Delta z) - T_2(x, -a + 2\Delta z)] + \kappa[4T_3(x, -a - \Delta z) - T_3(x, -a - 2\Delta z)]}{3(\kappa + \kappa_*)}. \tag{3.59}$$

3.3.1 *Marangoni Convection. The Case of a Symmetric System*

In this subsection, we consider the case where fluid 1 and fluid 3 are identical ($\nu_* = \chi_* = \kappa_* = \alpha_* = 1$) and the thicknesses of all the layers are equal ($a = a_* = 1$).

Stationary Flows

The results of nonlinear simulations justify the predictions of the linear theory described in the Sec. 3.2.1. Let us consider the system Fluorinert FC70 / silicone oil cs10 / Fluorinert FC70 (System 1 of Table 3.1). Stationary stream lines obtained in the case of free heat-insulated vertical boundaries in a cavity with $L = 3.2$ are presented in Fig. 3.19. The coordinates shown in this figure and other figures corresponding to stream lines and isotherms are $x + L/2$ and $z + a + a_*$. As expected, convective flows appear near the lower interface, because the ratio of thermal diffusivities $\chi < 1$. The field of the stream function satisfies the symmetry condition

$$\psi_m(x, z) = -\psi_m(-x, z), \quad m = 1, 2, 3. \tag{3.60}$$

The results of simulations in the same system with a larger aspect ratio L have been presented by Géoris and Legros (1996).

Quite opposite, in the system air / silicone oil 10cs / air (system 4 of Table 3.1, $\chi \gg 1$) the convection is more intensive near the upper interface (see Fig. 3.20).

In a closed cavity, the convective flows are influenced by the boundary conditions on the lateral walls. Both well-conducting (3.52) and heat-insulated (3.53) boundary conditions permit the solution (3.14) – (3.17) corresponding to the mechanical equilibrium state. In the region $M < M_c$, this state is stable, and the convective motion is absent. In the region $M > M_c$, a new solution of the boundary value problem appears ("perfect bifurcation") which corresponds to a convective flow.

However, in reality the rigid lateral walls of the cavity have a finite heat conductivity that is different from the heat conductivities of fluids. When the whole

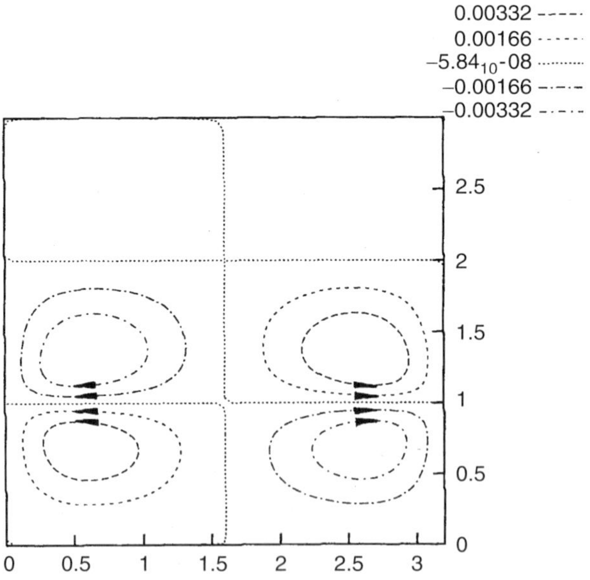

FIGURE 3.19. Stream lines for system 1 of Table 3.1 ($\chi < 1$); $M = 3000$.

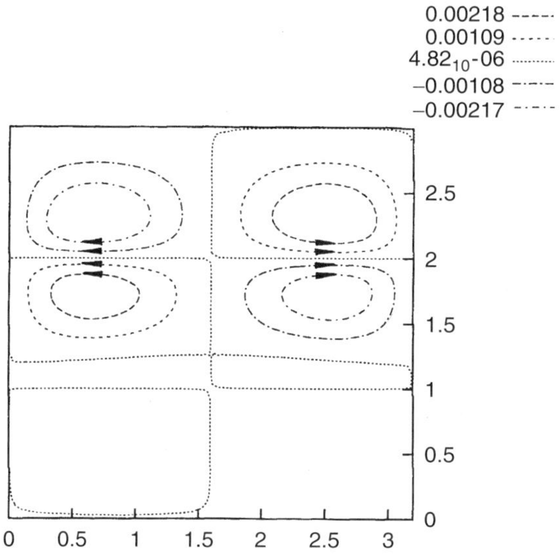

FIGURE 3.20. Stream lines for system 4 of Table 3.1 ($\chi > 1$); $M = 3000$.

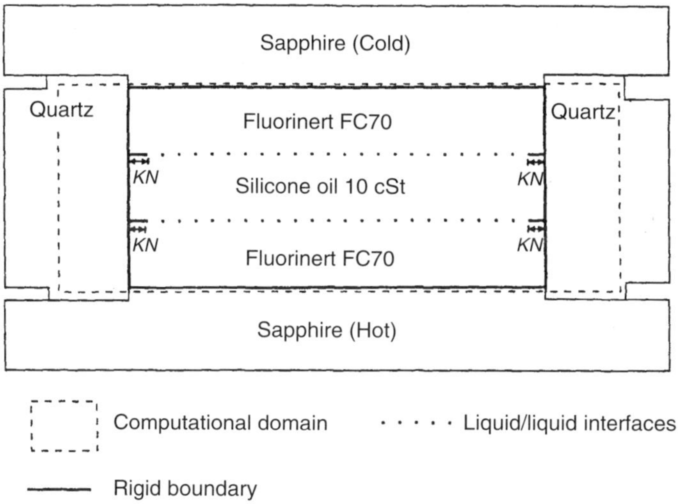

FIGURE 3.21. Computational domain.

system consisting of three fluid layers and lateral rigid walls is heated from below, the temperature field is different from the solution (3.15) – (3.17). It depends on the horizontal coordinates; therefore, *for any* $M \neq 0$, a thermocapillary convection appears. Simulations corresponding to this case have been performed by Géoris *et al.* (1999). The computational domain is shown in Fig. 3.21. In order to model antiwetting knife edges used in the experiments (see Sec. 3.4), some additional rows of nodes (KN in Fig. 3.21) with rigid boundary conditions have been used. The evolution equations have been discretized on a uniform rectangular grid and solved using the ADI scheme (Peyret and Taylor, 1990).

Indeed, for arbitrary weak heating, a thermocapillary flow appears close to the lateral walls (regime 1; see Fig. 3.22(a)). When the Marangoni number passes through the critical value $M = M_c$ corresponding to the onset of the Marangoni convection in the cavity with the heat-insulated lateral boundary, the intensity of the flow changes smoothly ("imperfect" bifurcation; see Fig. 3.23). The boundary temperature disturbance generates a system of convective rolls that fills the cavity (regime 2; see Figs. 3.22(b), and 3.22(c)). For larger values of the Marangoni number, the instability-generated convective rolls located in the center of the cavity suppress the lateral wall-induced rolls (regime 3; see Fig. 3.22(d)).

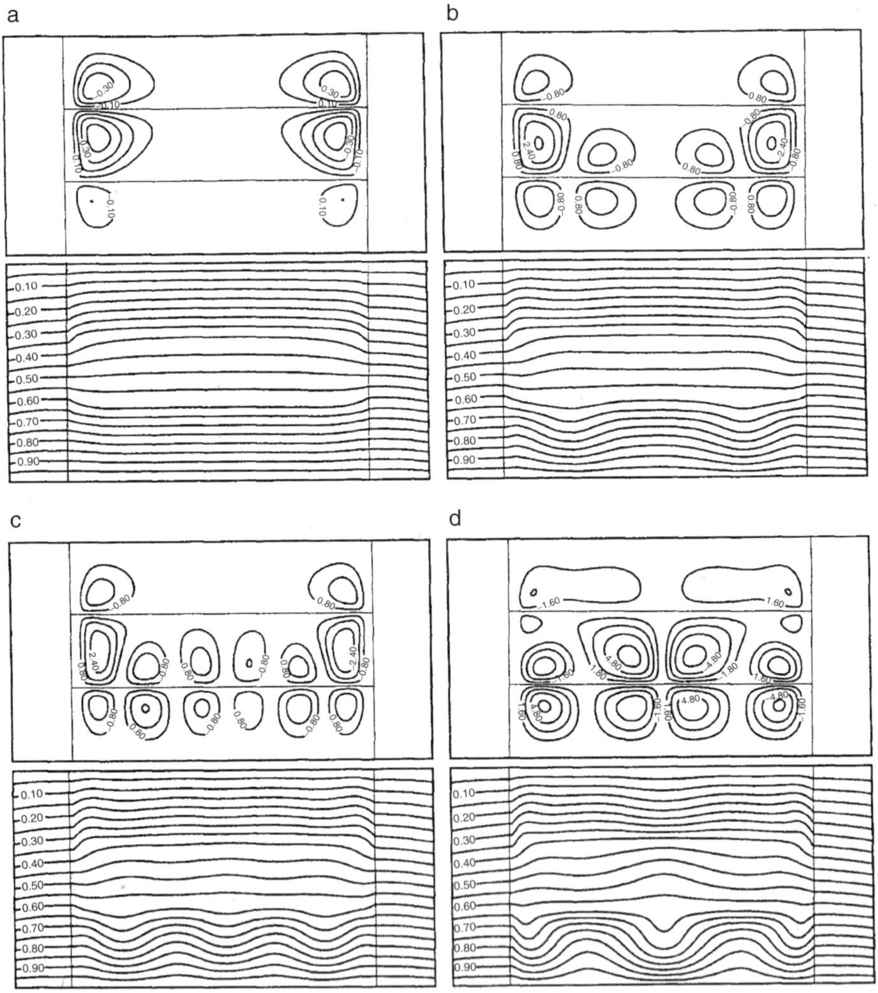

FIGURE 3.22. Stream lines and isotherms: (a) regime 1; (b) regime 2; (c) transition between regimes 2 and 3; (d) regime 3.

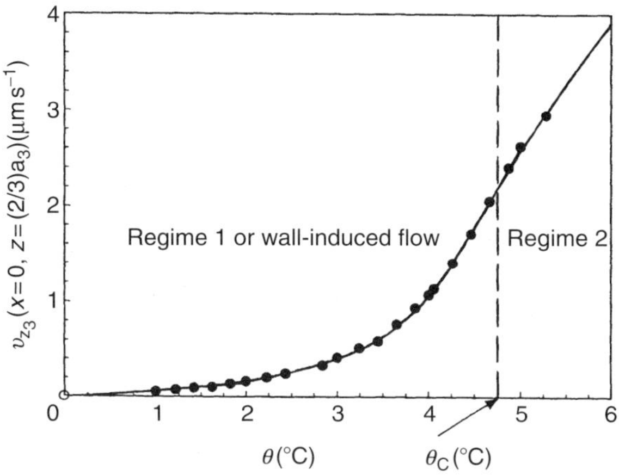

FIGURE 3.23. Dependence of the motion intensity on the temperature drop.

Oscillatory Flows

Let us describe now results of nonlinear simulations of oscillations. First, let us consider the model system with the set of parameters $\eta = \nu = 0.5$, $\kappa = \chi = P = a = 1$. As was discussed in the previous section, the mechanical equilibrium is unstable with respect to oscillatory disturbances; the linear stability curve is shown in Fig. 3.1 (line 5).

The nonlinear simulations have been performed for $L = 3$, $M = 12,000$ in the region with free heat-insulated vertical boundaries (Simanovskii *et al.*, 1992; Géoris *et al.*, 1993). The calculated flows correspond to a spatially periodic motion with period L in a system infinite in the horizontal direction. Let us describe qualitatively the flow evolution during the period of oscillations $0 < t < \tau$ (see Figs. 3.24 and 3.25). For any values of t, the fields of stream function and temperature satisfy the symmetry conditions

$$\psi_m(x, z, t) = -\psi_m(-x, z, t), \; T_m(x, z, t) = T_m(-x, z, t), \; m = 1, 2, 3. \tag{3.61}$$

We start from the state where an intensive thermocapillary convection takes place mainly in the top layer and in the middle layer, and the fluid in the bottom layer is almost stagnant (Figs. 3.24(a_1) and 3.25(a_2)). The fluid motion in the middle layer induces two weak vortices in the bottom layer. Because the descending flow in the center of the middle layer is more intensive than the ascending flow in the bottom layer, a minimum of the temperature distribution appears in the central point of the lower interface (Fig. 3.25(b_2)). As the result, the thermocapillary stresses on the lower interface generate a new four-vortex structure, which consists of two vortices in the bottom layer and two vortices in the middle layer (see Fig. 3.24(b_1)). A "two-storied" structure appears in the middle layer. Eventually, the new vortices oust the former ones in the middle layer. An intensive motion developing in the middle layer induces a motion in the opposite direction in the top layer

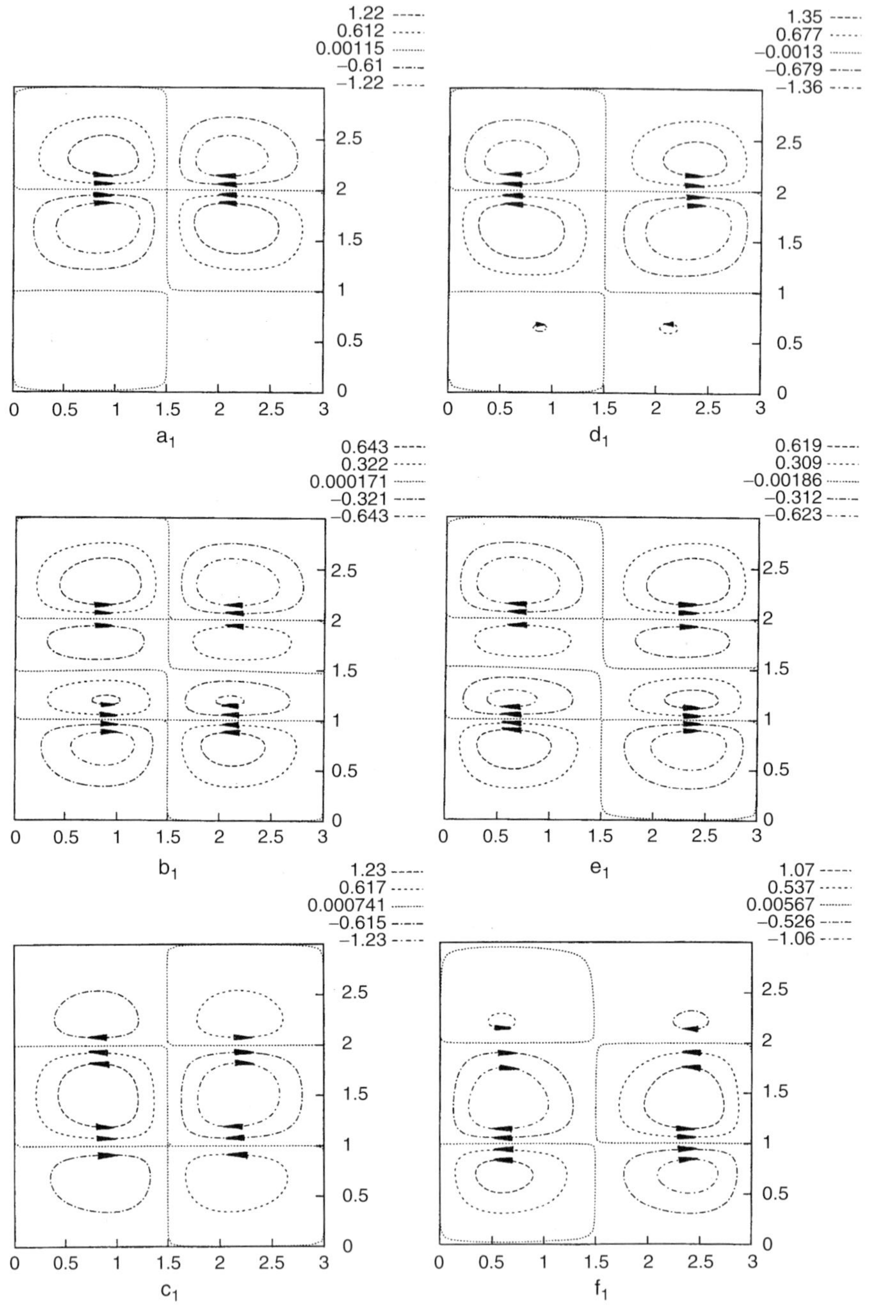

FIGURE 3.24. Stream lines (a_1) - (f_1) for the periodic oscillatory motion in the model system at $M = 6000$, $L = 3$.

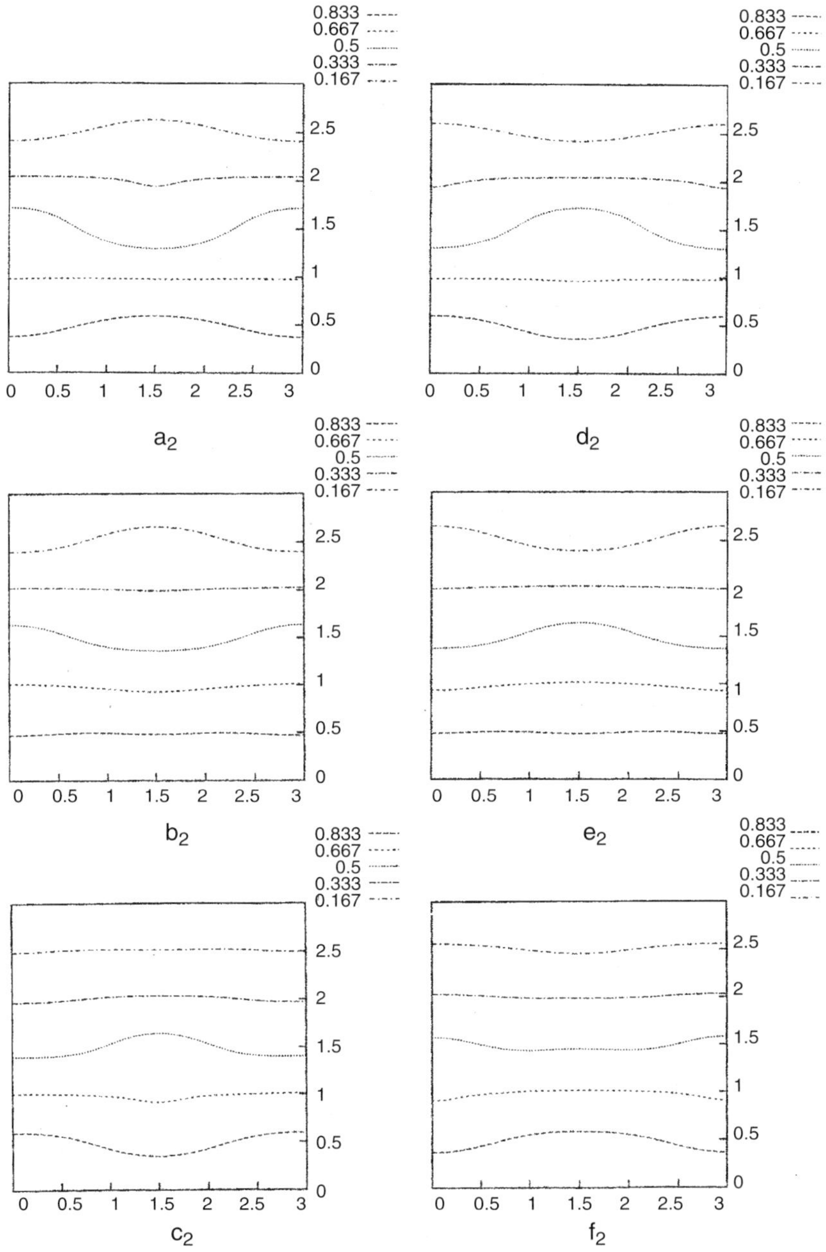

FIGURE 3.25. Isotherms (a_2) - (f_2) for the periodic oscillatory motion in the model system at $M = 6000$, $L = 3$.

(Fig. 3.24(c_1)) and diminishes the temperature in the center of the lower interface (Fig. 3.25(c_2)). Because of the latter phenomenon, the motion in the bottom layer slows down (Figs. 3.24(d_1) and 3.25(d_2)) and changes its direction (Figs. 3.24(e_1) and 3.25(e_2)). A "two-storied" structure appears again in the middle layer. Note that after half a period, $\tau/2$, the structure coincides with the initial one, but it is shifted on the distance $L/2$ in the horizontal direction:

$$\psi_m(x, z, t + \tau/2) = \psi_m(x + L/2, z, t), \quad T_m(x, z, t + \tau/2) = T_m(x + L/2, z, t), \tag{3.62}$$

$$m = 1, 2, 3.$$

The new growing vortices in the middle layer suppress the upper pair of vortices (Fig. 3.24(f_1)), enhance the temperature near the side walls on the upper interface (Fig. 3.25(f_2)), and diminish the temperature in the center of the lower interface. That is why the flow in the bottom layer is suppressed, and, finally, the structure returns to the configuration of Figs. 3.24(a_1) and 3.25(a_2).

Exactly the same type of oscillations is observed in the real system n-octane / methanol / n-octane (system 2 in Table 3.1) in the interval $M_c < M < M_*$, $M_c \approx 4300$ and $M_* \approx 20,450$ (Simanovskii *et al.*, 2004). The corresponding flow patterns shown in Figs. 3.26 and 3.27 are almost identical to those discussed above.

Let us define the following integral quantities:

$$S_{l1}(t) = \int_{-L/2}^{0} dx \int_{0}^{1} dz \psi_1(x, z, t), \ S_{r1}(t) = \int_{0}^{L/2} dx \int_{0}^{1} dz \psi_1(x, z, t); \tag{3.63}$$

$$S_{l2}(t) = \int_{-L/2}^{0} dx \int_{-a}^{0} dz \psi_2(x, z, t), \ S_{r2}(t) = \int_{0}^{L/2} dx \int_{-a}^{0} dz \psi_2(x, z, t); \tag{3.64}$$

$$S_{l3}(t) = \int_{-L/2}^{0} dx \int_{-a-a_*}^{-a} dz \psi_3(x, z, t), \ S_{r3}(t) = \int_{0}^{L/2} dx \int_{-a-a_*}^{-a} dz \psi_3(x, z, t). \tag{3.65}$$

The time evolution of the quantities $S_{lm}(t)$, $m = 1, 2, 3$, is shown in Fig. 3.28. The oscillations are almost harmonic; the largest amplitude of oscillations is observed in the middle layer. For symmetric oscillations satisfying conditions (3.61), the integral quantities satisfy the relations:

$$S_{l1}(t) = -S_{r1}(t), \ S_{l2}(t) = -S_{r2}(t), \ S_{l3}(t) = -S_{r3}(t) \tag{3.66}$$

(see Fig. 3.29(a)). The phase trajectory presented in Fig. 3.29(b) shows that the oscillations in the top layer and in the middle layer take place almost out of phase. The phase trajectory given in Fig. 3.29(c) demonstrates a significant phase delay of the oscillations in the top layer with respect to the oscillations in the bottom layer. The period of oscillations τ decreases with the growth of the Marangoni number M (see line 1 of Fig. 3.30).

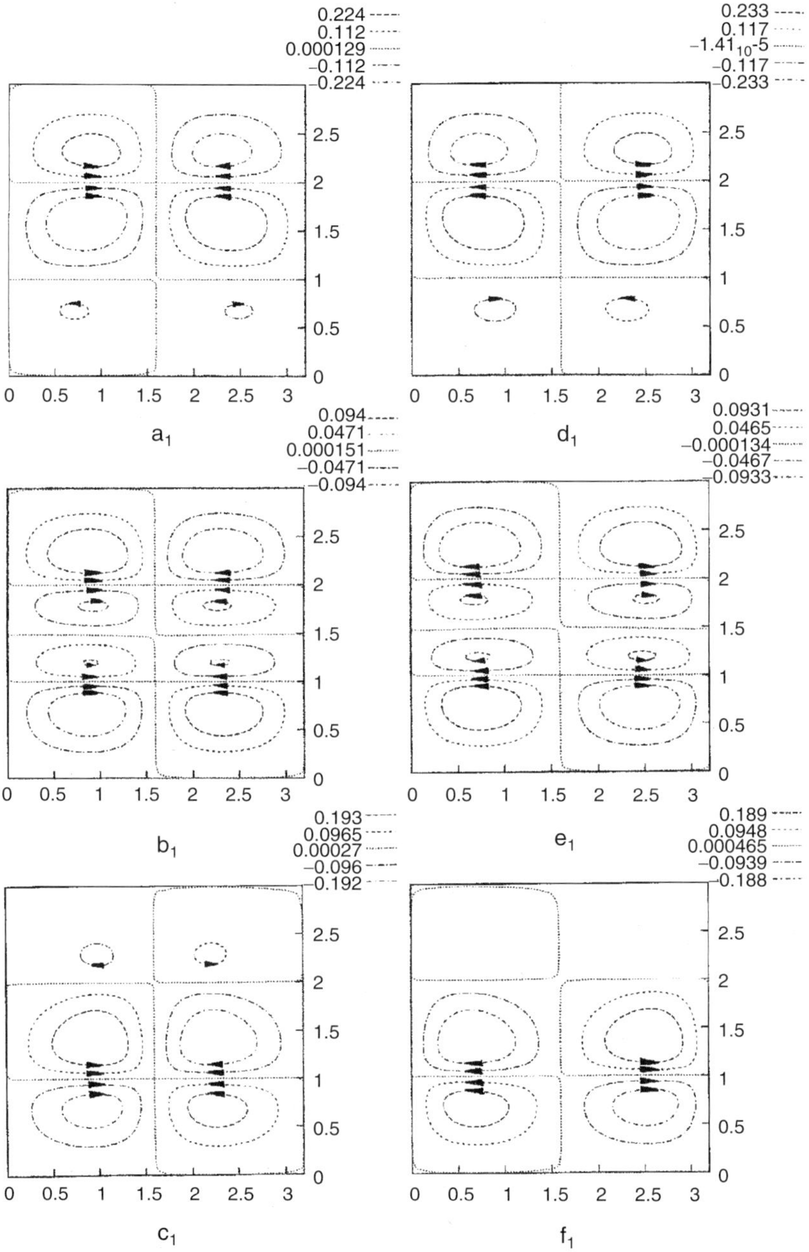

FIGURE 3.26. Stream lines (a_1) - (f_1) for the periodic oscillatory motion in the system n-octane / methanol / n-octane at $M = 6000$, $L = 3.2$.

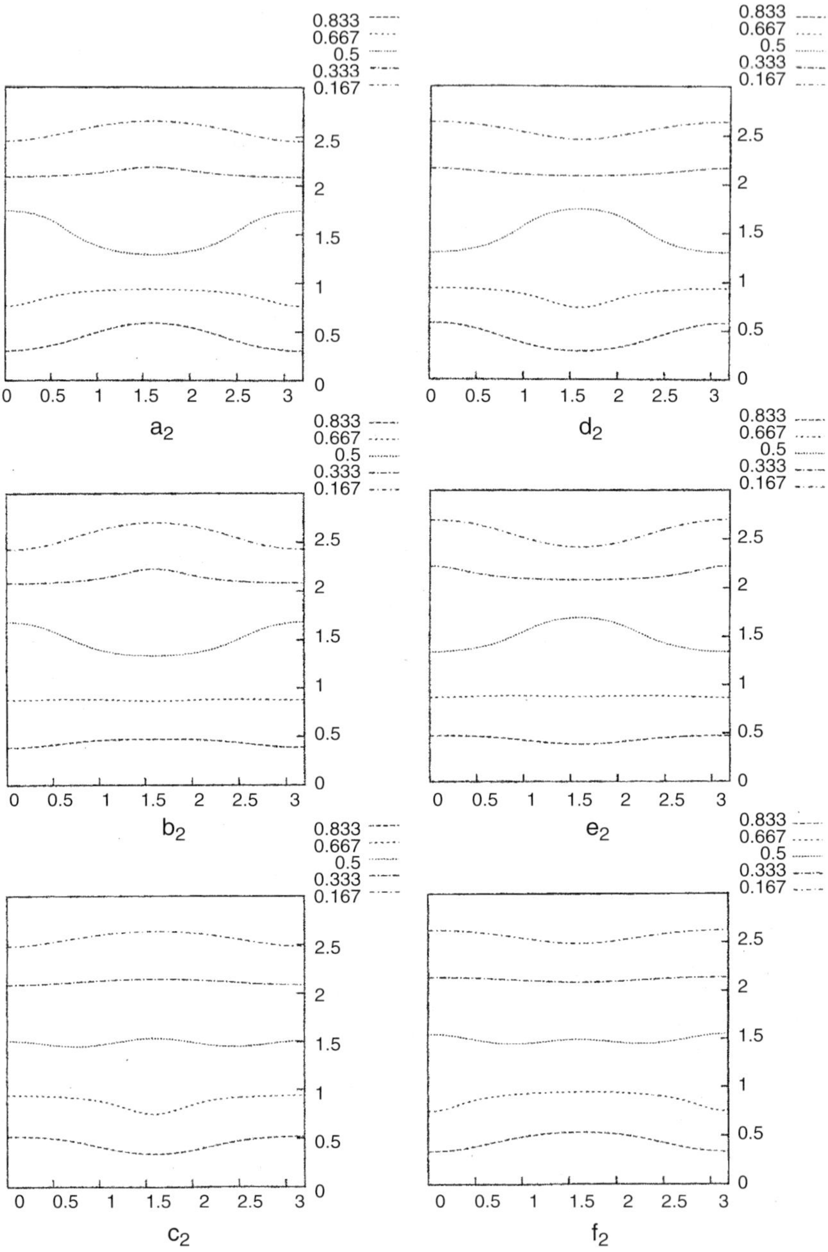

FIGURE 3.27. Isotherms (a_2) - (f_2) for the periodic oscillatory motion in the system n-octane / methanol / n-octane at $M = 6000$, $L = 3.2$.

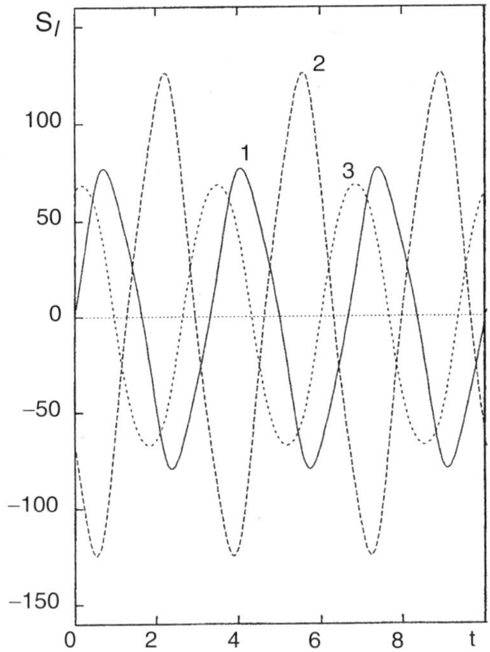

FIGURE 3.28. Dependence of S_{lm} $(m = 1, 2, 3)$ on t for the symmetric flow at $M = 6000$, $L = 3.2$.

If $M > M_*$, the symmetric oscillatory motion is unstable. A new, asymmetric, motion is developed through a period-doubling bifurcation. For this motion, the symmetry conditions (3.61) and (3.66) are violated. The time evolution of quantities $S_{l,m}$ $(m = 1, 2, 3)$ for the asymmetric motion is shown in Fig. 3.31. Note that this evolution is essentially nonsinusoidal. The phase trajectories of asymmetric periodic oscillations in variables (S_{1l}, S_{1r}), (S_{1l}, S_{2l}), and (S_{1l}, S_{3l}) are shown in Fig. 3.32. The period of oscillations decreases with M (line 2 of Fig. 3.30).

For larger values of M, the motion becomes aperiodic in time (see Fig. 3.33).

The numerical simulation of the oscillatory Marangoni convection in the same system but between lateral quartz walls has been fulfilled by Legros and Géoris (1997).

3.3.2 Marangoni Convection. The Case of a Nonsymmetric System

Symmetric systems considered in the previous subsection cannot exist under the action of gravity because of the Rayleigh-Taylor instability. We will discuss results of the numerical simulation of asymmetric systems.

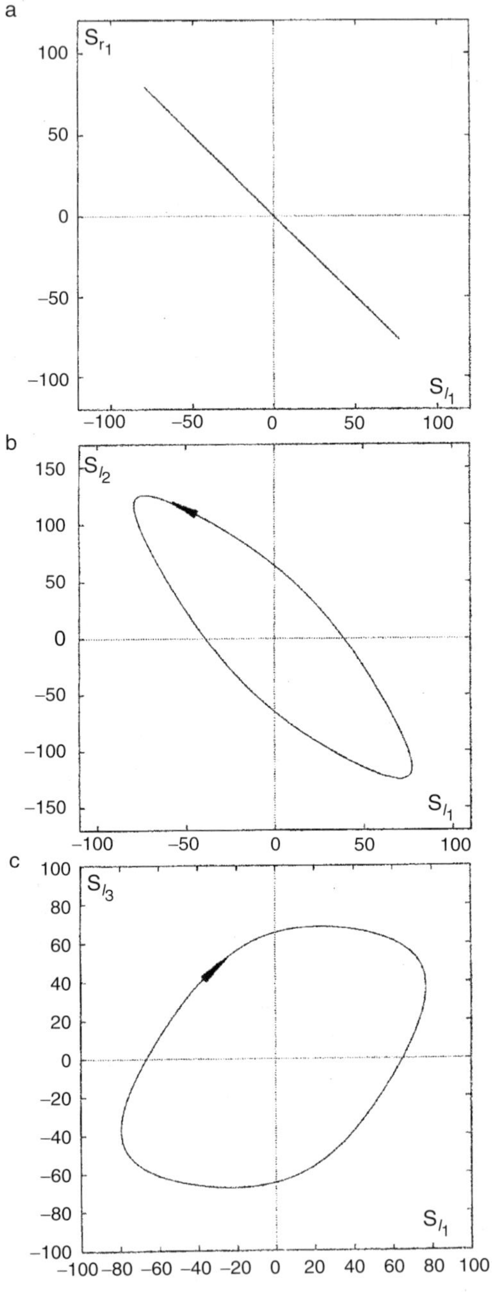

FIGURE 3.29. Phase trajectories in variables (a) S_{l1}, S_{r1}, (b) S_{l1}, S_{l2}, and (c) S_{l1}, S_{l3} for the periodic symmetric oscillatory motion in the system n-octane / methanol / n-octane at $M = 6000$, $L = 3.2$.

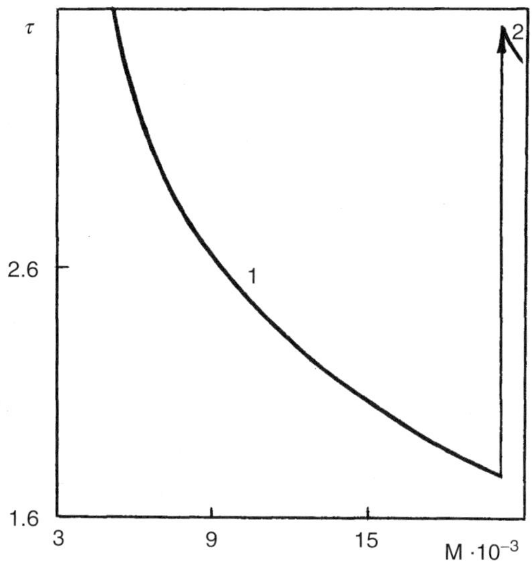

FIGURE 3.30. Dependence of the oscillations period τ on the Marangoni number M for the system n-octane / methanol / n-octane; $L = 3.2$.

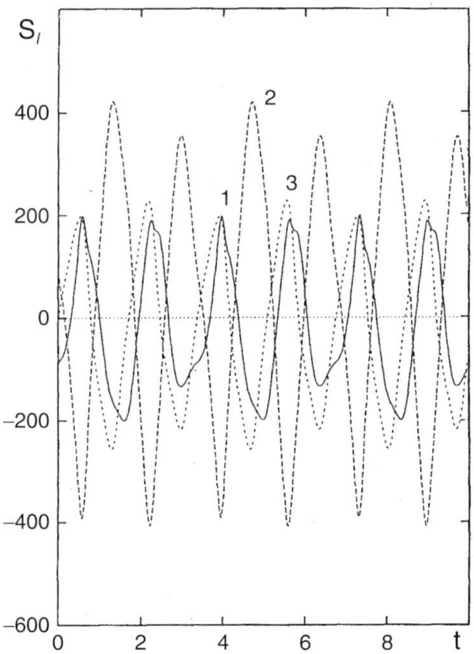

FIGURE 3.31. Dependence of S_{lm} ($m = 1, 2, 3$) on t for the asymmetric flow at $M = 20600$, $L = 3.2$.

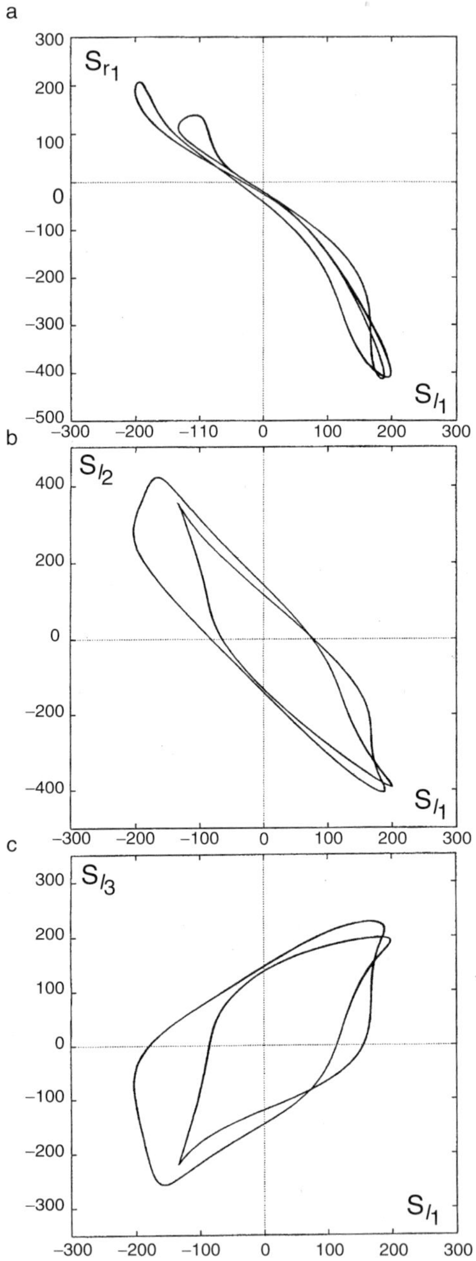

FIGURE 3.32. Phase trajectories in variables (a) S_{l1}, S_{r1}, (b) S_{l1}, S_{l2}, and (c) S_{l1}, S_{l3} for the oscillatory motion after the period-doubling bifurcation in the system n-octane / methanol / n-octane ($M = 20, 600$, $L = 3.2$).

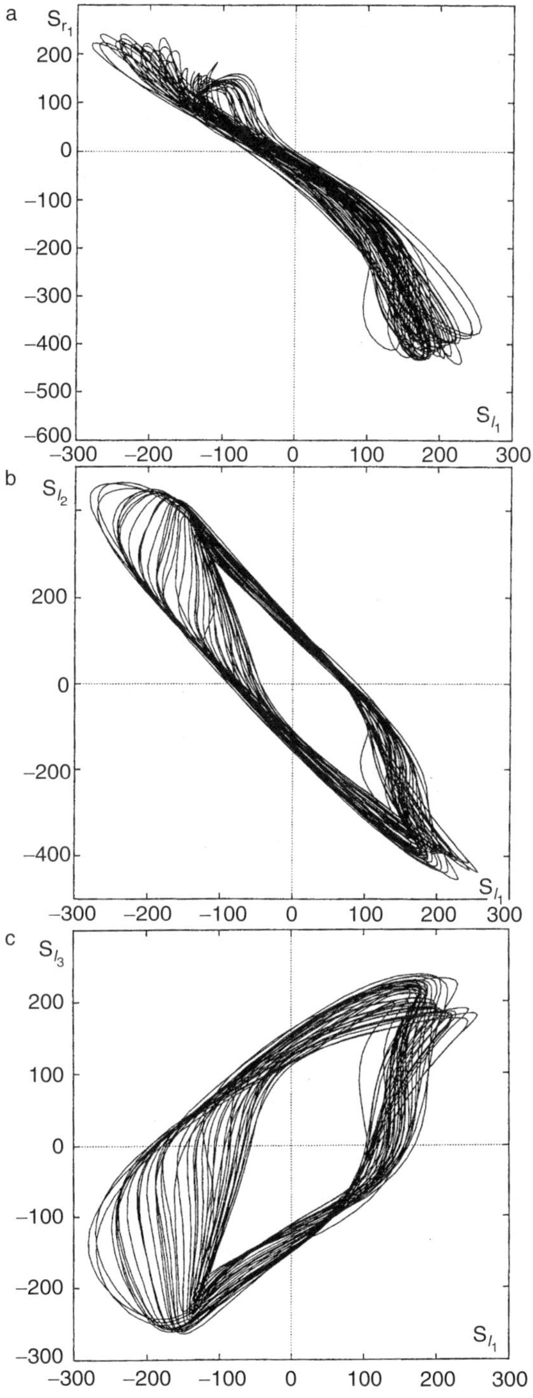

FIGURE 3.33. Phase trajectories in variables (a) S_{l1}, S_{r1}, (b) S_{l1}, S_{l2}, and (c) S_{l1}, S_{l3} for the aperiodic oscillatory motion in the system n-octane / methanol / n-octane ($M = 20, 750$, $L = 3.2$).

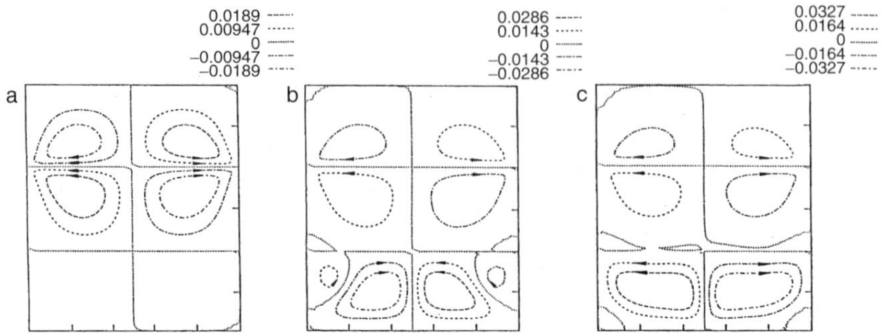

FIGURE 3.34. Stream lines for the system air / ethylene glycol / Fluorinert FC75 ($a = a_* = 1$); (a) $G = 0$, $M = 50,000$; (b) $G = 75$, $M = 25,000$; (c) $G = 100$, $M = 25,000$.

Stationary Motions

Let us consider the system air / ethylene glycol / Fluorinert FC75 (system 3 of Table 3.1) heated from below with $a = a_* = 1$. The simulation has been fulfilled for rigid well-conducting boundaries (3.51) and (3.52) with $L = 2.5$.

The linear stability theory predicts the appearance of the stationary convection near the upper interface (see Sec. 3.2.1). This prediction has been justified by the numerical simulation (see Fig. 3.34(a)). One could expect that two different kinds of thermocapillary motion are possible in a closed cavity: (i) the motion with the temperature near the lateral walls lower than that near the symmetry plane $x = 0$ and (ii) the motion with the opposite way of the temperature-field distribution. Let us define the first case as the case of "cold corners" and the second case as the case of "hot corners". It turns out, however, that only the flow with "cold corners" at the upper interface is observed in numerics. One can assume that this type of motion is selected due to the positive feedback for a thermocapillary flow in the case of a "cold corner" (see Canright, 1994). A similar motion has been obtained also in the case $a = 1$, $a_* = 4$ (see Fig. 3.35(a)).

Oscillatory Motions

Let us consider now the model system, in which all the parameters of fluids are equal except dynamic and kinematic viscosities: $\kappa = \kappa_* = \chi = \chi_* = a = a_* = \bar{a} = 1$, $\eta = \nu = 0.5$, and $\eta_* = \nu_* = 0.25$. It means that the bottom layer has the highest viscosity. We choose $P = 1$ and $L = 2.5$ (Simanovskii, 1996). The simulations have been performed with the rigid well-conducting boundary conditions (3.51) and (3.52).

The linear stability theory for infinite layers predicts an oscillatory instability. Periodic in time oscillations, which satisfy the symmetry conditions (3.61), appear above the instability threshold. The dependences of the integral quantity (3.63) on time for different values of M are shown in Fig. 3.36. Near the threshold, the oscillations have a rather simple, almost sinusoidal form (see Fig. 3.36, line 1), but the mean value of $S_{l1}(t)$ is different from zero. With the increase in the

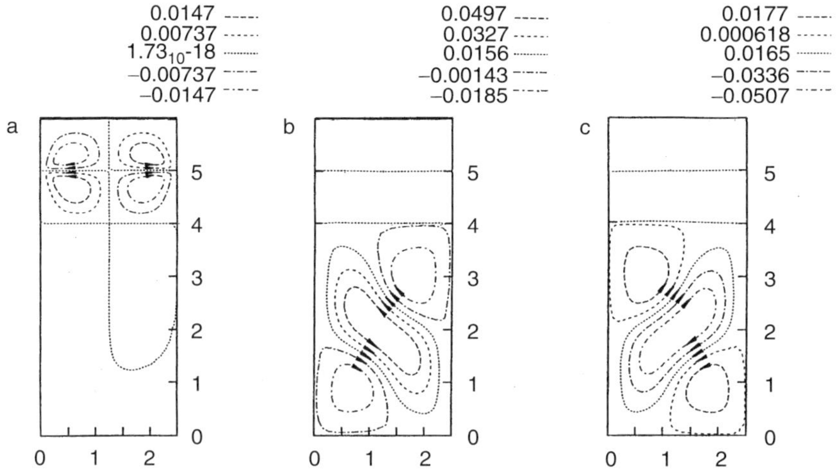

FIGURE 3.35. Stream lines ($a = 1$, $a_* = 4$) for (a) steady thermocapillary convection ($G = 0$, $M = 60,000$); (b), (c) oscillatory buoyancy convection ($G = 5$, $M = 0$).

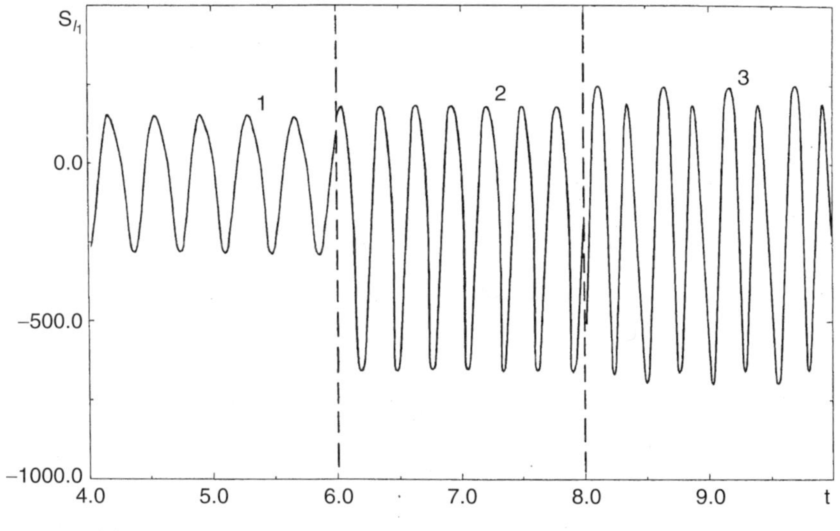

FIGURE 3.36. Oscillations of $S_{l1}(t)$ for the model system (line 1: $M = 16,000$, line 2: $M = 27,500$, line 3: $M = 31,200$).

Marangoni number, the amplitude of oscillations grows (see Fig. 3.36, line 2), and their period τ decreases (see Fig. 3.37, line 1).

The main stages of the flow evolution (see Fig. 3.38) coincide with those found for oscillations in infinite layers studied in the case of symmetric systems (cf. Figs. 3.24 – 3.27). Note, however, that in the case of infinite layers, the "cold

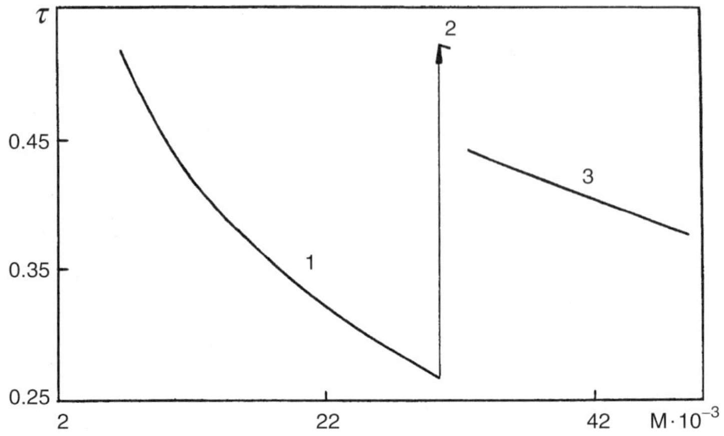

FIGURE 3.37. Dependence of the oscillation period τ on M for characteristic types of oscillation.

corner" configuration and the "hot corner" configuration are indistinguishable, and relations (3.62) are satisfied. In the closed cavity with rigid lateral boundaries, the intensity of flow in a "hot corner" configuration of Fig. 3.38(e) is much lower than that in a "cold corner" configuration of Fig. 3.38(a).

The effect of the "cold corner" is especially strong for the vorticity field. In Fig. 3.39, the time evolution of the vorticity on the interfaces in the points $(x, y) = (L/2 - \Delta x, 0)$ and $(x, y) = (L/2 - \Delta x, -a)$, where Δx is the mesh size for the horizontal coordinate, is presented (recall that in the case of well-conducting vertical boundaries, the vorticity is not defined in the corner points (Moffat, 1964)). One can see that the sign of the vorticity is positive during almost the whole period of oscillations. This sign of the vorticity corresponds to a "cold corner" motion. Let us note that the second harmonics caused by the nonlinearity are much stronger in the oscillations of the local value of vorticity than in the oscillations of the integral characteristics.

As $M \approx 30,500$, the symmetric oscillatory periodic motion becomes unstable with respect to disturbances violating the symmetry conditions (3.62). In order to characterize the violation of the symmetry, let us define

$$S_+ = S_{l1} + S_{r1}, \quad S_- = S_{l1} - S_{r1}. \tag{3.67}$$

For the symmetric motion, which satisfies conditions (3.62), $S_- = 2S_{l1}$ is nonzero and oscillates, whereas $S_+ = 0$. For the asymmetric motion, both variables S_- and S_+ are nonzero and oscillate. Generally, we could expect that the oscillations of S_+ would arise with a frequency incommensurable to the oscillation frequency of S_-. However, it turns out that a synchronization takes place: The oscillation frequency of S_- is exactly one-half of the oscillation frequency of S_+. Hence, one observes a *period-doubling* (subharmonic bifurcation) of the limit cycle (see Fig. 3.40(a)). The typical dependence $S_{l1}(t)$ is shown in Fig. 3.36, line 3.

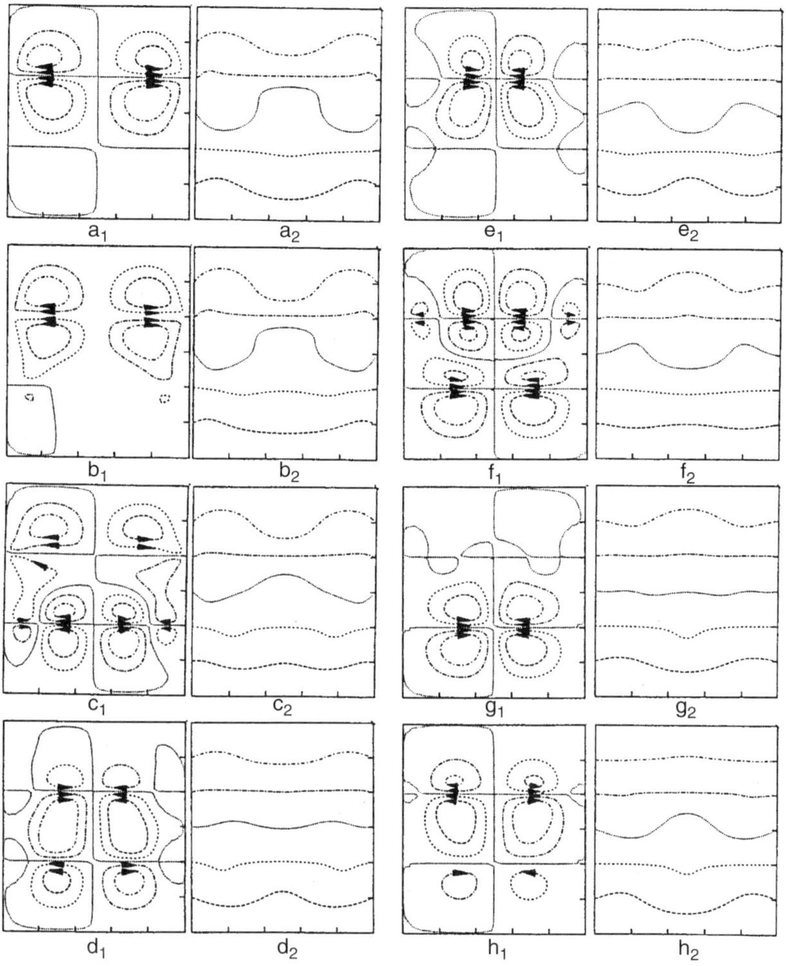

FIGURE 3.38. Stream lines (a_1) - (h_1) and isotherms (a_2) - (h_2) during the period of oscillations; $M = 27,500$. Time interval between neighboring pictures is $\tau/8$.

The time evolution of the stream-function field during a half of the period is shown in Fig. 3.41. The general evolution of flows is similar to one shown in Fig. 3.38. Some pictures look similar (e.g., Fig. 3.38(c) and Fig. 3.41(e), Fig. 3.38(d) and Fig. 3.41(g), Fig. 3.38(e) and Fig. 3.41(h)). However, the violation of the symmetry condition (3.62) is obvious during the main part of the period. In Figs. 3.41(i) and 3.41(j) the motion in the middle layer is dominated by one strong vortex. In Figs. 3.41(a), 3.41(k), and 3.41(l), a one-vortex structure is observed in the top layer.

Comparing Figs. 3.41(a) and 3.41(l) one can see that

$$\psi_m(x, z, t + \tau/2) = -\psi_m(-x, z, t), \quad T_m(x, z, t + \tau/2) = T_m(-x, z, t). \quad (3.68)$$

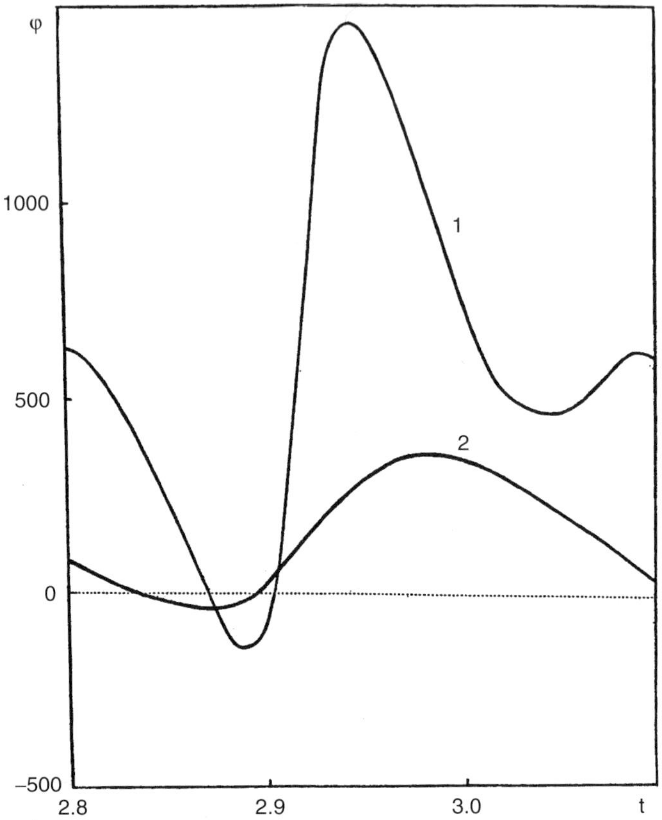

FIGURE 3.39. Oscillations of the vorticity during the period near the right corner on the upper interface (line 1) and on the lower interface (line 2); $M = 27,500$.

After the period-doubling bifurcation, with the increase of M the period of oscillations decreases. For larger values of M, the periodicity of the motion is destroyed; the phase trajectory is not closed (see Fig. 3.40(b)).

The appearance of asymmetric oscillations described above is connected with the competition of one- and two-vortex structures. In infinite layers, these modes would correspond to different values of the wave number. The nonlinear interaction of both modes generates complicated regimes of flow.

For $M > 32,000$, a new periodic regime is established. The oscillations are strongly nonsinusoidal from the very beginning (see Fig. 3.42). The amplitude increases and the period decreases with the increase in the parameter M (see Fig. 3.37, line 3). These oscillations satisfy the same symmetry property (3.68) as the ones described above (see Fig. 3.40(c)), but the evolution of stream lines and isotherms is completely different (see Fig. 3.43). As a rule, there is only one main vortex in each layer; two-vortex structures are absent. During one-half of the period, the vortex moves from the left to the right, then it is reflected by the

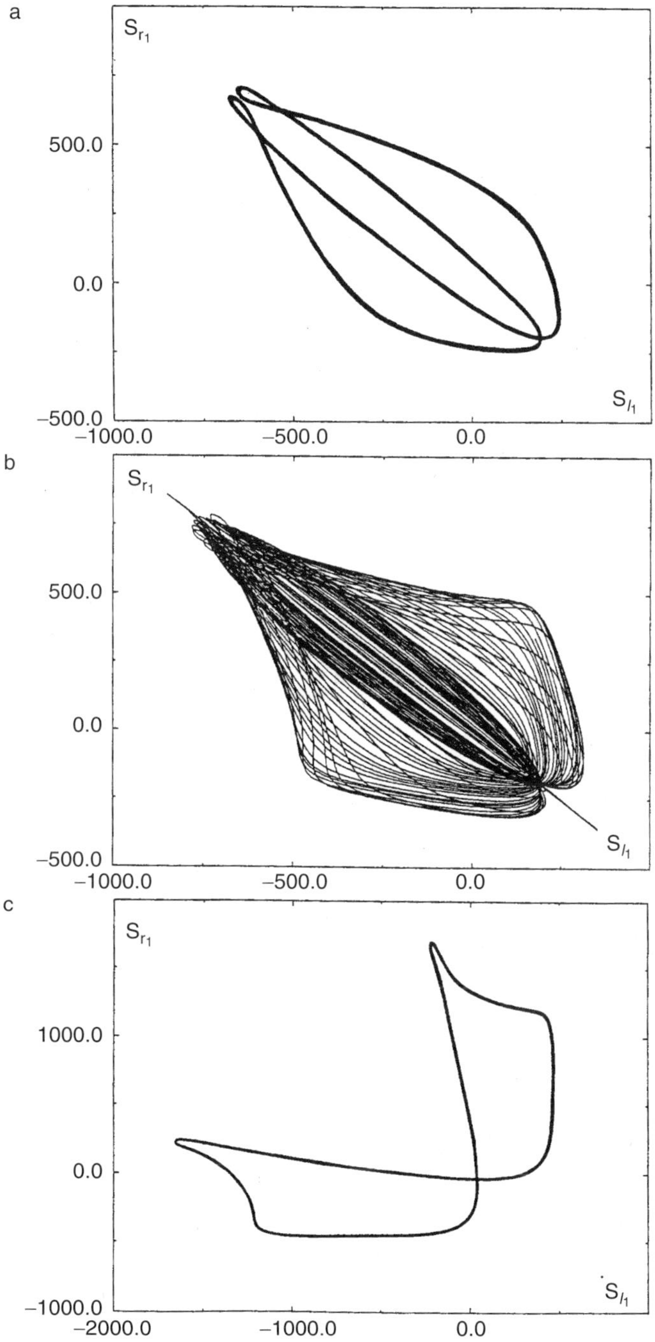

FIGURE 3.40. Phase trajectories of (a) the periodic motion after the period-doubling bifurcation ($M = 31,200$); (b) the nonperiodic motion ($M = 32,000$); (c) the periodic motion ($M = 47,500$).

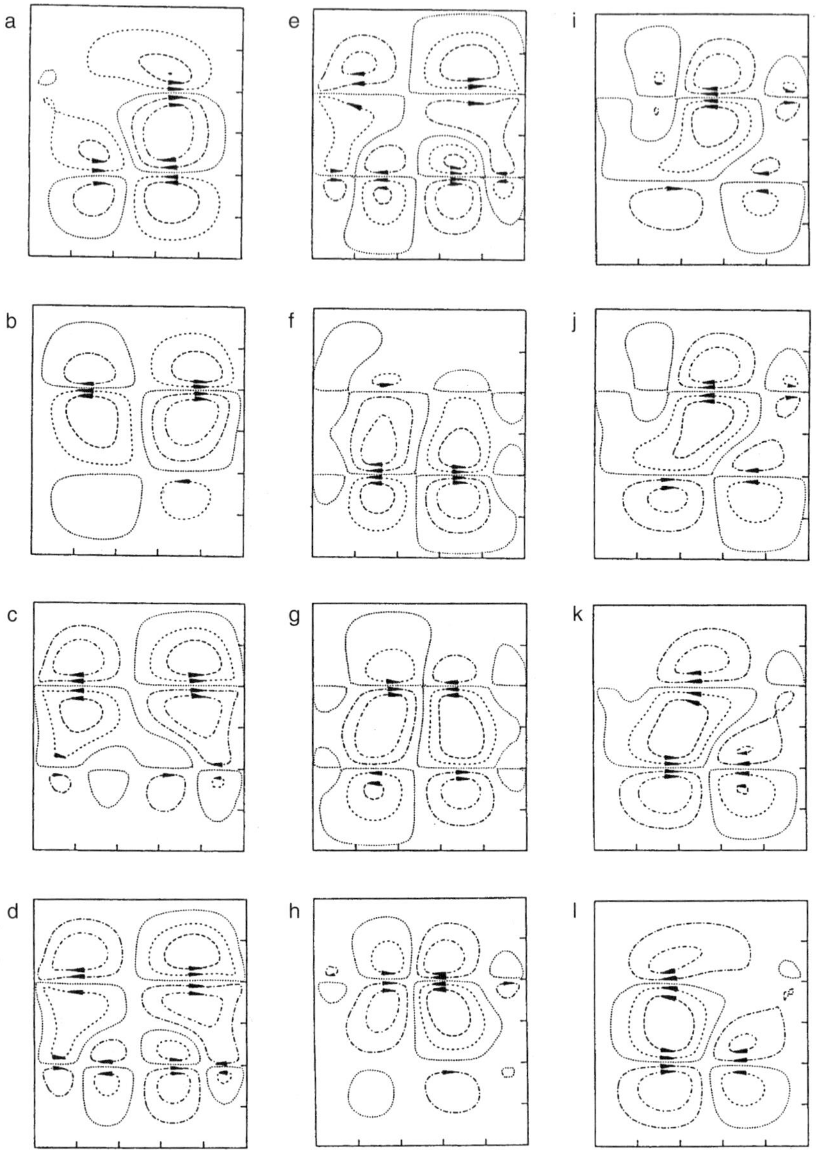

FIGURE 3.41. Stream lines (a) - (l) for the periodic oscillatory motion at $M = 31,200$ during half of the period.

lateral wall and starts to move in the opposite direction. Recall that Figs. 3.43(a) and 3.43(h) are connected by the transformation (3.68).

For $M \approx 49,960$, oscillations become unstable, and the steady motion satisfying the symmetry condition (3.61) is established. Thus, the oscillatory regimes are

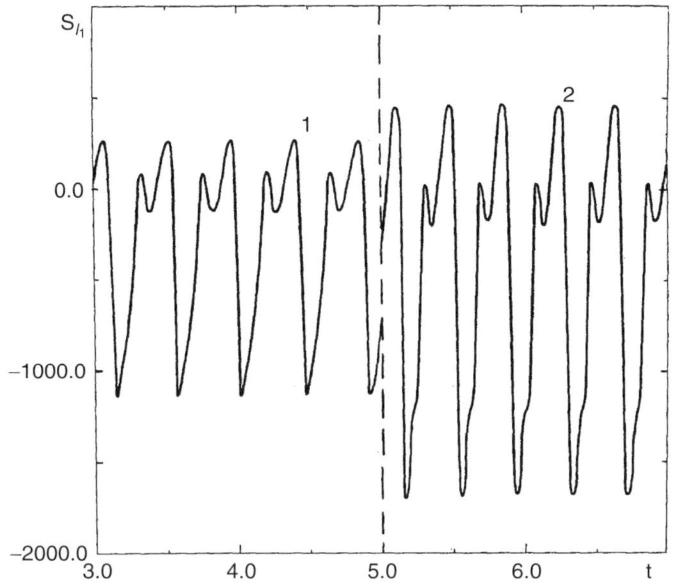

FIGURE 3.42. Oscillations of $S_{l1}(t)$ (line 1: $M = 32,350$, line 2: $M = 47,500$).

observed in the finite interval of M bounded both from below (by the mechanical equilibrium state) and from above (by the steady motion).

3.3.3 Rayleigh Convection

In the present subsection, we describe results of the nonlinear simulation of pure Rayleigh convection ($G \neq 0$, $M = 0$) in a three-layer system (Simanovskii, 1997; Nepomnyashchy and Simanovskii, 1999). We assume that the system is heated from below ($s = 1$). The computations have been fulfilled for a closed cavity with rigid well-conducting lateral boundaries (boundary conditions (3.51) and (3.52)).

As in the case of a two-layer system, we define "local" Rayleigh numbers determined by the parameters of each layer:

$$R_m = \frac{g\beta_m |A_m| a_m^4}{\nu_m \chi_m}, \quad m = 1, 2, 3. \tag{3.69}$$

Values of R_m are not independent and are expressed via dimensionless parameters introduced in Chapter 1 in the following way:

$$R_1 = \frac{GP}{1 + \kappa a + \kappa_* a_*}, \quad R_2 = \frac{GP}{1 + \kappa a + \kappa_* a_*} \frac{\kappa \nu \chi a^4}{\beta},$$
$$R_3 = \frac{GP}{1 + \kappa a + \kappa_* a_*} \frac{\kappa_* \nu_* \chi_* a_*^4}{\beta_*}. \tag{3.70}$$

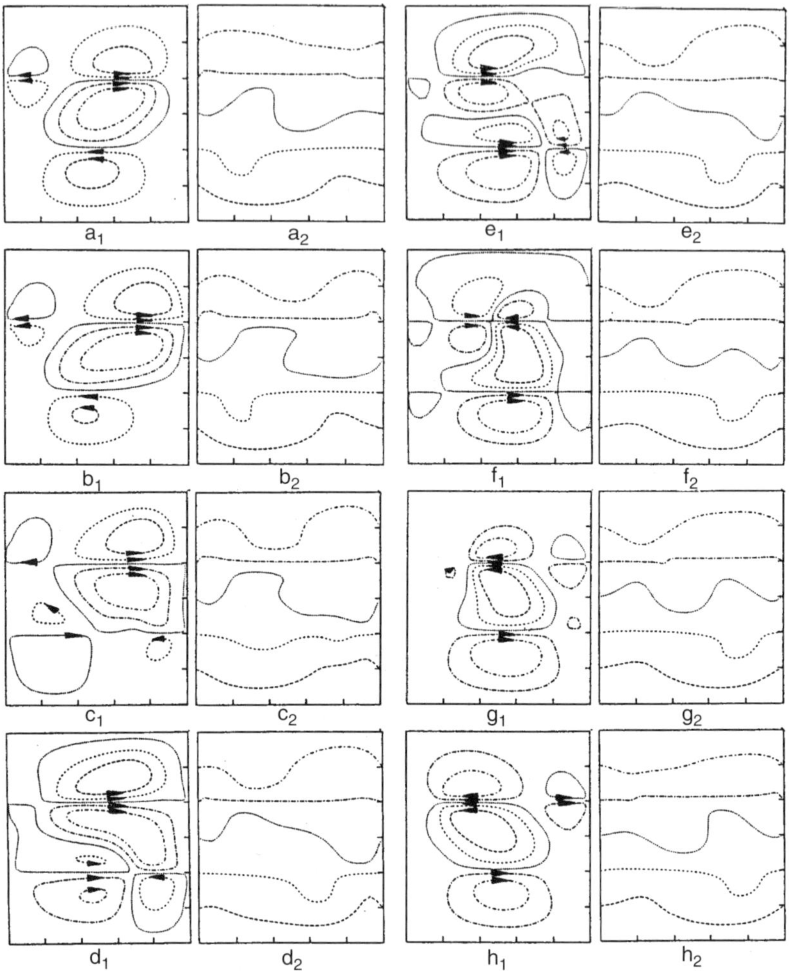

FIGURE 3.43. Stream lines (a_1) - (h_1) and isotherms (a_2) - (h_2) during half of the period; $M = 47,500$.

The ratios

$$\frac{R_2}{R_1} = \frac{\kappa \nu \chi a^4}{\beta}, \quad \frac{R_3}{R_1} = \frac{\kappa_* \nu_* \chi_* a_*^4}{\beta_*} \tag{3.71}$$

depend on the physical properties of fluids and on the relative thicknesses of layers, and they do not change as the heating intensity is changed.

Let us consider several examples.

Model System I

As the first example, let us consider the model system where all the parameters of the fluids are equal except dynamic and kinematic viscosities: $\kappa = \kappa_* = \chi =$

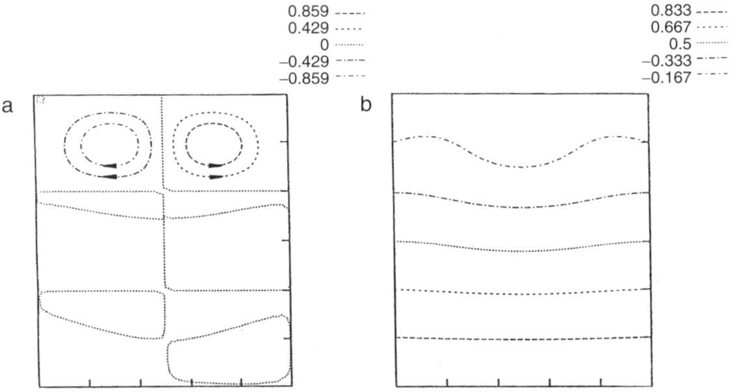

FIGURE 3.44. Stream lines (a) and isotherms (b) for $G = 6000$, $M = 0$.

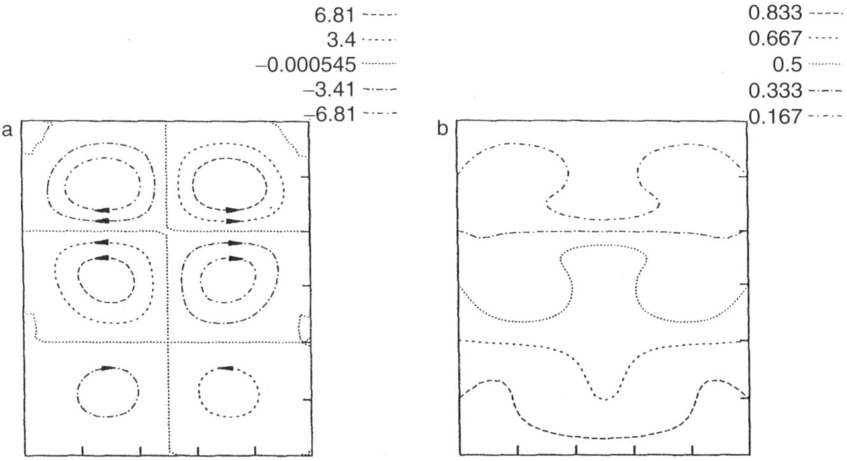

FIGURE 3.45. Stream lines (a) and isotherms (b) for $G = 28,000$, $M = 0$.

$\chi_* = \beta = \beta_* = 1$, $\eta = \nu = 0.5$, and $\eta_* = \nu_* = 0.25$. We choose $P = 1$, $a = a_* = 1$, and $L = 2.5$.

The ratios of the local Rayleigh numbers (3.71) are equal to

$$R_2/R_1 = 0.5, \;\; R_3/R_1 = 0.25.$$

Because the local Rayleigh number of the top layer exceeds those of other layers, the convective motion first arises in the top layer (see Fig. 3.44). When the Grashof number grows, the convection develops also in the middle layer and in the bottom layer (see Fig. 3.45). The motions satisfy the symmetry condition (3.61). The amplitude curves are presented in Fig. 3.46.

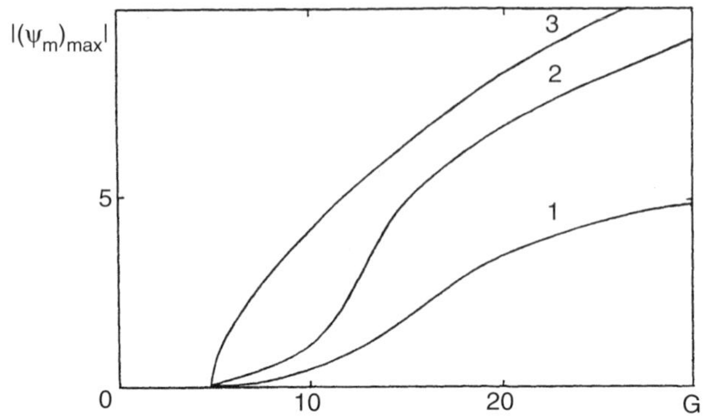

FIGURE 3.46. Dependence of the maximum value of the stream function modulus $|\psi_m|_{max}$ ($m = 1, 2, 3$) on the Grashof number.

Model System II

Let us consider now another model system characterized by the following set of parameters: $\eta = 0.2$, $\nu = 1$, $\kappa = 0.1$, $\chi = 0.1$, $\beta = 0.01$, $\eta_* = 0.04$, $\nu_* = 1$, $\kappa_* = 0.1$, $\chi_* = 0.07$, and $\beta_* = 0.01$. We fix $a = a_* = 1$, $P = 1$, and $L = 2.5$. The value of the equilibrium temperature gradient A_1 in the top layer is significantly higher than those in the middle layer, A_2, and in the bottom layer, A_3 (see Eq. (3.32)). Let us emphasize that the top fluid has the maximum viscosity and thermal diffusivity among all the three fluids, whereas the bottom fluid has the minimum viscosity and thermal diffusivity. In contradistinction to the previous example, the local Rayleigh numbers are close:

$$R_2/R_1 = 1, \quad R_3/R_1 = 0.7.$$

It turns out that in the subcritical region of the Grashof number $G < G_c = 1500 \pm 10$, the initial disturbances decay in an oscillatory manner. When the Grashof number increases, the mechanical equilibrium state becomes unstable, and regular oscillations develop in the system. The stream function and the temperature fields have the symmetry properties (3.61). The change of the stream-line patterns during one-half of the period is shown in Fig. 3.47.

The motion in the middle layer generates flows both in the top layer and in the bottom layer, but in a different way. First, let us discuss the interaction between the convective flows in the middle layer and in the top layer. The viscosity and the thermal diffusivity of the middle fluid is large in comparison with that of the top fluid. At the same time, the temperature gradient in the middle layer is relatively small in comparison with that of the top layer. Therefore, the convection in the middle layer generates a rather large velocity gradient in the top fluid near the interface, due to the tangential stresses, whereas the influence of the temperature disturbances in the middle layer on the temperature field in the upper layer is relatively weak. That is why the motion in the middle layer generates the motion

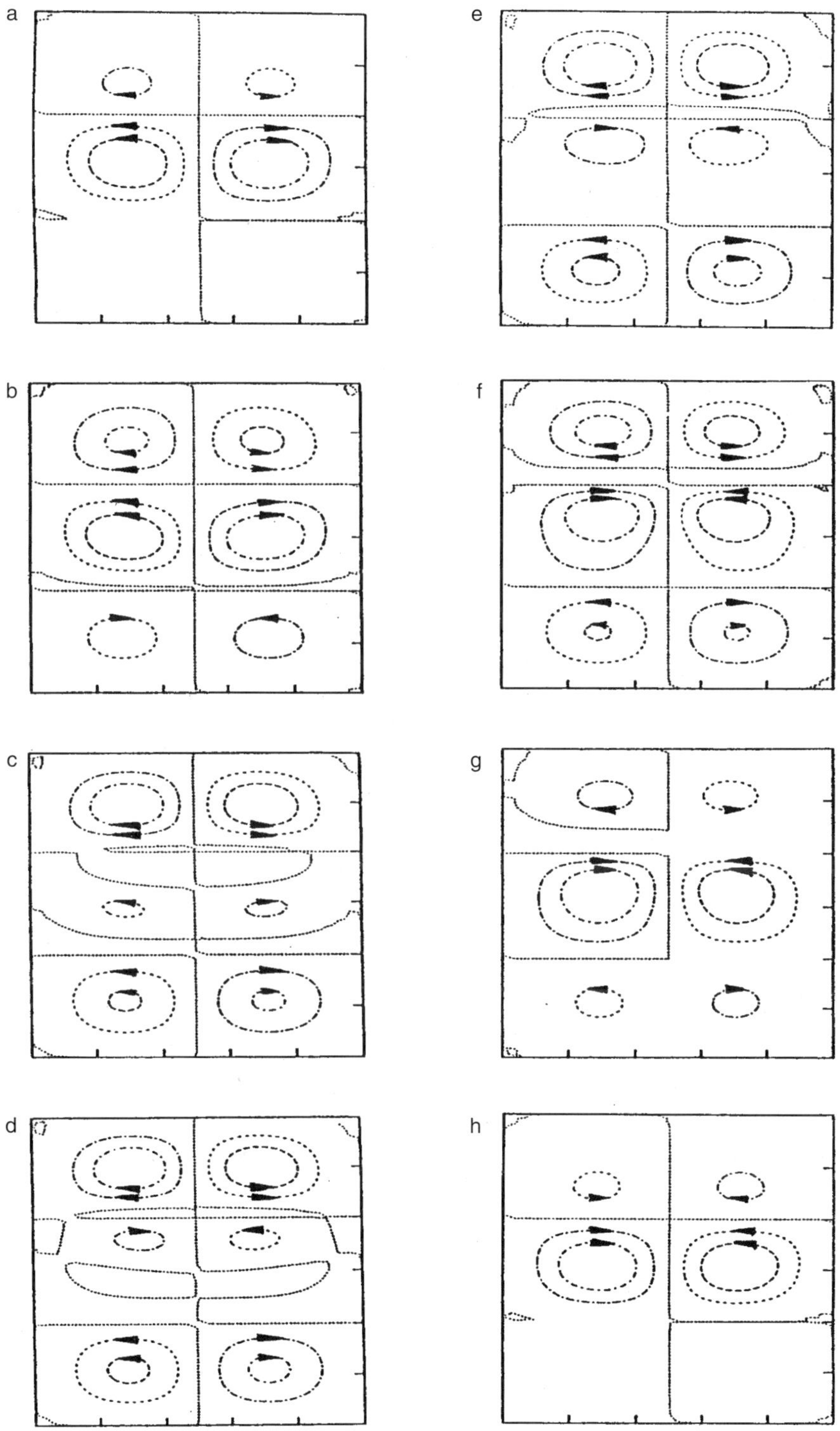

FIGURE 3.47. (a) - (h) Stream lines for the periodic oscillatory motion at $G = 1565$ during half of the period.

in the top layer, mainly by the tangential stresses, in such a way that the fluid rotation in the top layer occurs *in the opposite direction* compared with the one in the middle layer (see Figs. 3.47(a), and 3.47(b)). Since the temperature gradient in the top layer is much higher than that in the middle layer, the temperature disturbance in the top layer caused by the convective flow strongly influences the temperature field in the middle layer, generating vortices rotating *in the same direction* as those in the top layer (see Fig. 3.47(c)). As the result, the rotation direction in the middle layer is changed (see Figs. 3.47(d), and 3.47(e)). Then the motion in the middle layer suppresses the motion in the top layer by tangential stresses (Figs. 3.47(f) / 3.47(h)). Thus, there is a *negative feedback* on the upper interface leading to oscillations.

On the lower interface, the influence of the motion in the middle layer on the motion in the bottom layer by tangential stresses is weak, whereas the influence through the temperature field is strong, because of the differences of physical parameters. That is why the fluid rotation induced in the bottom layer is in the same direction as in the middle layer (see Figs. 3.47(b), and 3.47(c)). The action of the flow in the bottom layer on the motion in the middle layer is mainly due to tangential stresses. Again, there is a negative feedback that acts in favor of oscillations.

The oscillations are only obtained in a narrow region, close to the critical value of the Grashof number ($G_c < G < G_*$, $G_* \approx 1600$). With the increase in G, the evolution of the oscillations leads to the appearance of a steady motion, as shown in Fig. 3.48. Note that the main vortices in the middle layer and in the top layer rotate in the opposite directions (i.e. the viscous coupling prevails), while the main vortices in the middle layer and in the bottom layer rotate in the same directions (i.e. the thermal coupling prevails).

The System Air/Ethylene Glycol/Fluorinert FC75

Let us consider now the system air/ethylene glycol/Fluorinert FC75 (system 3 of Table 3.1). The simulations have been fulfilled with $a = a_* = 1$, $L = 2.5$.

For this system, the local Rayleigh numbers differ considerably:

$$R_2/R_1 = 3.5, \quad R_3/R_1 = 1.7 \times 10^3.$$

Because of this, an intensive convective motion arises only in the bottom layer ($G_c \approx 2.5$); in the other layers, weak induced motions exist (Fig. 3.49). It is interesting that the structure of the motion changes with the increase in G. For relatively small values of G (see Fig. 3.49(a)), the fluid goes down in the middle of the bottom layer (structure I). When G increases, the lateral vortices become stronger and suppress the central vortices, so that structure I is smoothly transformed into a steady four-vortex structure II (see Fig. 3.49(b)). For larger values of G (see Fig. 3.49(c)), the fluid goes up in the middle of the bottom layer (structure III). There is a hysteretic transition between structures II and III: in the interval of the Grashof numbers $30 < G < 80$, the stability regions of these structures are overlapped. For all of the configurations, the fields of stream function and temperature

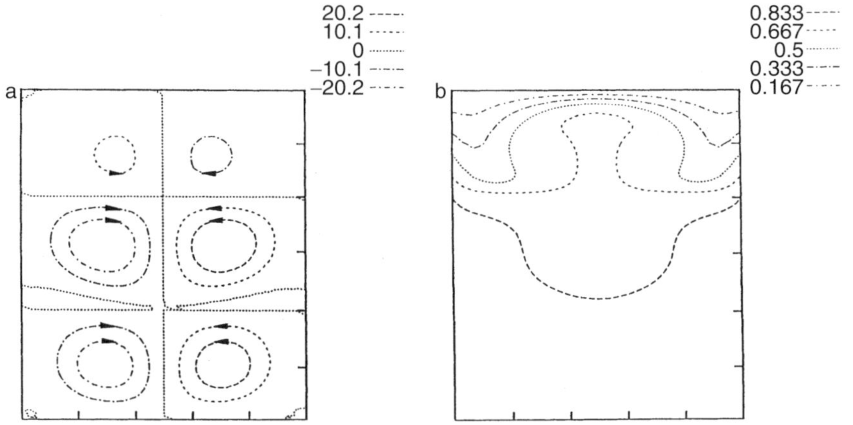

FIGURE 3.48. (a) Stream lines and (b) isotherms for the steady motion at $G = 1700$.

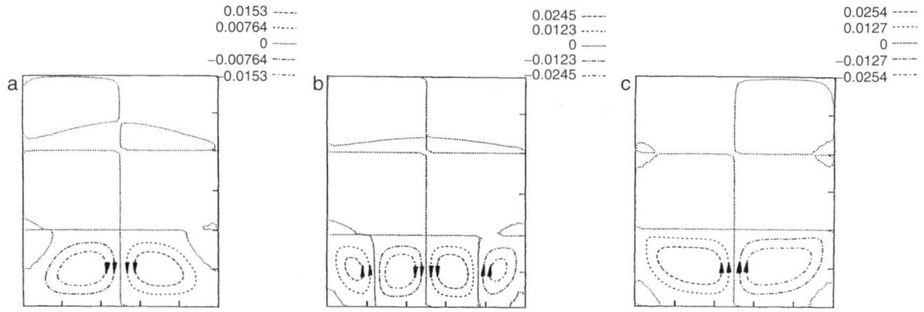

FIGURE 3.49. Stream lines for $M = 0$; (a) $G = 20$; (b) $G = 75$; (c) $G = 100$.

have the symmetry properties

$$\psi_m(x, z) = -\psi_m(-x, z), \ T_m(x, z) = T_m(-x, z), \ m = 1, 2, 3. \tag{3.72}$$

At larger values of G, nonlinear oscillations take place (see Figs. 3.35(b), and 3.35(c); $a = 1$, $a_* = 4$).

The System Air/Water/Mercury

The last example is the system air/water/mercury (system 5 of Table 3.1). The ratios of the local Rayleigh numbers are equal to

$$R_2/R_1 = 4.66, \ R_3/R_2 = 0.0785.$$

Because of this, the steady convective motion appears first of all in the middle layer (see Fig. 3.50). At the larger values of the Grashof number ($G > 2100$), the intensity of the motion in the top layer becomes higher than that of the motion in the other layers.

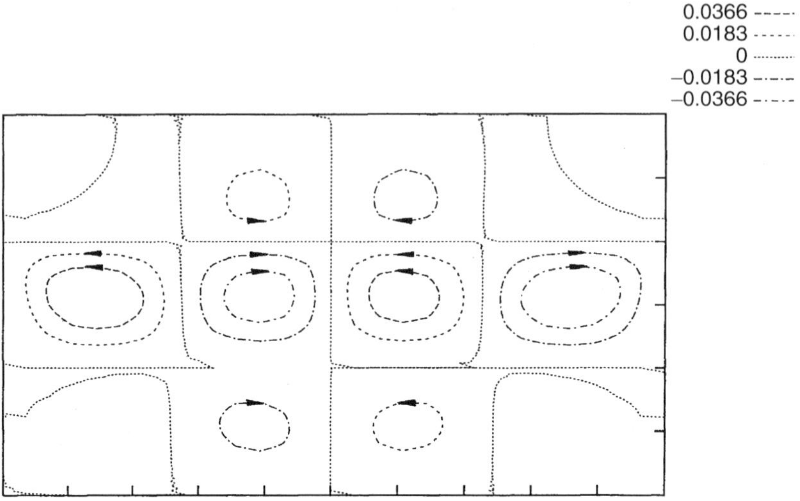

FIGURE 3.50. Stream lines for the steady motion at $M = 0$, $G = 1500$.

3.3.4 *Mixed Rayleigh-Marangoni Convection*

In this section, we present some examples of convective flows produced by the simultaneous action of buoyancy and thermocapillary effect.

Model System

First, we consider the model system with all the parameters of the fluids equal, except dynamic and kinematic viscosities: $\kappa = \kappa_* = \chi = \chi_* = \beta = \beta_* = 1$, $\eta = \nu = 0.5$, and $\eta_* = \nu_* = 0.25$. We choose $P = 1$, $a = a_* = 1$, and $L = 2.5$.

As was described above, thermocapillary oscillations have been found in this system in the case $M \neq 0$, $G = 0$, whereas the pure buoyancy convection (in the case $G \neq 0$, $M = 0$) is stationary. The diagram of regimes observed under the combined action of buoyancy and thermocapillary effect is shown in Fig. 3.51. The preferred kind of oscillatory regimes turned out to be similar to one shown in Fig. 3.40(c). However, there exists also another type of oscillation that does not satisfy condition (3.68) (see Fig. 3.52).

The System Air/Ethylene Glycol/Fluorinert FC75

As was shown in the previous sections, the Marangoni convection in the above-mentioned system (system 3 of Table 3.1) is produced mainly by the upper interface, whereas the Rayleigh convection is located mainly in the bottom layer. That is why one can expect that the interaction between the surface-tension-driven flow (in the top layer and in the middle layer) and the buoyancy-driven flow (in the bottom layer) will be negligible. Nevertheless, a certain kind of indirect interaction between the flows takes place. In the case $a = a_* = 1$, the *structure* of the buoyancy convective motion in the bottom layer is actually influenced by the

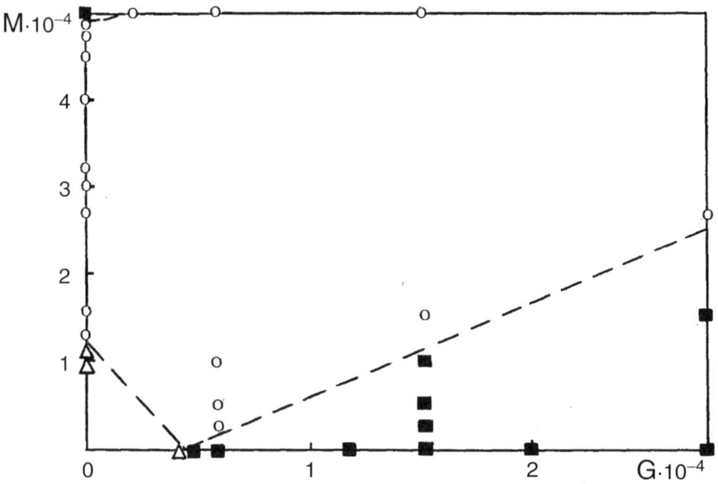

FIGURE 3.51. The diagram of regimes (triangles: equilibrium, filled squares: steady state, circles: oscillations). The dashed lines separate the regions of different regimes.

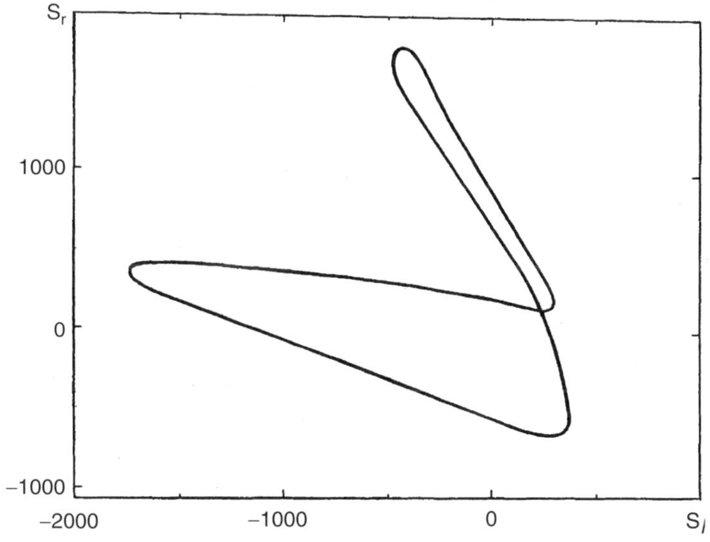

FIGURE 3.52. Phase trajectory of the periodic motion ($G = 15,000$, $M = 16,000$).

thermocapillary convection in the other layers. Indeed, the viscous stresses produced on the lower interface by the motion in the middle layer support structure I, which replaces structures II and III (see Fig. 3.34(b)). For larger values of G, structure III survives as $M \neq 0$ (see Fig. 3.34(c)). In the latter case, some small vortices appear separating the convection in the bottom layer, which is produced mainly by the buoyancy, and the convection in the middle layer and in the top layer, which is mainly of the thermocapillary origin.

In the case $a = 1$, $a_* = 4$, we observe a weak interaction between the steady thermocapillary motion around the upper interface and an oscillatory buoyancy-induced motion in the bottom layer. Under the combined action of both instability mechanisms, an oscillatory motion develops in the system, which satisfies relation (3.72). The evolution of the stream-line patterns during one-half of the period is shown in Fig. 3.53. The most intensive motion of the thermogravitational origin takes place in the bottom layer. The thermocapillary motion in the top layer and in the middle layer is weakly influenced by the buoyancy convection in the bottom layer.

3.3.5 *Anticonvection*

In the present subsection, we consider a three-layer system heated uniformly *from above* ($s = -1$). As was shown in Section 2.3, a specific, non-Rayleigh mechanism of instability can act in this case. It leads to the developement of an "anti-convective" flow.

Model System

First, we will study the appearance of the anticonvection in the model system II described in Sec. 3.2.3. The simulations have been done for $a = a_* = 1$, $P = 1$, and $L = 2.5$.

Recall the physical mechanism of instability (see also Sec. 2.3). Let us consider a velocity disturbance such that a fluid element in the top layer moves down toward the upper interface. Since the thermal diffusivity of the top fluid is small ($\chi \ll 1$, $\chi_* \ll 1$), the temperature of the element changes slowly and remains higher than that of the surrounding fluid for a long time. Because the heat expan-

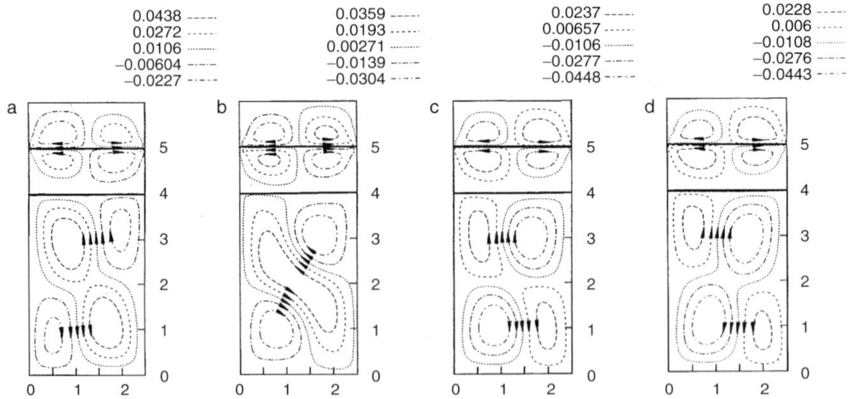

FIGURE 3.53. Stream lines (a) - (d) for the periodic oscillatory motion during half of the period ($G = 5$, $M = 60000$, $a = 1$, $a_* = 4$).

sion coefficient of the top fluid is small ($\beta \ll 1$, $\beta_* \ll 1$), the buoyancy does not prevent the motion of the relatively hot fluid element toward the interface. This element enhances the temperature on the interface, so that a "hot spot" is generated. Due to the heat conductivity, a temperature inhomogeneity reaches the middle layer, where the heat expansion coefficient of the fluid is relatively large. In the middle layer, a convective motion is produced that includes an ascending flow below the hot spot and an advective motion along the interface directed from the hot spot to the periphery. Because of the large thermal diffusivity of the middle fluid, this flow does not destroy the hot spot. At the same time, the interfacial viscous stresses caused by this flow generate a descending flow in the top layer. The latter flow brings the new portion of the hot fluid toward the interface, and the anticonvective motion can be sustained.

When the Grashof number is small enough, the system relaxes to the mechanical equilibrium state: All the disturbances decay. The mechanical equilibrium state becomes unstable for the values of the Grashof number above $G_c \approx 2900$, and a steady motion is obtained in the system. Typical stream lines are shown in Fig. 3.54(a). The fields of the stream function and of the temperature have the symmetry properties (3.61). One can see that the motion takes place mainly near the active upper interface. The intensity of the anticonvective motion is relatively large in the layer with the higher heat expansion coefficient (i.e., in the layer below the active interface).

The anticonvection has to be distinguished from the Marangoni convection that also appears in this model system near the upper interface by heating from above (see Fig. 3.54(b)). In contradistinction to the case of anticonvection, the intensities of the Marangoni convection in both the top and middle layers are of the same order. The directions of rotation coincide for anticonvective and thermocapillary motions. When both mechanisms of instability act simultaneously, the intensity of the motion increases in both fluids, but its increase in the middle layer is much stronger (see Fig. 3.54(c)). Direct and indirect interaction of anticonvective and thermocapillary mechanisms of instability was considered by Simanovskii (2000).

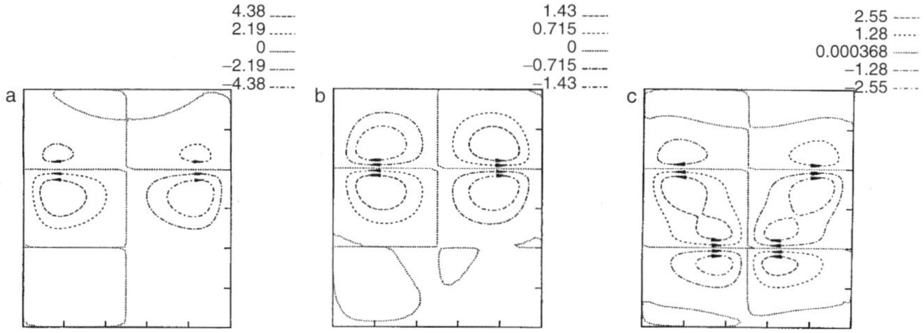

FIGURE 3.54. Stream lines for the model system; (a) $M = 0$, $G = 4500$, $\bar{a} = 1$; (b) $M = 12,000$, $G = 4500$, $\bar{a} = 1$; (c) $M = 25,000$, $G = 4500$, $\bar{a} = 200$.

The System Air/Water/Mercury

The system air/water/mercury (system 6 of Table 3.1) gives an example of a real system where the anticonvection takes place by heating from above ($s = -1$). In contrast to the model system described in the previous subsection, the active interface generating the anticonvection is the lower interface, and the most intense motion takes place in the bottom fluid (see Fig. 3.55).

3.4 Space Experiments

This section presents results of two microgravity experiments devoted to the Marangoni-Bénard instability in three-layer systems. The first experiment (Experiment I), which was performed in June 1994 on the IML2 mission of Spacelab on-board the US Space Shuttle, deals with the stationary convection in the system Fluorinert FC70 / silicone oil 10 cSt / Fluorinert FC70 (system 1 of Table 3.1). The second experiment (Experiment II) was done on the LMS mission of Spacelab in 1996. It was devoted to the investigation of the oscillatory Marangoni instability in the system n-octane - methanol - n-octane (system 2 of Table 3.1).

3.4.1 Experiment Description

Fluid Container

The experimental setup used in both experiments is represented in Fig. 3.56. The fluid container was made of a rectangular quartz frame closed on two sides by

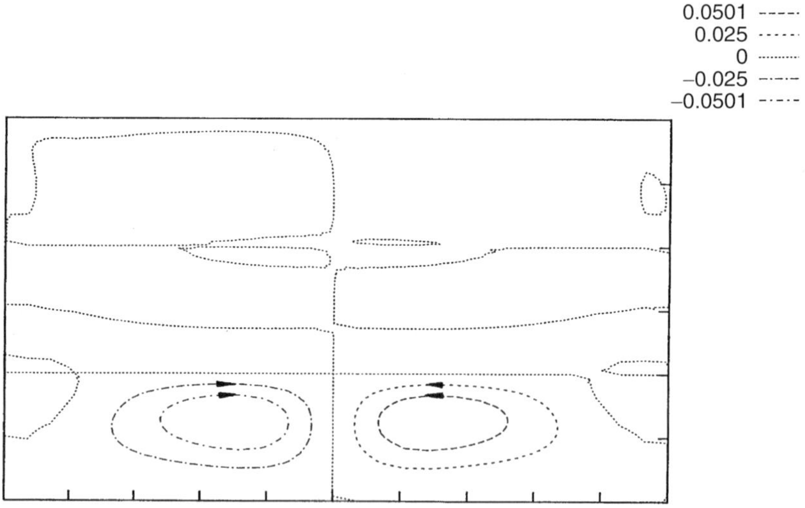

FIGURE 3.55. Stream lines for the steady motion in the air/water/mercury system at $M = 0$, $G = 40,000$.

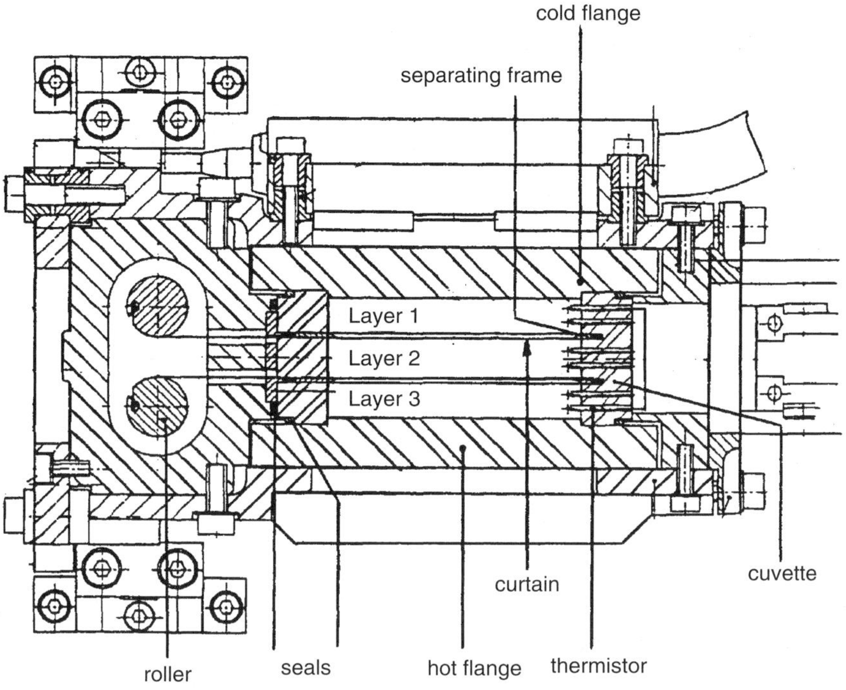

FIGURE 3.56. The experimental setup.

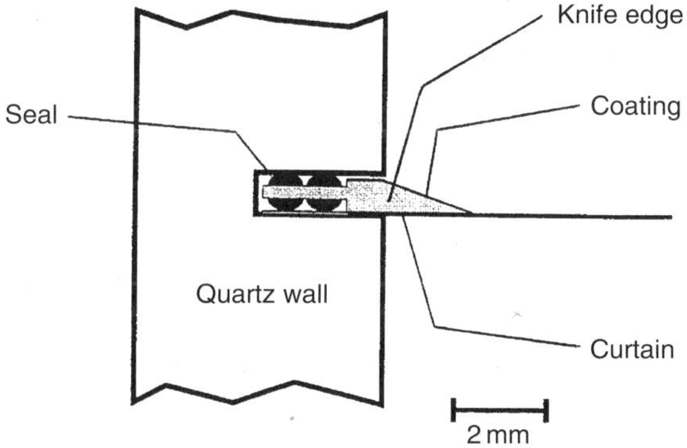

FIGURE 3.57. Close-up of the antiwetting knife edge.

two sapphire windows, each being 1 cm thick. Each liquid layer was 8 mm high, 50 mm long, and 35 mm wide. Initially, the layers were separated by two 50-μm-thick stainless-steel curtains to prevent mixing before the beginning of the experiment. The boundaries between the layers were kept leak tight by Viton seals molded on an antiwetting frame forming a knife edge, as depicted in Fig. 3.57. Because the knife edge penetrated 2 mm into the fluid cell, the effective lengths of the free interfaces were 46 mm rather than 50 mm. At the beginning of the experiment, the curtains were slowly pulled out and gently wound (0.25 mm s^{-1}) on the roller (Fig. 3.56), creating two free liquid/liquid interfaces subjected to Marangoni effect. Heating and cooling were achieved with an electrical resistances and Peltier elements located along the edges of the sapphire windows.

Stability of the Three-layer Configuration

In microgravity, the static shapes of connected liquid phases are governed by the interfacial forces only. In the absence of walls, a three-layer configuration such as shown in Fig. 3.56 would be unstable. Indeed, the surface-tension forces would act to minimize the surface with respect to the volume of the layers, curving the free interfaces and leading presumably to connected spherical phases. In a finite container, the situation is different because the layers will remain flat, provided that their edges are properly anchored on the walls. Here, anchorage was achieved using two stainless-steel frames. In Experiment I, the frames were coated on one side with a Teflon film that prevented creeping of the silicone oil along the quartz wall. Indeed, unlike quartz and stainless steel, Teflon is wetted preferentially by Fluorinert rather than by silicone oil. Thus, the Teflon-coated sides of the frames are oriented toward the Fluorinert layers. In Experiment II, the frames were coated on one side with a polypropylen film and with a nylon film on the other side. Nylon is wetted preferentially by methanol rather than by n-octane, whereas polypropylen is wetted by n-octane rather than by methanol.

In addition to controlling the wall wetting, the experimental setup must be designed to satisfy another specific requirement: The interfacial tension has to counterbalance effectively the liquid density difference. This condition has to be fulfilled because of the low residual gravity level in the Space Shuttle.

In the simplified case of the Rayleigh-Taylor instability, the residual gravity being normal to the interfaces, the surface tension will stabilize the arrangement for sufficiently short wavelengths. The critical wavelength λ_m for the onset of the Rayleigh-Taylor instability for two liquids of infinite lateral extent is given by the capillary length (Chandrasekhar 1961):

$$\lambda_m = 2\pi \left(\frac{\sigma}{g(\rho_2 - \rho_1)} \right)^{1/2} . \tag{3.73}$$

The maximum tolerable gravity level g^* can be estimated from equation (3.73) knowing that the 46-mm length of the free available surface has to be smaller than λ_m to exclude the appearance of the Rayleigh-Taylor instability. It was found that

$$g^* \approx 1.30 \times 10^{-2} g_0, \text{ where } g_0 = 9.81 \text{ m s}^{-2}.$$

In the Space Shuttle, the mean microgravity level is lower than $10^{-4} g_0$. The Rayleigh-Taylor instability mechanism was inactive for the present geometry and indeed was not observed.

Flow Visualization

The liquids were seeded with silver-coated ceramic spheres whose diameters range between 80 and 100 μm. The flow was visualized using a He-Ne laser light sheet oriented parallel to the longest side of the test container and located 10 mm from the front wall. To avoid sedimentation due to the residual acceleration, the densities of the particles matched the densities of the liquids to within a few percent.

Thermal Measurements

Up to nine thermistors, penetrating 5 mm into the liquid bulk, were inserted along the lateral wall of the quartz frame to record the temperatures near the lateral walls, throughout the experiment. The locations of the thermistors are given in Fig. 3.58. The instrumentation had to obey several external constraints such that, in order to change the temperature difference across the three-layer system, the reference point on the hot side was varied by steps of one degree. The time span allowed for experiments was rather short and it did not allow very accurate measurements of the critical temperature difference at the onset of convection.

3.4.2 *Experiment I*

Experiment I was devoted to the investigation of convection in the system Fluorinert FC70 / silicone oil 10 cSt / Fluorinert FC70 (system 1 of Table 3.1) with equal thicknesses of layers. As was shown in Secs. 3.1 and 3.2, the convection in this system is stationary.

Thermal Data

The measurements were performed by means of six thermistors. Their location and the recording of the thermistors signals during the experiment are shown in Fig. 3.59.

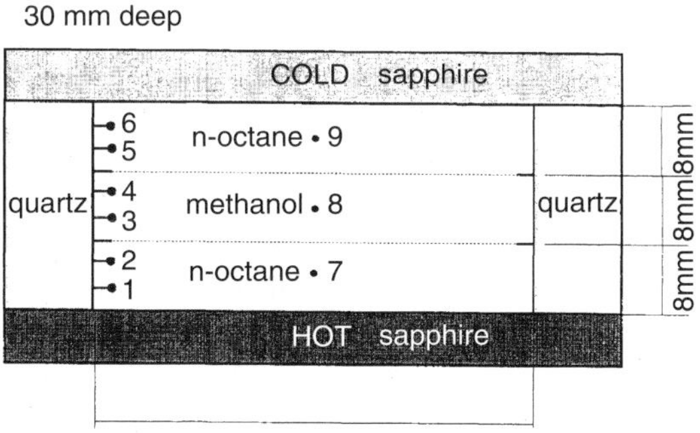

FIGURE 3.58. The locations of the thermistors.

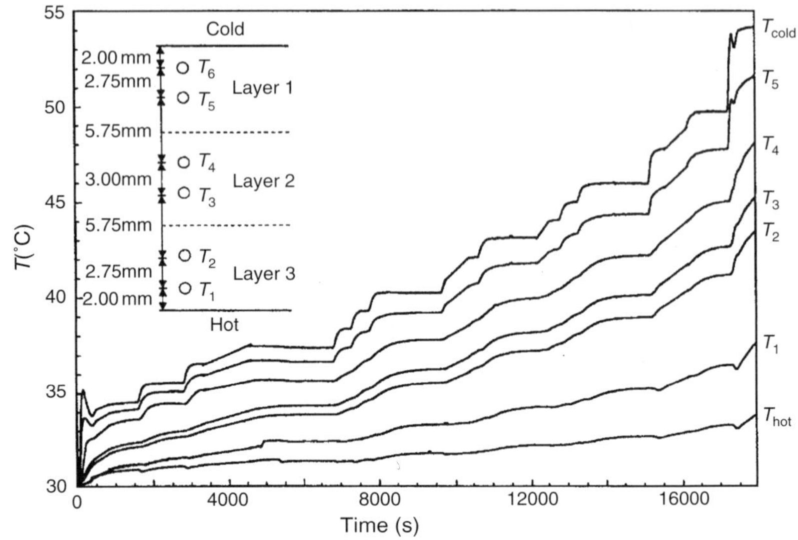

FIGURE 3.59. The locations and the recording of the thermistors.

The Prandtl numbers of Fluorinert FC70 and silicone oil 10 cSt are 390 and 95, respectively. For high Prandtl number fluids, beyond the onset of convection, the velocity field is slaved to the thermal field. The intensity of the flow is reflected by the temperature recordings at the thermistors. From these signals, three different regimes (see Sec. 3.2.1) are identified depending on the temperature difference imposed between the two sapphire conductive plates. Since the distance between the thermistors above and beneath any interface is 5.75 mm (Fig. 3.59), we define the following quantity Δ_2 from the thermistors signals to compare the temperature gradients across the interfaces:

$$\Delta_2 = \frac{T_4 - T_5}{5.75} - \frac{T_2 - T_3}{5.75}. \tag{3.74}$$

To compare experimental data with numerical data of Sec. 3.2.1, Δ_2 is plotted versus the temperature difference Θ (Θ^{exp} and Θ^{num}) across the three layers (see Fig. 3.60). The general tendency is very similar for both the experimental and the numerical approaches. The reason for choosing equation (3.74) is that in regime 1, Δ_2 has to be close to zero because of the symmetry of the configuration and because the isotherms are not perturbed: $T_4 - T_5$ is more or less equal to $T_2 - T_3$ in the range of the imposed Θ. When Θ is increased, the hotter lower interface pushes the liquids away from the walls, bringing hot Fluorinert and cold silicone oil to the lower interface near the lateral wall. However, in the top layer, the flow is very slow and the thermal field is undistorted from the conductive state. This means that $T_2 - T_3$ increases, whereas $T_4 - T_5$ hardly changes: Consequently Δ_2 becomes negative. In the region $\Delta_2 < 0$, the experimental curve reaches its minimum value earlier than the numerical curve, indicating that Θ_c^{exp} is smaller

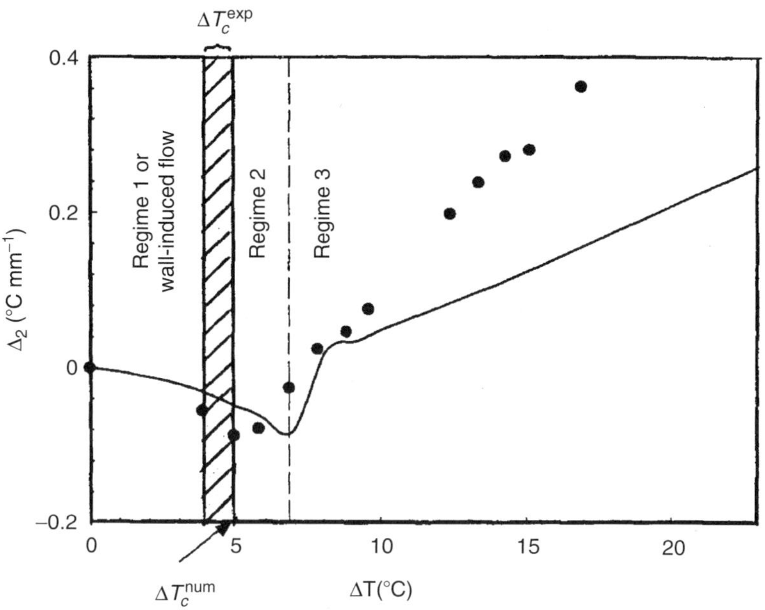

FIGURE 3.60. Comparison between experimental (circles) and numerical (solid line) values of Δ_2 at quasi-steady state.

than $\Theta_c^{num} = 4.75°C$. As Θ is increased still further, the Marangoni convection becomes more and more important, reducing the relative thermal influence of the walls on the flow. This means that the rotation direction of the vortices at the lateral walls is no longer determined by the wall lateral gradient effect but is essentially dictated by the Marangoni-Bénard instability. When the lower interface is pushing the liquids toward the wall, the colder Fluorinert and the hotter silicone oil move away from that interface. However, $T_2 - T_3$ now decreases and, on the other hand, $T_4 - T_5$ increases so that Δ_2 increases toward zero and becomes positive.

This behavior is well reproduced by both the numerical and the experimental results. Generally, the numerical curve matches quite well the experimental points for $\Theta < 10°C$. Beyond this value of Θ, the recorded experimental values of Δ_2 are significantly larger than their numerical counterparts, indicating that the real flow strength is larger than the simulated one (Fig. 3.60). The transition between regime 2 and regime 3 corresponding to the reverse of the rotation of the lateral vortices occurs for $\Theta_c^{exp} = 6.9°C$, whereas it occurs for $\Theta_c^{num} = 7.6°C$ in the numerical simulations.

Flow Patterns

The flow was visualized using a light sheet parallel to the widest side of the experimental cell; it was 1 mm thick and located 10 mm from the front window inside the liquids. The images were digitized and the velocity of the tracer particles was measured using a gray-level correlation algorithm. The flow was resolved in 1.2-mm^2 windows. To improve the signal-to-noise ratio and to increase the number of velocity vectors computed, the velocity maps were time-averaged. The average velocity vector $\bar{v}(x, y)$ was computed as

$$\bar{v}(x, y) = \frac{1}{N}\sum_{t=t_0}^{t_1} v(t)(x, y),$$

where N was the number of samples acquired between t_0 and t_1, t_0 was the start of the sampling sequence, and t_1 was the end of the sampling sequence.

No convection was found inside the fluid cell for time $t < 850$ s, for which the temperature differences imposed across the three layers were smaller than $\Theta^{\text{exp}} = 3.8°\text{C}$. The Marangoni-Bénard instability set in for Θ^{exp} between 3.8°C and 4.9°C.

From $t = 2860$ s ($\Theta^{\text{exp}} = 4.9°\text{C}$) onward, an organized flow was observed in layers 2 and 3 with two rolls located in the central region of the cell. Layer 1 was at rest (Fig. 3.61(a)). In the middle of the cell, the flow in layer 3 was rising toward layer 2. Note that the same flow pattern was observed in the numerical simulations (regime 2). Near the lateral walls, the flow seemed to be very weak. Actually, for the Marangoni-Bénard instability, convection typically starts near the lateral walls (Koschmieder, 1992). It is very likely that, in that region, the largest components of the velocity vectors were perpendicular to the visualization

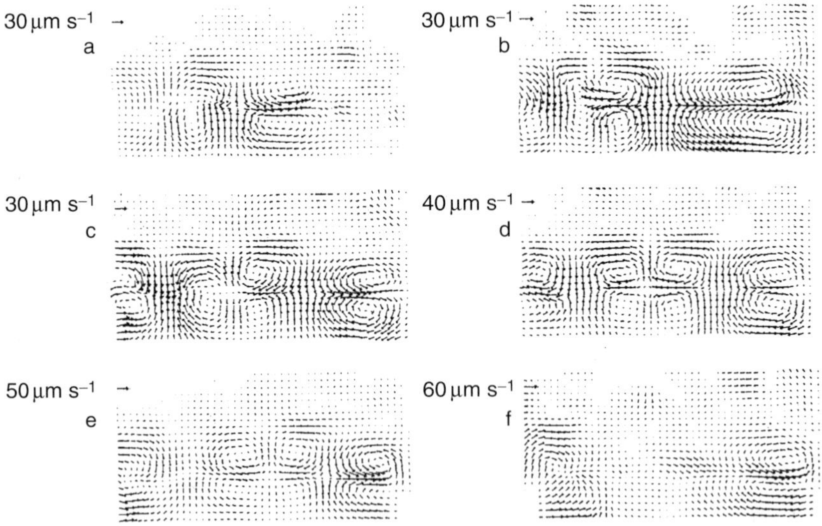

FIGURE 3.61. Experimental velocity fields; (a) $\Theta^{\text{exp}} = 5.5°\text{C}$; (b) $\Theta^{\text{exp}} = 6.7°\text{C}$; (c) $\Theta^{\text{exp}} = 7.8°\text{C}$; (d) $\Theta^{\text{exp}} = 8.7°\text{C}$; (e) $\Theta^{\text{exp}} = 9.7°\text{C}$; (f) $\Theta^{\text{exp}} = 12.4°\text{C}$.

plane, and thus no measurements were possible. As Θ^{exp} was increased, four convective rolls became clearly visible in layers 2 and 3 (Fig. 3.61(b)).

The roll at the right-hand side, initially much larger than the other ones, shrinked progressively (Figs. 3.61(c), and 3.61(d)). The pattern became more regular as the different rolls became of the same size (Figs. 3.61(e)). The long exposure time picture given in Fig. 3.62 is a visualization of the stream lines in regime 3. Whereas convection is intense in layers 2 and 3, it is very weak in layer 1. At the walls, the direction of rotation is opposite to the ones of regimes 1 and 2. That coincides with the prediction of the numerical scheme. Near the upper interface, close to the lateral walls, the two residual vortices are clearly apparent.

As Θ is further increased, the v_y velocity component perpendicular to the light sheet is dominant in the central region of layers 2 and 3, and the structure cannot be understood using two-dimensional visualization (Fig. 3.61(f)). Vortices are, however, still visible near the lateral walls.

Experimental velocity v_z profiles (taken on the height $(2/3)a_3$ above the bottom) are shown in Fig. 3.63. For the profile that corresponds to regime 2 (circles), the flow is rising up in the center region and very weak near the walls. The profile is shifted to the left when Θ^{exp} is increased. The magnitude of the velocity does not increase as Θ^{exp} is increased. This is probably an artifact: Due to the three-dimensional nature of the convection cells, the light sheet is not necessarily being located exactly in a plane where v_y velocity component vanishes.

FIGURE 3.62. Long exposure time picture of the convective flow at $\Theta^{\mathrm{exp}} = 9.7°\mathrm{C}$.

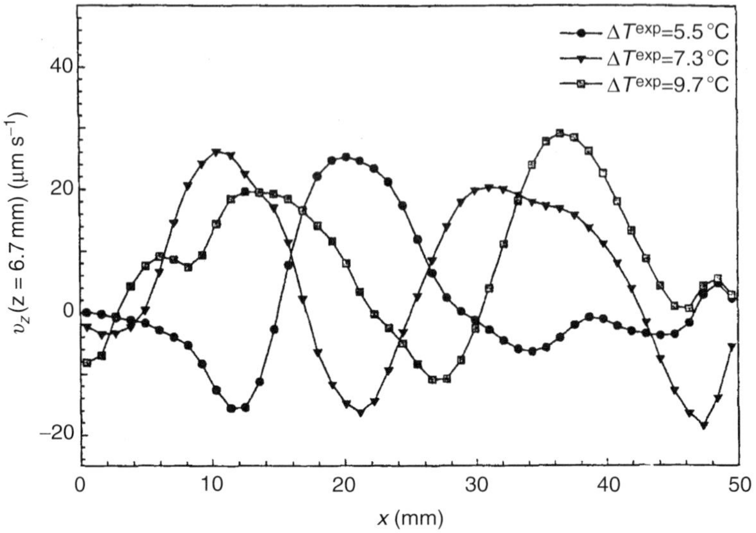

FIGURE 3.63. Experimental velocity profile.

3.4.3 Experiment II

In the present subsection, we shall follow a recent study by Simanovskii *et al.* (2003). For the experiment II, n-octane was selected for the exterior layers ($m = 1, 3$), and methanol was selected for the central layer ($m = 2$). The physical properties of the liquids are given in Table 1. One can see that the relevant parameters of liquids differ slightly, and χ belongs to a certain interval around $\chi = 1$, where the oscillations are expected.

The locations of the thermistors are given in Fig. 3.58, and the recordings of the signals during the experiment are shown in Fig. 3.64. The typical fragments of the phase trajectories are presented in Fig. 3.65.

When the temperature difference across the three layers, Θ, is rather small, all of the disturbances decay, and the system, in practice, keeps the mechanical equilibrium state. With the increase of Θ, the mechanical equilibrium state becomes unstable, and the oscillatory instability develops in the system. Oscillations appear when the experimental time $t \geq t_c = 11,700$ s and further during all of the experiment ($t = 47,500$ s) oscillations have an irregular character (see Figs. 3.64, and 3.65). Note, that both interfaces play an active role in the oscillatory process. The critical Marangoni number determined in the experiment $M_c^e = 4000$. (It is interesting to compare this value with the critical Marangoni number determined from the linear analysis, $M_c^t = 4335.4$. The agreement thus appears quite satisfactory).

Consider the evolution of temperature recordings in more detail. The evolution of temperature recordings inside the methanol layer and the corresponding frequency spectrum are shown in Fig. 3.66. At the onset of convection, the oscil-

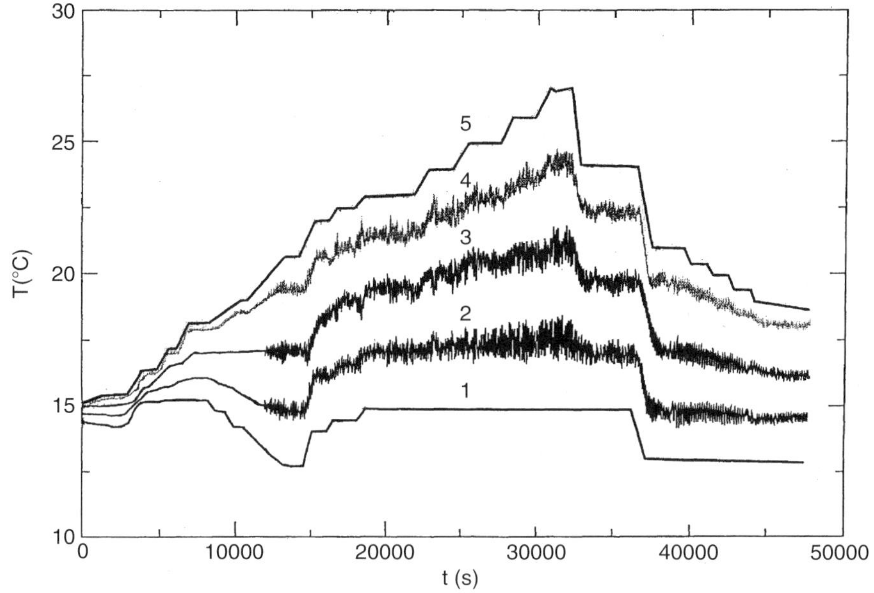

FIGURE 3.64. Recordings of the thermistors' signals.

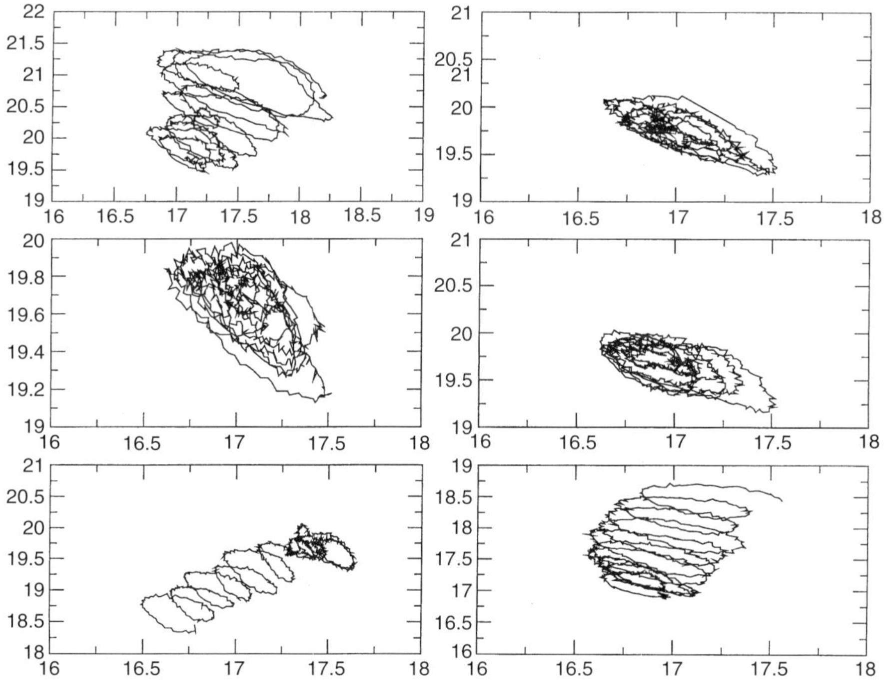

FIGURE 3.65. Phase trajectories. Each fragment corresponds to the time interval $\Delta t = 1000$s.

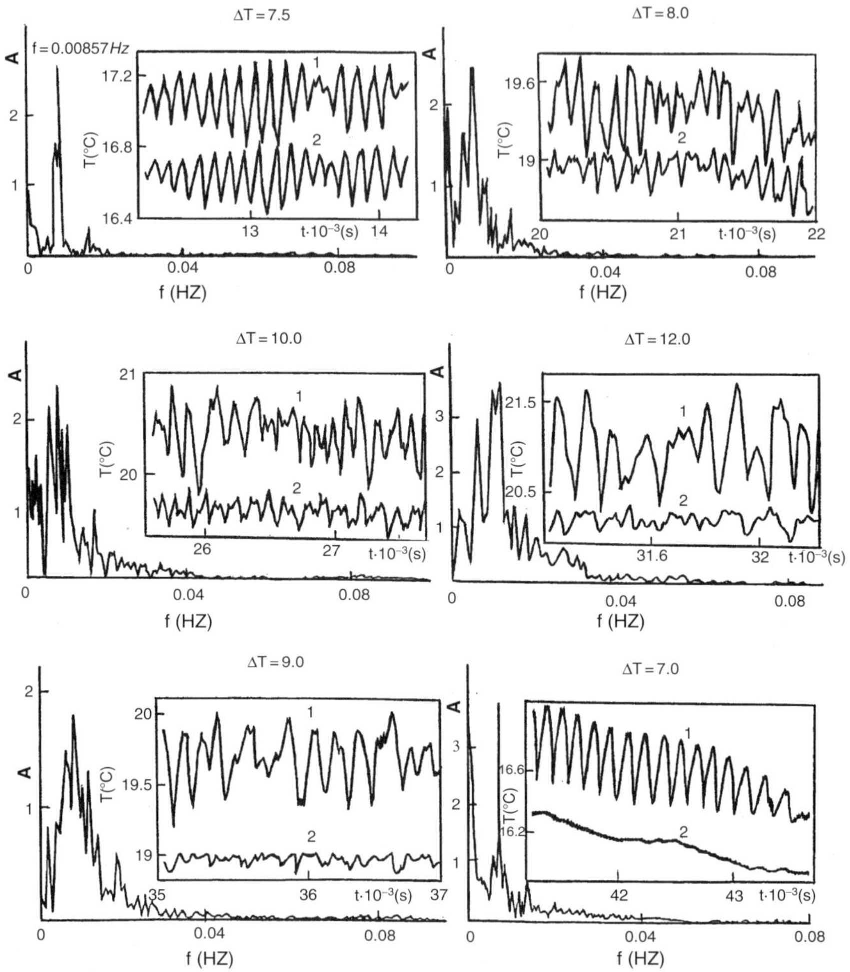

FIGURE 3.66. The evolution of temperature recordings and the corresponding frequency spectrum.

lation pattern is simple, although the amplitude is not regular. The amplitudes of the temperature oscillations recorded on the side thermistors and on the central one are of the same order of magnitude. The frequency spectrum is composed of a dominant peak at $f = 0.00857$ Hz. When Θ is increased, the oscillation pattern becomes more and more complex and the spectrum becomes composed of multiple peaks. After the onset, the main frequency tends to increase as Θ increases. When Θ is decreased back to the threshold value, the oscillations slow down and become very regular, as at the beginning of the experiment. As the experiment goes on, the amplitude of the signal recorded with the side thermistors is decreasing continuously. The frequency of oscillations tends to increase as Θ increases.

The experimental results confirm the existence of theoretically predicted oscillatory instability, essentially connected with the interaction of the interfaces. The oscillations are robust and tend to accelerate when the thermal constraint is increased.

4

Bénard Problem in Multilayer Systems with Deformable Interfaces

The present chapter is devoted to the investigation of phenomena related to the deformations of interfaces. As it was explained in Chapter 1, these phenomena take place at finite values of the Galileo number Ga; thus, the Grashof number G has to be set equal to zero. In Section 4.1, we recall the formulation of the problem formerly presented in Section 1.2.4. Section 4.2 contains the results of the linear stability analysis. In Section 4.3 we derive and simulate the long-wave amplitude equations that describe the interaction between the deformations of both interfaces.

4.1 Formulation of the Problem

The nonlinear equations governing the convection in a three-layer system have been given in Chapter 3 (see equations (3.1) – (3.3)). The boundary conditions on the rigid boundaries, (1.28) and (1.29), are not changed. However, now we use the boundary conditions appropriate for the deformable interfaces. At $z = h$, they can be written in the form

$$p_1 - p_2 + \frac{W_0}{R}(1 - \delta_\alpha T_1) + Ga\,\delta h$$
$$= \left[\left(\frac{\partial v_{1i}}{\partial x_k} + \frac{\partial v_{1k}}{\partial x_i}\right) - \eta^{-1}\left(\frac{\partial v_{2i}}{\partial x_k} + \frac{\partial v_{2k}}{\partial x_i}\right)\right] n_i n_k, \tag{4.1}$$

$$\left[\left(\frac{\partial v_{1i}}{\partial x_k} + \frac{\partial v_{1k}}{\partial x_i}\right) - \eta^{-1}\left(\frac{\partial v_{2i}}{\partial x_k} + \frac{\partial v_{2k}}{\partial x_i}\right)\right] \tau_i^{(l)} n_k - \frac{M}{P}\tau_i^{(l)}\frac{\partial T_1}{\partial x_i} = 0, \; l = 1, 2, \tag{4.2}$$

$$\mathbf{v}_1 = \mathbf{v}_2, \tag{4.3}$$

$$\frac{\partial h}{\partial t} + v_{1x}\frac{\partial h}{\partial x} + v_{1y}\frac{\partial h}{\partial y} = v_{1z}, \tag{4.4}$$

$$T_1 = T_2, \tag{4.5}$$

$$\left(\frac{\partial T_1}{\partial x_i} - \kappa^{-1}\frac{\partial T_2}{\partial x_i}\right) n_i = 0, \tag{4.6}$$

and at $z = -a + h_*$ they are

$$p_2 - p_3 + \frac{W_{*0}}{R_*}(1 - \delta_{\alpha *} T_1) + Ga\,\delta_* h_*$$

$$= \left[\eta^{-1}\left(\frac{\partial v_{2i}}{\partial x_k} + \frac{\partial v_{2k}}{\partial x_i}\right) - \eta_*^{-1}\left(\frac{\partial v_{3i}}{\partial x_k} + \frac{\partial v_{3k}}{\partial x_i}\right)\right] n_{*i} n_{*k}, \tag{4.7}$$

$$\left[\eta^{-1}\left(\frac{\partial v_{2i}}{\partial x_k} + \frac{\partial v_{2k}}{\partial x_i}\right) - \eta_*^{-1}\left(\frac{\partial v_{3i}}{\partial x_k} + \frac{\partial v_{3k}}{\partial x_i}\right)\right] \tau_{*i}^{(l)} n_{*k} - \frac{\bar{\alpha} M}{P}\tau_{*i}^{(l)}\frac{\partial T_3}{\partial x_i} = 0, \tag{4.8}$$

$$l = 1, 2,$$

$$\mathbf{v}_2 = \mathbf{v}_3, \tag{4.9}$$

$$\frac{\partial h_*}{\partial t} + v_{3x}\frac{\partial h_*}{\partial x} + v_{3y}\frac{\partial h_*}{\partial y} = v_{3z}, \tag{4.10}$$

$$T_2 = T_3, \tag{4.11}$$

$$\left(\kappa^{-1}\frac{\partial T_2}{\partial x_i} - \kappa_*^{-1}\frac{\partial T_3}{\partial x_i}\right) n_{*i} = 0, \tag{4.12}$$

where $M = \alpha\theta a_1/\eta_1\chi_1$ is the Marangoni number, $Ga = ga_1^3/\nu_1^2$ is the Galileo number, $W_0 = \sigma_0 a_1/\eta_1\nu_1$, $\delta_\alpha = \alpha\theta/\sigma_0$, $\delta = \rho^{-1} - 1$, $W_{*0} = \sigma_{*0}a_1/\eta_1\nu_1$, $\delta_{\alpha_*} = \alpha_*\theta/\sigma_{*0}$, $\delta_* = \rho_*^{-1} - \rho^{-1}$, and $\bar{\alpha} = \alpha_*/\alpha$; R and R_* are the radii of curvature, $\mathbf{n}$ and $\mathbf{n}_*$ are the normal vectors, and $\boldsymbol{\tau}^{(l)}$ and $\boldsymbol{\tau}_*^{(l)}$ are the tangential vectors of the upper and lower interfaces, respectively; p_m is the difference between the overall pressure and the hydrostatic pressure.

As was discussed in Chapter 1, the conditions of the deformability of interfaces ($Ga\delta = O(1)$, $Ga\delta_* = O(1)$) are incompatible with the assumption $G \neq 1$ in the framework of the Boussinesq approximation, unless $\delta \ll 1$ and $\delta_* \ll 1$. Therefore, the boundary conditions (4.1) – (4.12) can be used only if $G = 0$ or $\delta \ll 1, \delta_* \ll 1$.

4.2 Linear Stability Analysis

We investigate the linear stability of the mechanical equilibrium state, described by the solution (3.14) – (3.17) with respect to infinitesimal disturbances

$$(\mathbf{v}_m(z),\ T_m(z),\ p_m(z),\ h,\ h_*)\exp(\lambda t + ikx),\ m = 1, 2, 3,$$

which include the deformations of interfaces. As in the previous section, we introduce the stream function ψ_m, $m = 1, 2, 3$:

$$v_{mx} = \psi_m', \; v_{mz} = -ik\psi_m.$$

Eliminating p_m, $m = 1, 2, 3$, we obtain the following eigenvalue problem:

$$\lambda(\psi_m'' - k^2\psi_m) = c_m(\psi_m'''' - 2k^2\psi_m'' + k^4\psi_m) + ikGb_m T_m, \tag{4.13}$$

$$\lambda T_m = \frac{d_m}{P}(T_m'' - k^2 T_m) + ikA_m\psi_m; \tag{4.14}$$

$$\psi_1 = \psi_1' = T_1 = 0 \text{ at } z = 1, \tag{4.15}$$

$$\psi_3 = \psi_3' = T_3 = 0 \text{ at } z = -a - a_*, \tag{4.16}$$

$$\psi_1''' - \eta^{-1}\psi_2''' + [\lambda\delta - 3k^2(1 - \eta^{-1})]\psi_1' + ik[Ga\delta + k^2 + Wk^2]h = 0, \tag{4.17}$$

$$\psi_1'' + k^2\psi_1 - \eta^{-1}(\psi_2'' + k^2\psi_2) - ikMP^{-1}(T_1 + A_1 h) = 0, \tag{4.18}$$

$$\psi_1 = \psi_2, \tag{4.19}$$

$$\psi_1' = \psi_2', \tag{4.20}$$

$$\lambda h = -ik\psi_1, \tag{4.21}$$

$$T_1 - T_2 = (A_2 - A_1)h, \tag{4.22}$$

$$T_1' - \kappa^{-1}T_2' = 0 \text{ at } z = 0 \tag{4.23}$$

and

$$\eta^{-1}\psi_2''' - \eta_*^{-1}\psi_3''' + [\lambda\delta_* - 3k^2(\eta^{-1} - \eta_*^{-1})]\psi_3' + ik[Ga\delta_* + k^2 W_*]h_* = 0, \tag{4.24}$$

$$\eta^{-1}(\psi_2'' + k^2\psi_2) - \eta_*^{-1}(\psi_3'' + k^2\psi_3) - ik\bar{a}MP^{-1}(T_3 + A_3 h_*) = 0, \tag{4.25}$$

$$\psi_2 = \psi_3, \tag{4.26}$$

$$\psi_2' = \psi_3', \tag{4.27}$$

$$\lambda h_* = -ik\psi_3, \tag{4.28}$$

$$T_2 - T_3 = (A_3 - A_2)h_*, \tag{4.29}$$

$$\kappa^{-1}T_2' - \kappa_*^{-1}T_3' = 0 \text{ at } z = -a. \tag{4.30}$$

Here, primes denote differentiation with respect to z, A_m is the equilibrium temperature gradient in the mth fluid

$$A_1 = -\frac{s}{1 + \kappa a + \kappa_* a_*}, \; A_2 = -\frac{s\kappa}{1 + \kappa a + \kappa_* a_*}, \; A_3 = -\frac{s\kappa_*}{1 + \kappa a + \kappa_* a_*}, \tag{4.31}$$

and

$$W = W_0\left(1 - \delta_a \frac{s}{1+\kappa a+\kappa_* a_*}\right), \quad W_* = W_{*0}\left[1 - \delta_{a*}\frac{s(1+\kappa a)}{1+\kappa a+\kappa_* a_*}\right]. \tag{4.32}$$

The mechanical equilibrium state is unstable if the real part of an eigenvalue $\lambda(M, G, k)$ is positive for some k.

4.2.1 Long-Wave Asymptotics

General Formulas

As it was discussed in Sec. 2.4.4, the deformations of the interfaces are especially important in the limit of long waves. We consider deformational modes with the following asymptotics as $k \to 0$:

$$\lambda = \lambda^{(2)}k^2 + \cdots, \quad h = h^{(0)} + \cdots, \quad h_* = h_*^{(0)} + \cdots,$$

$$\psi_m = k\psi_m^{(1)} + \cdots, \quad T_m = T_m^{(0)} + \cdots, \quad m = 1, 2, 3.$$

Solving the problem for $T_m^{(0)}(z)$, $m = 1, 2, 3$, we find that

$$T_1^{(0)} = -s\frac{(1-\kappa)h^{(0)} + (\kappa-\kappa_*)h_*^{(0)}}{(1+\kappa a+\kappa_* a_*)^2}(z-1),$$

$$T_2^{(0)} = \frac{s}{(1+\kappa a+\kappa_* a_*)^2}\left\{-\kappa\left[(1-\kappa)h^{(0)} + (\kappa-\kappa_*)h_*^{(0)}\right]z \right.$$
$$\left. -(1-\kappa)(\kappa a+\kappa_* a_*)h^{(0)} + (\kappa-\kappa_*)h_*^{(0)}\right\},$$

$$T_3^{(0)} = -s\kappa_*\frac{(1-\kappa)h^{(0)} + (\kappa-\kappa_*)h_*^{(0)}}{(1+\kappa a+\kappa_* a_*)^2}(z+a+a_*).$$

The stream function $\psi_m^{(1)}(z)$, $m = 1, 2, 3$, has the following representation:

$$\psi_1^{(1)} = E_1(z-1)^3 + D_1(z-1)^2, \tag{4.33}$$

$$\psi_3^{(1)} = E_3(z+a+a_*)^3 + D_3(z+a+a_*)^2, \tag{4.34}$$

$$\psi_2^{(1)} = E_2 z^3 + D_2 z^2 + F_2 z + G_2, \tag{4.35}$$

where E_1, D_1, E_3, D_3, E_2, D_2, F_2, and G_2 are constants. Substituting expressions (4.33) – (4.35) into the boundary conditions, one can find the explicit formulas for all of these constants as functions of $h^{(0)}$ and $h_*^{(0)}$, which have been obtained by Nepomnyashchy and Simanovskii (1997).

Using these formulas, we find expressions for $\psi_1^{(1)}(0)$ and $\psi_3^{(1)}(-a)$, which have the following structure:

$$\psi_1^{(1)}(0) = i(\alpha_{11}Ga\delta + \beta_{11}sM)h^{(0)} + i(\alpha_{12}Ga\delta_* + \beta_{12}sM)h_*^{(0)}, \quad (4.36)$$

$$\psi_3^{(1)}(-a) = i(\alpha_{21}Ga\delta + \beta_{21}sM)h^{(0)} + i(\alpha_{22}Ga\delta_* + \beta_{22}sM)h_*^{(0)}, \quad (4.37)$$

where α_{ij} and β_{ij}, $i, j = 1, 2$, are real numbers.

Finally, we get the quadratic equation for $\lambda^{(2)}$:

$$\lambda^{(2)2} - (A_{11} + A_{22})\lambda^{(2)} + A_{11}A_{22} - A_{21}A_{12} = 0, \quad (4.38)$$

where

$$A_{ij} = \alpha_{ij}Ga\delta + \beta_{ij}sM, \; i, j = 1, 2.$$

Equation (4.38) determines two kinds of instability boundaries:

(i) monotonic instability: $A_{11}A_{22} - A_{21}A_{12} = 0$;
(ii) oscillatory instability: $A_{11} + A_{22} = 0$, $(A_{11} - A_{22})^2 + 4A_{21}A_{12} < 0$.

One can see that the system cannot be unstable in a monotonic and oscillatory manner at the same parameter values.

The Limit of Small a_*

The cumbersome general expressions are simplified in the limit $a_* \ll 1$, when the third layer is thin.

First, let us assume that $M = O(1)$. In the lowest order in a_*, the coefficients E_1 and D_1 in equation (4.33) are described by

$$E_1 = \frac{\Delta_E}{\Delta}, \; D_1 = \frac{\Delta_D}{\Delta}, \quad (4.39)$$

where

$$\Delta = \eta^2 a^4 + 4\eta a^3 + 6\eta a^2 + 4\eta a + 1, \quad (4.40)$$

$$\Delta_E = -\frac{1}{6} iGa\delta h^{(0)}(\eta^2 a^4 + 4\eta a^3 + 3\eta a^2) - Q(\eta a^2 + \eta a), \quad (4.41)$$

$$\Delta_D = -\frac{1}{2} iGa\delta h^{(0)}(\eta^2 a^4 + 2\eta a^3 + \eta a^2) + \frac{1}{2} Q(\eta^2 a^4 + 3\eta a^3 - 2\eta a), \quad (4.42)$$

$$Q = -\frac{isM}{P}\frac{\kappa(1 + a_h^{(0)}) + (\kappa_* - \kappa)h_*^{(0)}}{(1 + \kappa a)^2}. \quad (4.43)$$

Hence,

$$\psi_1^{(1)}(0) = \frac{h^{(0)}}{\Delta}\left[-\frac{1}{3} iGa\delta\eta a^3(1 + \eta a) + \frac{isM}{P}\frac{\kappa(1 + a)}{(1 + \kappa a)^2}\frac{1}{2}\eta a^2(1 - \eta a^2)\right]$$
$$+ \frac{h_*^{(0)}}{\Delta}\frac{isM}{P}\frac{\kappa_* - \kappa}{(1 + \kappa a)^2}\frac{1}{2}\eta a^2(1 - \eta a^2). \quad (4.44)$$

Similarly, we find that

$$\psi_3^{(1)}(-a) = \frac{\eta_* a_*^2}{\Delta}\left\{h^{(0)}\left[iGa\delta\left(-\frac{1}{2}\eta a^3 - \eta a^2 - \frac{1}{2}a\right)\right.\right.$$
$$\left. + \frac{isM}{P}\frac{\kappa(1+a)}{(1+\kappa a)^2}\left(-\eta a^3 - \frac{3}{2}\eta a^2 + \frac{1}{2}\right)\right]$$
$$\left. + h_*^{(0)}\frac{isM}{P}\left[\frac{\kappa_* - \kappa}{(1+\kappa a)^2}\left(-\eta a^3 - \frac{3}{2}\eta a^2 + \frac{1}{2}\right) + \frac{\bar{\alpha}\kappa_*}{2(1+\kappa a)}\Delta\right]\right\}. \tag{4.45}$$

Finally, we obtain the following quadratic equation for $\lambda^{(2)}$:

$$\lambda^{(2)2} + (B_{11} + a_*^2 B_{22})\lambda^{(2)} + a_*^2(B_{11}B_{22} - B_{12}B_{21}) = 0, \tag{4.46}$$

where

$$B_{11} = \frac{1}{\Delta}\left[\frac{1}{3}Ga\delta\eta a^3(1+\eta a) - \frac{sM}{P}\frac{\kappa(1+a)}{(1+\kappa a)^2}\frac{1}{2}\eta a^2(1-\eta a^2)\right] + O(a_*), \tag{4.47}$$

$$B_{12} = -\frac{1}{\Delta}\frac{sM}{P}\frac{\kappa_* - \kappa}{(1+\kappa a)^2}\frac{1}{2}\eta a^2(1-\eta a^2) + O(a_*), \tag{4.48}$$

$$B_{21} = \frac{\eta_*}{\Delta}\left[Ga\delta\left(\frac{1}{2}\eta a^3 + \eta a^2 + \frac{1}{2}a\right)\right.$$
$$\left. + \frac{sM}{P}\frac{\kappa(1+a)}{(1+\kappa a)^2}\left(\eta a^3 + \frac{3}{2}\eta a^2 - \frac{1}{2}\right)\right] + O(a_*), \tag{4.49}$$

$$B_{22} = \frac{\eta_*}{\Delta}\frac{sM}{P}\left[\frac{(\kappa_* - \kappa)}{(1+\kappa a)^2}\left(\eta a^3 + \frac{3}{2}\eta a^2 - \frac{1}{2}\right) - \frac{\bar{\alpha}\kappa_*}{2(1+\kappa a)}\Delta\right] + O(a_*). \tag{4.50}$$

Monotonic instability. The condition determining the monotonic instability boundary $\lambda^{(2)} = 0$ is

$$D = B_{11}B_{22} - B_{12}B_{21} = 0.$$

Using expressions (4.47) – (4.50), we find two boundaries:

$$M \equiv M_1 = \frac{sGa\delta PaK(1+\kappa a)}{3\bar{\alpha}\kappa\kappa_*(1+a)(1-\eta a^2)} + O(a_*), \tag{4.51}$$

$$K = \kappa + \kappa_*[2\bar{\alpha}(1+\eta a)(1+\kappa a) - 1],$$

$$M \equiv M_2 = O(a_*). \tag{4.52}$$

In order to determine M_2, we have to assume that $M = O(a_*)$ and find expressions similar to expressions (4.47) – (4.50). The final result is

$$M = M_2 = \frac{4sGa\delta_* Pa_*(1+\kappa a)^2(1+\eta a)}{3K} + O(a_*^2). \tag{4.53}$$

If $\kappa = \kappa_*$, the disturbances of both interfaces do not interact to the lowest order in a_*. The monotonic mode with the critical value M_1 (mode 1) corresponds to the deformation $h^{(0)}$ of the upper interface, whereas the monotonic mode with the critical value M_2 (mode 2) corresponds to the deformation $h_*^{(0)}$ of the lower interface.

If $\kappa \neq \kappa_*$, the disturbances interact because of the boundary condition (4.29). For monotonic mode 1, the disturbances $h^{(0)}$ and $h_*^{(0)}$ are of the same order. For monotonic mode 2, $h^{(0)} = O(a_* h_*^{(0)})$.

It is necessary to emphasize that the above-mentioned boundaries (4.51) and (4.53) do not always separate stable and unstable regions. Another possibility is the separation of two unstable regions having different numbers of positive eigenvalues.

In order to distinguish these two cases, let us consider both roots of equation (4.46). If $|B_{11}|$ is not small, they are

$$\lambda_+^{(2)} = -B_{11} + o(1), \ \lambda_-^{(2)} = -a_*^2 \frac{D}{B_{11}} + o(a_*^2). \tag{4.54}$$

One can see that expressions (4.51) and (4.53) describe the stability boundary only if $B_{11} > 0$.

Oscillatory instability. To the lowest order in a_*, the oscillatory instability boundary $\mathrm{Re}\lambda^{(2)} = 0$, $\mathrm{Im}\lambda^{(2)} \equiv \omega^{(2)} \neq 0$ is determined by two relations:

$$B_{11} = 0 \tag{4.55}$$

and

$$\omega^{(2)2} = -B_{12} B_{21} a_*^2 > 0. \tag{4.56}$$

The first criterion (4.55) gives the formula

$$M = M_0 = \frac{2}{3} s G a \delta P \frac{a(1+\eta a)(1+\kappa a)^2}{\kappa(1+a)(1-\eta a^2)}, \tag{4.57}$$

which coincides exactly with the threshold of the *monotonic* instability of a *two-layer* system with $a_* = 0$, $\bar{a} = 0$, and $h_* = 0$ (Simanovskii and Nepomnyashchy, 1993). Substituting expressions (4.47) – (4.50) into equation (4.56), we obtain the following formula, which determines the frequency of oscillations:

$$\omega^{(2)2} = \frac{(G a \delta)^2 a^4 (1+\eta a)\eta\eta_*}{18 \Delta \kappa (1+a)} \frac{\kappa_* - \kappa}{1-\eta a^2} a_*^2. \tag{4.58}$$

Using formulas (4.57) and (4.58), we come to the conclusion that the oscillatory instability can arise in two cases:

$$\kappa - \kappa_* < 0, \ 1 - \eta a^2 > 0, \ s > 0 \tag{4.59}$$

(heating from below) and

$$\kappa - \kappa_* > 0, \ 1 - \eta a^2 < 0, \ s < 0 \tag{4.60}$$

(heating from above).

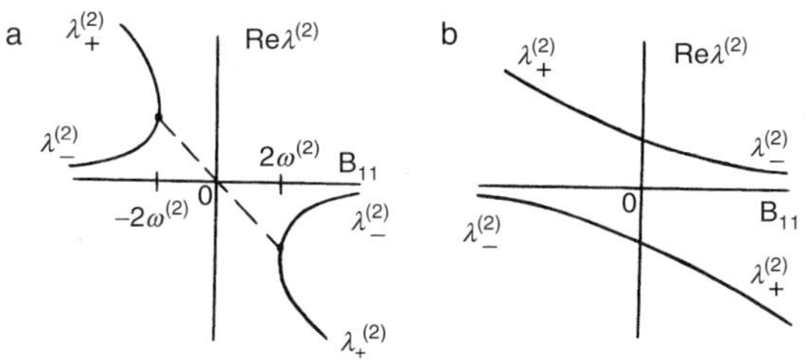

FIGURE 4.1. Dependence of Re ($\lambda^{(2)}$) on B_{11} (a) in the case $B_{12}B_{21} < 0$ and (b) in the case $B_{12}B_{21} > 0$.

In order to estimate the size of the oscillatory instability region, we consider the vicinity of the point $M = M_0$, where $B_{11} = O(a_*)$: $B_{11} = Ba_*$. Solving equation (4.46), we find that

$$\lambda^{(2)} = -\left[\frac{1}{2}B \pm \left(\frac{1}{4}B^2 + B_{12}B_{21}\right)^{1/2}\right]a_* + O(a_*^2).$$

If $-B_{12}B_{21} = \omega^{(2)2} > 0$, there are oscillatory disturbances only in the narrow region of M, where $|B| < 2\omega^{(2)}$ or $|B_{11}| < 2\omega^{(2)}a_*$ (see Fig. 4.1(a)). For $|B_{11}| > 2\omega^{(2)}a_*$, the eigenvalues are real. They tend to the values (4.54), when $|B_{11}| \gg O(a_*)$.

Note that in the case when $B_{12}B_{21} > 0$, the system always has a monotonic instability near the point $B_{11} = 0$ (see Fig. 4.1(b)). The real eigenvalues never coincide and tend to the values (4.54) when $|B_{11}| \gg O(a_*)$.

The configuration of stable and unstable regions is determined by the signs of the three parameters: $\kappa - \kappa_*$, $1 - \eta a^2$, and K. All of the possibilities are shown in Fig. 4.2.

The system is linearly stable in cases (a) and (c) of Fig. 4.2 if $sM < 0$ (for heating from above).

Oscillatory instability is absent in cases (a), (d), and (f) of Fig. 4.2. In case (a), monotonic instability appears only for heating from below, whereas in cases (d) and (f), it appears for both ways of heating.

Oscillatory instability takes place when heating from below in cases (c) and (e) of Fig. 4.2. In case (c), there is a linear stability window between the regions of monotonic instability and oscillatory instability. In case (e), monotonic instability takes place in the finite region $sM_1 < sM < sM_2$ for heating from above.

4.2.2 *Neutral Stability Curves*

In the previous subsection, we investigated the instability in the long-wave limit. Here, we present the results of the numerical calculation of the whole neutral stability curves, in order to clarify the questions under what conditions the minimum of the stability curve $M(k)$ indeed corresponds to the limit $k \to 0$. These results have been obtained by L. M. Braverman (unpublished).

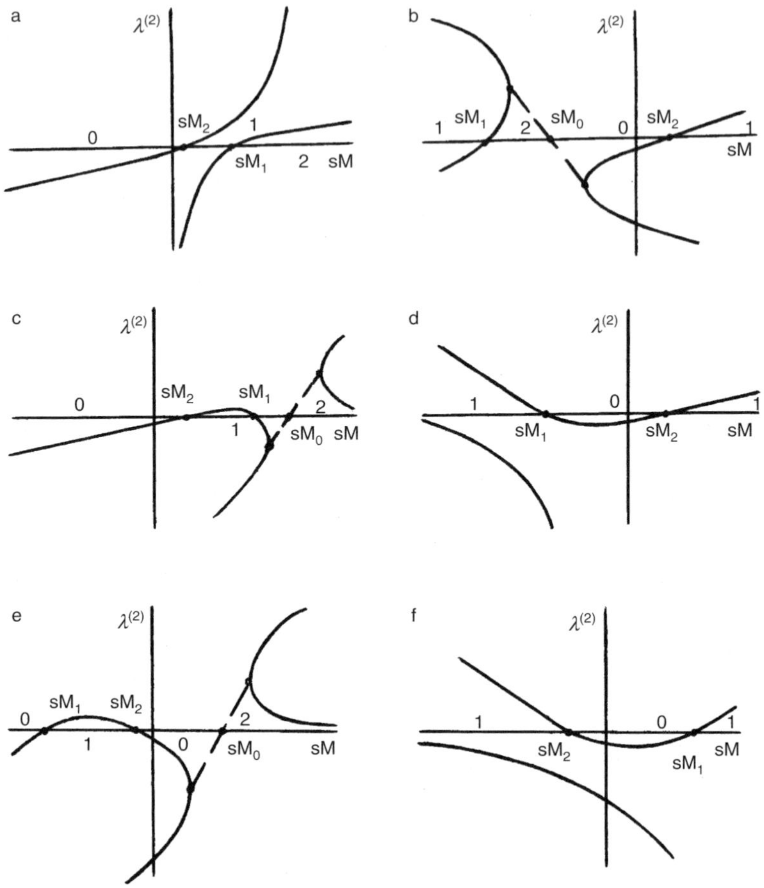

FIGURE 4.2. Configurations of stable and unstable regions: (a) $\kappa - \kappa_* > 0$, $1 - \eta a^2 > 0$, $K > 0$; (b) $\kappa - \kappa_* > 0$, $1 - \eta a^2 < 0$, $K > 0$; (c) $\kappa - \kappa_* < 0$, $1 - \eta a^2 > 0$, $K > 0$; (d) $\kappa - \kappa_* < 0$, $1 - \eta a^2 < 0$, $K > 0$; (e) $\kappa - \kappa_* < 0$, $1 - \eta a^2 > 0$, $K < 0$; (f) $\kappa - \kappa_* < 0$, $1 - \eta a^2 < 0$, $K < 0$.

The calculations have been done for the system air/ethylene glycol/Fluorinert FC-75 (system 3 of Table 3.1). According to the criterion obtained in the previous subsection, the system belongs to class (e) if $a < 1/\sqrt{\eta} \approx 31.6$; thus there is a long-wave oscillatory instability by heating from below, at least in the limit of small a_*. The theoretical prediction (4.57) for the critical Marangoni number in the limit $k \to 0$, $a_* \to 0$ is $M \approx 19,853$, whereas the leading order asymptotics for the frequency of oscillations ω is $\omega \approx 7.05 a_* k^2$. This prediction was used for testing the numerical code. Table 4.1 contains the results of the comparison of the numerical data fullfiled with $k = 0.001$ and the theoretical predictions. The coincidence is rather good.

TABLE 4.1. Comparison of numerical data and theoretical predictions.

	Numerics	Numerics	Theory
a_*	M	ω/k^2	ω/k^2
10^{-3}	20, 800	7.1×10^{-3}	7.05×10^{-3}
5×10^{-4}	19, 969	3.54×10^{-3}	3.52×10^{-3}
10^{-4}	19, 877	7.05×10^{-4}	7.05×10^{-4}

The leading order expressions for M and ω do not contain the parameters W and W_*, which characterize the interfacial tension on two interfaces. The interfacial tension suppresses the deformational instability mode at finite values of the wave number k, but it hardly influences the short-wave (Pearson) instability mode. Note that for a chosen system of fluids and a fixed value of the gravity acceleration, the parameters W and Ga are not independent: Both of them are determined by the dimensional top layer thickness a_1. Specifically, for the above-mentioned three-fluid system under the normal gravity $W = 1.85 \times 10^6 a_1$, $Ga = 4.36 \times 10^4 a_1^3$, where a_1 is measured in centimeters. For instance, for $a = 1$mm, $W = 1.85 \times 10^5$, $Ga = 43.6$; for $a = 0.1$mm, $W = 1.85 \times 10^4$, $Ga = 4.36 \times 10^{-2}$. Note that $Ga = 6.9 \times 10^{-15} W^3$. Under the microgravity conditions, the values of Ga will be lower for the same W. Note that $W_*/W = \sigma_{0*}/\sigma$; for the above-mentioned system of fluids, $W_*/W = 0.479$.

Let us discuss the results of calculations. Four typical situations determined by the location of the minimum value of the Marangoni number have been found: (a) the short-wave monotonic instability (the critical wave number k_m is about 1); (b) the short-wave oscillatory instability; (c) the oscillatory instability with a small but nonzero k_o; (d) the oscillatory instability with $k_o = 0$. The corresponding neutral curves are shown in Figs. 4.3 – 4.6, the critical values of Marangoni number M, wave number k, and frequency ω are presented in Table 4.2. Note that in case (d) (long-wave oscillatory instability), the critical frequency $\omega(k_o)$ is equal to zero because $k_o = 0$, and it is not shown in Table 4.2.

In all the above-mentioned cases, we observe some competition between the short-wave, nondeformational, instabilities and long-wave, deformational, instabilities. If a_* is not small, a short-wave monotonic instability takes place for relatively large W and Ga (type (a); see Table 4.2). The neutral curve corresponding to the case $a = a_* = 1$, $W = 10^5$, $Ga = 6.9$ is shown in Fig. 4.3(a). The minimum of the monotonic neutral curve $M_m \approx 4200$ is achieved at $k_m \approx 1.46$. The oscillatory neutral curve has a minimum at very small value of the wave number $k_o \approx 8.4 \times 10^{-3}$ (see Fig. 4.3(b)); the corresponding minimum value of the Marangoni number $M_o \approx 25,000$ is essentially higher than M_m. The dependence of the frequency of oscillations on the wave number is shown in Fig. 4.3(c). The oscillatory neutral curve merges the monotonic one in the codimension-2 point where the frequency of oscillations becomes zero.

If W and G are relatively large but a_* is small, the minimum of the neutral curve corresponds to a short-wave oscillatory instability (type (b); see Table 4.2).

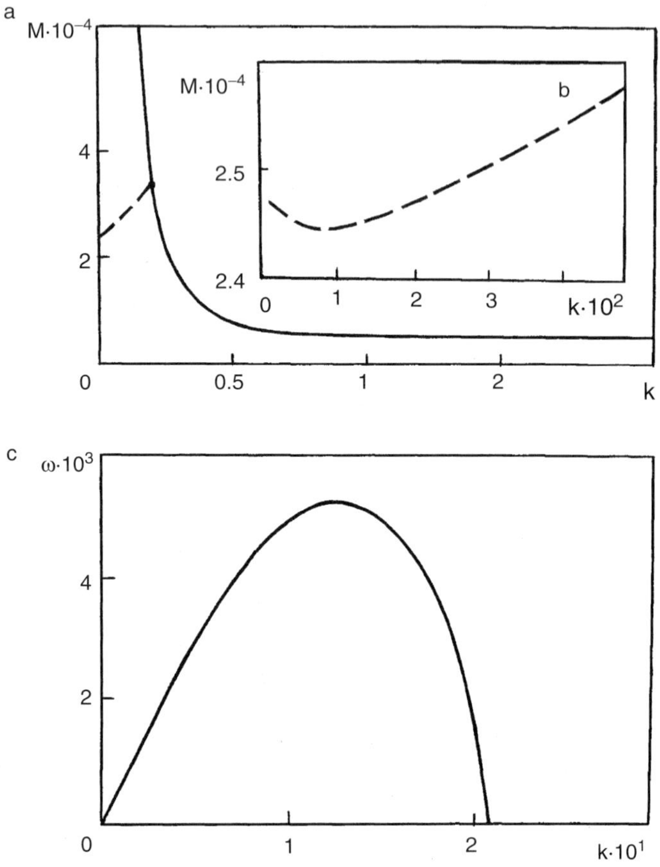

FIGURE 4.3. The case $a = a_* = 1$, $W = 10^5$, $Ga = 6.9$. (a) The oscillatory and monotonic neutral curves; (b) the oscillatory neutral curve in the long-wave region; (c) dependence of the frequency of oscillations ω on the wave number k.

A typical neutral curve calculated for $a = 1$, $a_* = 10^{-3}$, $W = 10^6$, $Ga = 6.9 \times 10^3$ is shown in Fig. 4.4.

For smaller values of W and Ga, the oscillatory instability becomes more "dangerous" than the monotonic one even when a_* is not small. For instance, in the case $a = a_* = 1$, $W = 5 \times 10^4$, $Ga = 0.86$ (see Fig. 4.5), $M_m \approx 4200$, $k_m \approx 1.47$, whereas $M_o \approx 3100$, $k_o \approx 7.9 \times 10^{-3}$ (type (c)).

In the case $a = 1$, $a_* = 0.1$, $W = 5 \times 10^4$, $Ga = 0.86$ (see Fig. 4.6), the oscillatory instability curve has two minima. The lowest one, $M_o \approx 3000$, is located at $k_o = 0$, and it is more "dangerous" than both of the other short-wave minima, that of the oscillatory instability curve and that of the monotonic instability curve (type (d) instability).

The transition between the short-wave oscillatory instability ($W > W_c$, type (b)) and long-wave oscillatory instability ($W < W_c$, type (d)) by changing W in

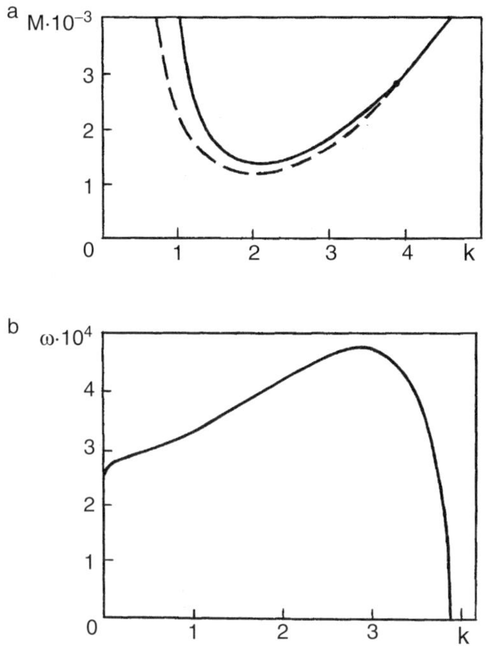

FIGURE 4.4. The case $a = 1$, $a_* = 10^{-3}$, $W = 10^6$, $Ga = 6.9 \times 10^3$. (a) The oscillatory and monotonic neutral curves; (b) dependence of the frequency of oscillations ω on the wave number k.

the system with $a = 1$, $a_* = 10^{-3}$ is shown in Fig. 4.7. Note that the threshold of the short-wave oscillatory instability (line 1) is almost independent of W, whereas for the long-wave oscillatory instability $M_o \sim G \sim W^3$.

Thus, the long-wave oscillatory instability studied analytically in Sec. 4.1.1 is indeed the most important type of the instabilities in the case of small a_* and $W < W_c$.

4.3 Nonlinear Theory

In the present section, we develop the nonlinear theory of long Marangoni waves in systems with two interfaces. We concentrate on the analytically tractable case, where the relative thickness of the bottom layer a_* is small (see Sec. 4.1.1).

4.3.1 Derivation of the Amplitude Equations

Here, we derive the system of amplitude equations governing the evolution of interfacial deformations.

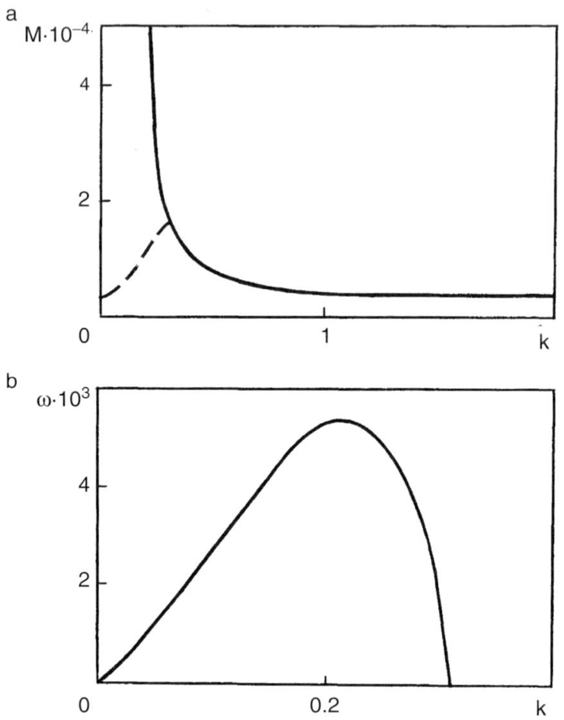

FIGURE 4.5. The case $a = a_* = 1$, $W = 5 \times 10^4$, $Ga = 0.86$. (a) The oscillatory and monotonic neutral curves; (b) dependence of the frequency of oscillations ω on the wave number k.

Let us remember the main results of the weakly nonlinear theory of the monotonic deformational instability in the case of a sole interface between two fluids (this corresponds to $h_* = a_* = \bar{a} = 0$; see Sec. 2.4.4).

The deflection of the interface h is equivalent to the local change of the upper and lower layer thicknesses:

$$a_1' = 1 - h, \quad a_2' = a + h. \tag{4.61}$$

Using expressions (4.61), it is possible to recalculate the critical Marangoni number M_c as a function of h. It is necessary to distinguish between the two cases: (A) $Q_1 = (dM_c/dh)_{h=0} \neq 0$, and (B) $Q_1 = 0$.

In case (A), one can expect that the instability is non-saturable; that is, it leads to blow-up, physically corresponding to the formation of dry spots (VanHook *et al.*, 1995). Assume that $Q_1 > 0$. Let us take a system at the critical value of the Marangoni number M_c $(h = 0)$ and consider a long-wave modulation of its local thickness $h(x, y)$. In the regions where the thickness decreases, the *local critical value* of the Marangoni number $M_c(h)$ decreases, and the instability is enhanced. The case of $Q_1 < 0$ can be considered in a similar way. The asymptotic analysis (Nepomnyashchy and Simanovskii, 1990b; Simanovskii and

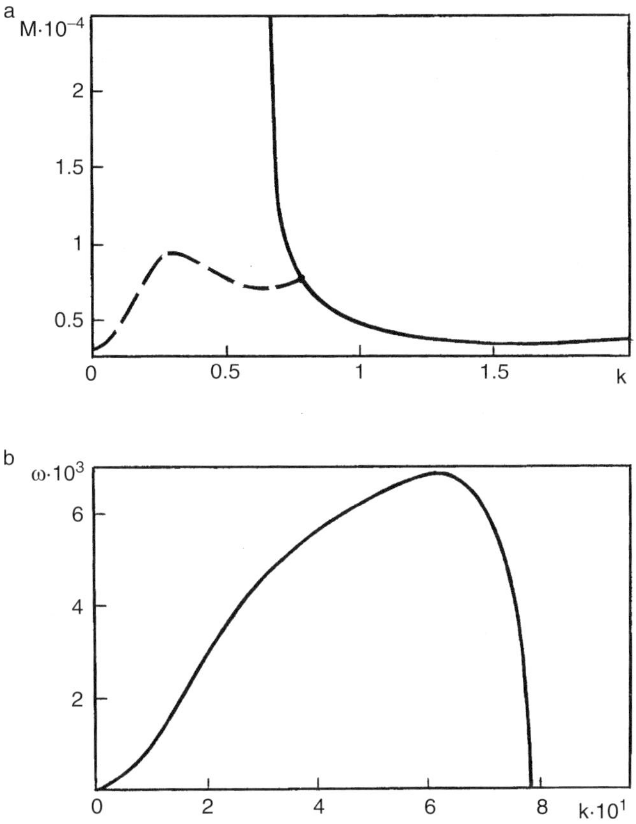

FIGURE 4.6. The case $a = 1$, $a_* = 0.1$, $W = 5 \times 10^4$, $Ga = 0.86$. (a) The oscillatory and monotonic neutral curves; (b) dependence of the frequency of oscillations ω on the wave number k.

Nepomnyashchy, 1993) shows that the nonlinear evolution of the interfacial deformation is governed to the leading order by the *Sivashinsky-Funada equation* (Sivashinsky, 1983; Funada, 1987):

$$\frac{\partial h^{(0)}}{\partial \tau} = -\Delta_{\perp}(A\Delta_{\perp}h^{(0)} + BM^{(1)}h^{(0)} + Ch^{(0)2}), \tag{4.62}$$

where $h^{(0)}$ is the leading term in the asymptotic expansion of $h = \epsilon h^{(0)} + \cdots$ in powers of a small parameter ϵ, $M^{(1)} = (M - M_c)/\epsilon$, $\tau = \epsilon^2 t$, and $\Delta_{\perp}$ is the scaled Laplacian operator proportional to $\partial^2/\partial x^2 + \partial^2/\partial y^2$; the coefficients A and B are positive, and the real constant C is proportional to $-Q_1$.

Equation (4.62) may be written in the form

$$\frac{\partial h^{(0)}}{\partial \tau} = \Delta_{\perp}\frac{\delta L}{\delta h^{(0)}}; \tag{4.63}$$

TABLE 4.2. Critical values of parameters for different types of instability.

a_* | W	10^5	$7 \cdot 10^4$	$6 \cdot 10^4$	$5 \cdot 10^4$
0.001	(b)	(b)	(d)	(d)
	$M_o = 4478$	$M_o = 4475$	$M_o = 4337$	$M_o = 2979$
	$k_o = 1.98$	$k_o = 1.97$	$k_o = 0$	$k_o = 0$
	$\omega_o = 3.69 \cdot 10^{-4}$	$\omega_o = 3.68 \cdot 10^{-4}$		
0.0025	(b)	(b)	(b)	(d)
	$M_o = 4361$	$M_o = 4356$	$M_o = 4354$	$M_o = 2549$
	$k_o = 1.94$	$k_o = 1.94$	$k_o = 1.94$	$k_o = 0$
	$\omega_o = 8.76 \cdot 10^{-4}$	$\omega_o = 8.67 \cdot 10^{-4}$	$\omega_o = 8.64 \cdot 10^{-4}$	
0.005	(b)	(b)	(b)	(d)
	$M_o = 4219$	$M_o = 4206$	$M_o = 4201$	$M_o = 2608$
	$k_o = 1.90$	$k_o = 1.90$	$k_o = 1.90$	$k_o = 0$
	$\omega_o = 1.68 \cdot 10^{-3}$	$\omega_o = 1.63 \cdot 10^{-3}$	$\omega_o = 1.62 \cdot 10^{-3}$	
0.05	(a)	(a)	(a)	(d)
	$M_m = 3801$	$M_m = 3819$	$M_m = 3829$	$M_o = 2979$
	$k_m = 1.80$	$k_m = 1.81$	$k_m = 1.82$	$k_o = 0$
0.1	(a)	(a)	(a)	(d)
	$M_m = 3448$	$M_m = 3454$	$M_m = 3458$	$M_o = 2979$
	$k_m = 1.63$	$k_m = 1.64$	$k_m = 1.64$	$k_o = 0$
0.25	(a)	(a)	(a)	(d)
	$M_m = 3118$	$M_m = 3118$	$M_m = 3119$	$M_o = 2807$
	$k_m = 1.41$	$k_m = 1.41$	$k_m = 1.42$	$k_o = 0$
0.5	(a)	(a)	(a)	(d)
	$M_m = 3186$	$M_m = 3185$	$M_m = 3184$	$M_o = 2734$
	$k_m = 1.29$	$k_m = 1.29$	$k_m = 1.29$	$k_o = 0$
0.75	(a)	(a)	(a)	(c)
	$M_m = 3581$	$M_m = 3579$	$M_m = 3578$	$M_o = 2857$
	$k_m = 1.30$	$k_m = 1.30$	$k_m = 1.30$	$k_o = 4.88 \cdot 10^{-3}$
				$\omega_o = 1.15 \cdot 10^{-5}$
1	(a)	(a)	(a)	(c)
	$M_m = 4215$	$M_m = 4213$	$M_m = 4212$	$M_o = 3090$
	$k_m = 1.47$	$k_m = 1.47$	$k_m = 1.47$	$k_o = 7.93 \cdot 10^{-3}$
				$\omega_o = 1.04 \cdot 10^{-4}$

here, the Lyapunov functional L is defined as

$$L(h^{(0)}) = \int dx\, dy \left[\frac{1}{2} A (\nabla_\perp h^{(0)})^2 - \frac{1}{2} B M^{(1)} h^{(0)2} - \frac{1}{3} C h^{(0)3} \right].$$

With the growth of time, the functional L decreases. It is not bounded from below, and the blow-up of solutions takes place in a finite time (Bernoff and Bertozzi, 1995).

However, the dependence of the critical Marangoni number on the thicknesses of the liquid layers may be *nonmonotonic* (Simanovskii and Nepomnyashchy, 1993). If the ratio of the thicknesses of the layers $a = a_c$ corresponds to the minimum value of the critical Marangoni number, so that $Q_1 = (dM_c/dh)_{h=0} = 0$, and $Q_2 = (d^2 M_c/dh^2)_{h=0} > 0$, both the increase and the decrease of this ratio weaken the instability. In the latter case, the coefficient C in equation (4.62)

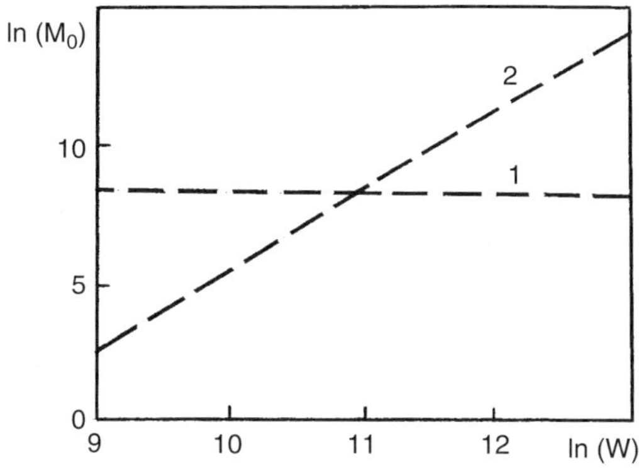

FIGURE 4.7. Transition between the short-wave oscillatory instability (line 2) and the long-wave oscillatory instability (line 1).

vanishes, and the cubic nonlinearity has to be incorporated. One obtains the following amplitude equation:

$$\frac{\partial h^{(0)}}{\partial \tau} = -\Delta_{\perp}\left(A\Delta_{\perp}h^{(0)} + BM^{(1)}h^{(0)} - Dh^{(0)3}\right), \tag{4.64}$$

where $h = \epsilon^{1/2}h^{(0)} + \cdots$; the coefficient D is positive if $Q_2 > 0$. Equation (4.64), which was first derived in the theory of phase transitions, is known as the *Cahn-Hilliard equation* (Langer, 1980). It describes the formation of two locally stable "phases" $h^{(0)} \approx \pm\left(BM^{(1)}/D\right)^{1/2}$ separated by kinks (Kawasaki and Ohta, 1982).

If a is close to a_c, $|a - a_c| \sim O(\epsilon^{1/2})$, the quadratic and the cubic nonlinearities are of the same order, and the problem is governed by the following amplitude equation:

$$\frac{\partial h^{(0)}}{\partial \tau} = -\Delta_{\perp}\left(A\Delta_{\perp}h^{(0)} + BM^{(1)}h^{(0)} + C^{(0)}h^{(0)2} - Dh^{(0)3}\right); \tag{4.65}$$

here, $C = C^{(0)}\epsilon^{1/2}$. The corresponding Lyapunov functional

$$L(h^{(0)}) = \int dx\, dy \left[\frac{1}{2}A(\nabla_{\perp}h^{(0)})^2 - \frac{1}{2}BM^{(1)}h^{(0)2} - \frac{1}{3}h^{(0)3} + \frac{1}{4}Dh^{(0)4}\right]$$

is bounded from below, so that the instability is saturable.

Now, we will consider the case of a system with two interfaces. As was shown in Sec. 4.1.1, there exist two stationary instability boundaries $M = M_1$ and $M = M_2$, and in some cases (see Eqs. (4.59) and (4.60)), also the oscillatory instability boundary $M = M_0$. If M_1 and M_2 are not close to each other, the deformations of both interfaces h and h_* near each instability threshold are mutually proportional, and the problem is governed by equation (4.62), equation (4.64), or (4.65). If M_1 and M_2 are close, and in the case of the oscillatory instability, the deformations of both interfaces can be considered as independent active variables. In the latter case, we can expect that the problem is governed by a system of *two coupled equations* for h and h_*.

We are going to derive the system of amplitude equations for an oscillatory instability in a specific case $a_* \ll 1$. Recall that the threshold Marangoni number of the oscillatory instability and the frequency of oscillations in the long wave limit $k \to 0$ are determined by formulas (4.57) and (4.58).

A uniform deflection of the upper interface (4.61) is equivalent to the following local change of the upper and middle layer thicknesses (in dimensional variables):

$$a_1' = a_1(1-h),\ a_2' = a_2 + ha_1.$$

The onset of convection instability is determined by the parameters

$$M' = \frac{\alpha\theta a_1'}{\eta_1\chi_1},\quad Ga' = \frac{ga_1'^3}{\nu_1^2},\quad a' = \frac{a_2'}{a_1'}.$$

According to equation (4.37), the threshold is governed by the formula

$$M_0' = \frac{2s\, Ga'\, \delta P(1+\eta a')(1+\kappa a')^2}{3\kappa(1+a')(1-\eta a'^2)} + o(1).$$

Returning to variables M, Ga, and δ, we rewrite the latter equation in the form

$$M_0 = \frac{2s\, Ga\delta P\,(1-h)(a+h)[(1-h)+\eta(a+h)][(1-h)+\kappa(a+h)]^2}{3\kappa(1+a)[(1-h)^2-\eta(a+h)^2]} + o(1). \tag{4.66}$$

Parameter h can vary within the interval $-a < h < 1$.

Let us present the explicit expressions for variables $Q_1 = (dM_0/dh)_{h=0}$ and $Q_2 = (d^2M_0/dh^2)_{h=0}$, which are important for the nonlinear analysis:

$$Q_1 = M_0\left[\frac{\eta(1+a)}{1+\eta a} + \frac{2\kappa(1+a)}{1+\kappa a} + \frac{1-2a}{a} + \frac{2\eta a(1+a)}{1-\eta a^2}\right]; \tag{4.67}$$

$$\begin{aligned} Q_2 = M_0\Bigg[&-\frac{\eta-1}{1+\eta a} - \frac{\kappa-1}{1+\kappa a} - \frac{1}{a} - \frac{1+2\eta a}{1-\eta a^2} \\ &+\frac{(\eta-1)(\kappa-1)}{(1+\eta a)(1+\kappa a)} + \frac{\eta-1}{a(1+\eta a)} + \frac{(\eta-1)(1+2\eta a)}{(1-\eta a^2)(1+\eta a)} \\ &+\frac{\kappa-1}{a(1+\kappa a)} + \frac{(\kappa-1)(1+2\eta a)}{(1+\kappa a)(1-\eta a^2)} + \frac{1+2\eta a}{a(1-\eta a^2)} \\ &+\frac{\eta}{1-\eta a^2} + \left(\frac{1+2\eta a}{1-\eta a^2}\right)^2\Bigg]. \end{aligned} \tag{4.68}$$

Cases (A) $Q_1 \neq 0$ and (B) $Q_1 = 0,\ Q_2 > 0$ are studied separately.

The Case $Q_1 \neq 0$

Let us consider the region near the threshold of oscillatory instability:

$$M = M_0 + ma_*,\ m = O(1),\ a_* \ll 1. \tag{4.69}$$

The linear theory predicts an oscillatory instability in the region of small wave numbers $k = O(a_*^{1/2})$. This instability is characterized by both growth rate and frequency of oscillations of order $O(a_*^2)$. From expression (4.66), we can expect that the deformations $h = O(a_*)$ are relevant in the case $Q_1 \neq 0$. Taking into account the scaling properties of eigenfunctions appearing in the linear theory, we introduce the following scaling of variables:

$$\bar{x} = a_*^{1/2}x,\ \bar{y} = a_*^{1/2}y,\ \bar{t} = a_*^2 t,\ h = a_*\bar{h},$$

$$T_j = T_j^0 + a_*\Theta_j,\ p_j = a_* P_j,\ \mathbf{v}_{j\perp} = a_*^{3/2}\mathbf{V}_{j\perp},\ v_{jz} = a_*^2 V_{jz}\ (j = 1, 2); \tag{4.70}$$

$$h_* = a_*^2\bar{h}_*,\ T_3 = s + a_*\Theta_3,\ p_3 = a_* P_3,$$

$$\mathbf{v}_{3\perp} = a_*^{5/2}\mathbf{V}_{3\perp},\ v_{3z} = a_*^4 V_{3z},$$

where $\mathbf{v}_{m\perp} \equiv (v_{mx}, v_{my})$, $m = 1, 2, 3$. In the region $-a - a_* \leq z \leq -a$, the variable

$$\bar{z} = (z + a)/a_* \tag{4.71}$$

is used.

The solution $f = (H, H_*, \Theta_m, P_m, \mathbf{V}_{m\perp}, V_{mz})\ (m = 1, 2, 3)$ is presented in the form of a series,

$$f = f^{(0)} + a_* f^{(1)} + \cdots. \tag{4.72}$$

We substitute the expansions (4.69) – (4.72) into the problem (3.1) – (3.13) and collect the terms of the same order in a_*. We obtain the amplitude equations for "active" variables H and H_* from the solvability conditions. The details of the derivation of the amplitude equations can be found in the paper of Kliakhandler *et al.* (1998).

In the zeroth order, the equations and boundary conditions are linear. We reproduce the results of the linear theory in the long-wave limit presented in Sec. 4.1.1, including expression (4.57) for the critical Marangoni number of the oscillatory instability, and obtain the equation

$$\frac{\partial \bar{h}_*^{(0)}}{\partial \bar{t}} = E\bar{\Delta}_\perp \bar{h}^{(0)}, \tag{4.73}$$

which describes in the leading order the time evolution of the deformation of the lower interface. Here,

$$E = \frac{Ga\,\delta}{6}\frac{\eta_* a}{1 - \eta a^2},\ \bar{\Delta}_\perp = \frac{\partial^2}{\partial \bar{x}^2} + \frac{\partial^2}{\partial \bar{y}^2}.$$

In the first order in a_* we obtain the nonlinear amplitude equation for the evolution of the deformation of the upper interface:

$$\frac{\partial \bar{h}^{(0)}}{\partial \bar{t}} = -\bar{\Delta}_\perp \left[A\bar{\Delta}_\perp \bar{h}^{(0)} + B(m - m_*)\bar{h}^{(0)} + C\bar{h}^{(0)2} + F\bar{h}_*^{(0)} \right], \tag{4.74}$$

where the coefficients A, B, and C are *exactly* the same as in equation (4.62); the quantity m_* describing the deviation of the critical Marangoni number from the value M_0 in the first order in a_* is given by the expressions

$$m_* = \frac{2s\,Ga\,\delta P(1+\kappa a)}{3(1-\eta a^2)\kappa(1+a)} \left[\frac{\kappa_*(1+\eta a)(-1+2\kappa+\kappa a)}{\kappa(1+a)} \right.$$
$$\left. + \frac{\eta_*(1+2\eta a+\eta a^2)(1+\kappa a^2)(1+\kappa a)}{\eta(1-\eta a^2)} \right], \tag{4.75}$$

and

$$F = \frac{sM_0(\kappa_* - \kappa)\eta a^2(1-\eta a^2)}{2P(1+\kappa a)^2\Delta}.$$

Thus, we obtained the *coupled* system of amplitude equations (4.73) and (4.74) describing the nonlinear evolution of long-wave deformations of both interfaces near the instability threshold in the case $a_* \ll 1$.

Let us note that equations (4.73) and (4.74) predict a linear oscillatory instability with the frequency $\omega = \sqrt{EF}k^2 + O(k^4)$ in the case of E and F having the same sign, which coincides with formula (4.58).

The Case $Q_1 = 0$

In the case $Q_1 = 0$, the nonlinear coefficient C in equation (4.74) vanishes. In order to obtain the nonlinear saturation, it is necessary to use a different scaling of functions:

$$h = a_*^{1/2}\bar{h},\ T_j = T_j^0 + a_*^{1/2}\Theta_j,\ p_j = a_*^{1/2}P_j,$$
$$\mathbf{v}_{j\perp} = a_*\mathbf{V}_{j\perp},\ v_{jz} = a_*^{3/2}V_{jz}\ (j=1,2); \tag{4.76}$$
$$h_* = a_*^{3/2}\bar{h}_*,\ T_3 = s + a_*^{1/2}\Theta_3,\ p_3 = a_*^{1/2}P_3,$$
$$\mathbf{v}_{3\perp} = a_*^2\mathbf{V}_{3\perp},\ v_{3z} = a_*^{7/2}V_{3z}.$$

The solution is expanded into series in powers of $a_*^{1/2}$. Finally, equation (4.73) is not changed, whereas equation (4.74) is replaced by the amplitude equation

$$\frac{\partial \bar{h}^{(0)}}{\partial \bar{t}} = -\bar{\Delta}_\perp \left[A\bar{\Delta}_\perp \bar{h}^{(0)} + B(m - m_*)\bar{h}^{(0)} - D\bar{h}^{(0)3} + F\bar{h}_*^{(0)} \right], \tag{4.77}$$

where the coefficient D is the same as in equation (4.64).

In the case of small but nonzero Q_1, where the coefficient C is not exactly equal to zero but small ($C = \bar{C}a_*^{1/2}$, $\bar{C} = O(1)$), the equation containing both quadratic and cubic nonlinearities is obtained (see Simanovskii and Nepomnyashchy, 1993):

$$\frac{\partial \bar{h}^{(0)}}{\partial \bar{t}} = -\bar{\Delta}_\perp \left[A\bar{\Delta}_\perp \bar{h}^{(0)} + B(m - m_*)\bar{h}^{(0)} + \bar{C}\bar{h}^{(0)2} - D\bar{h}^{(0)3} + F\bar{h}_*^{(0)} \right]. \tag{4.78}$$

Equations (4.74) and (4.77) can be considered as particular cases of equation (4.78).

By means of the transformation

$$\bar{h} = (FE/D^2)^{1/4}H, \; \bar{h}_* = (E^3/FD^2)^{1/4}H_*,$$

$$X = (FE/A^2)^{1/4}\bar{x}, \; Y = (FE/A^2)^{1/4}\bar{y}, \; \tau = (FE/A)\bar{t}, \tag{4.79}$$

equations (4.78) and (4.73) are rewritten in the form

$$\frac{\partial H}{\partial \tau} + \Delta_\perp(\Delta_\perp H + \mu H + \gamma H^2 - H^3 + H_*) = 0, \tag{4.80}$$

$$\frac{\partial H_*}{\partial \tau} - \Delta_\perp H = 0, \tag{4.81}$$

where

$$\Delta_\perp = \frac{\partial}{\partial X^2} + \frac{\partial}{\partial Y^2}, \; \mu = \frac{B(m - m_*)}{(FE)^{1/2}}, \; \gamma = \frac{\bar{C}}{(FED^2)^{1/4}}.$$

4.3.2 Traveling Wave Solutions

The trivial solution $H = 0$, $H_* = 0$ of the system (4.80), (4.81) is stable with respect to disturbances with the wave number k in the region $\mu < k^2$. On the neutral curve $\mu = \mu_0 = k^2$, an oscillatory instability with the frequency $\omega_0 = \pm k^2$ appears. In the present subsection, we consider small-amplitude solutions, which bifurcate on the neutral curve.

Bifurcation of Traveling Wave Solutions

In order to describe the traveling wave solutions near the neutral curve ($\mu - \mu_0 = O(\epsilon^2)$, $\epsilon \ll 1$), we introduce the timescales

$$\tau_0 = \tau, \; \tau_2 = \epsilon^2\tau, \; \ldots \tag{4.82}$$

and use the following expansions:

$$\mu = \mu_0 + \epsilon^2\mu_2, \; H = \epsilon H_1 + \epsilon^2 H_2 + \cdots, \; H_* = \epsilon H_{*1} + \epsilon^2 H_{*2} + \cdots \tag{4.83}$$

In the first order in ϵ, we obtain a linear eigenvalue problem. Let us consider a particular solution corresponding to a traveling wave that propagates along the X-axis in the positive direction and has the wave number k. We choose

$$H_1 = A(\tau_2)e^{i(kX-\omega_0\tau_0)} + \text{c.c.}, \ H_{*1} = B(\tau_2)e^{i(kX-\omega_0\tau_0)} + \text{c.c.}$$

(c.c. means complex conjugate) and find $\mu_0 = k^2$, $\omega_0 = k^2$, and $B = -iA$.
In the second order, the solution is

$$H_2 = A_2(\tau_2)e^{2i(kX-\omega_0\tau_0)} + \text{c.c.}, \ H_{*2} = B_2(\tau_2)e^{2i(kX-\omega_0\tau_0)} + \text{c.c.}$$

(we omit the general solution of the homogeneous system renormalizing the amplitude), where

$$A_2 = \frac{2\gamma A^2}{3(i + 2k^2)}, \quad B_2 = -2iA_2.$$

The solvability condition of the third-order equations determines the Landau equation for the amplitude evolution:

$$\frac{dA}{d\tau_2} = \frac{k^2}{2}\left\{\mu_2 A + \left[\frac{4\gamma^2(2k^2 - i)}{3(1 + 4k^4)} - 3\right]|A|^2 A\right\}. \tag{4.84}$$

The limit cycle corresponding to a traveling wave solution is $A = |A|\exp(i\omega_2\tau_2)$, where

$$\omega_2 = \frac{2\gamma^2 k^2}{3(1 + 4k^4)}, \tag{4.85}$$

$$|A|^2 = \mu_2\left[3 - \frac{8\gamma^2 k^2}{3(1 + 4k^4)}\right]^{-1}. \tag{4.86}$$

One can see that the bifurcation is supercritical for any values of k if $\gamma < 3/\sqrt{2}$. Otherwise, an interval of a subcritical bifurcation appears in a certain interval of k.

Interaction of Traveling Waves

In order to consider the interaction of traveling waves, we use the same scales (4.82) and expansions (4.83) as in the previous subsection, but in the leading order of the expansion, we choose the solution

$$H_1 = A^{(1)}(\tau_2)e^{i(\mathbf{k}^{(1)}\cdot\mathbf{X}-\omega_0\tau_0)} + A^{(2)}(\tau_2)e^{i(\mathbf{k}^{(2)}\cdot\mathbf{X}-\omega_0\tau_0)} + \text{c.c.},$$

$$H_{*1} = B^{(1)}(\tau_2)e^{i(\mathbf{k}^{(2)}\cdot\mathbf{X}-\omega_0\tau_0)} + B^{(2)}(\tau_2)e^{i(\mathbf{k}^{(2)}\cdot\mathbf{X}-\omega_0\tau_0)} + \text{c.c.},$$

$|\mathbf{k}^{(1)}| = |\mathbf{k}^{(2)}| = k$, $\omega_0 = k^2$, $B^{(1)} = -iA^{(1)}$, and $B^{(2)} = -iA^{(2)}$, corresponding to two traveling waves moving in different directions.

In the second order, the solution has the following structure:

$$
\begin{aligned}
H_2 &= A_2^{(1,1)}(\tau_2)e^{2i(\mathbf{k}^{(1)}\cdot\mathbf{X}-\omega_0\tau_0)} + A_2^{(2,2)}(\tau_2)e^{2i(\mathbf{k}^{(2)}\cdot\mathbf{X}-\omega_0\tau_0)} \\
&\quad + A_2^{(1,2)}(\tau_2)e^{2i(\mathbf{k}^{(1)}+\mathbf{k}^{(2)})\cdot\mathbf{X}-2\omega_0\tau_0)} + A_2^{(1,-2)}(\tau_2)e^{2i(\mathbf{k}^{(1)}-\mathbf{k}^{(2)})\cdot\mathbf{X}} + \text{c.c.}, \\
H_{*2} &= B_2^{(1,1)}(\tau_2)e^{2i(\mathbf{k}^{(1)}\cdot\mathbf{X}-\omega_0\tau_0)} + B_2^{(2,2)}(\tau_2)e^{2i(\mathbf{k}^{(2)}\cdot\mathbf{X}-\omega_0\tau_0)} \\
&\quad + B_2^{(1,2)}(\tau_2)e^{2i(\mathbf{k}^{(1)}+\mathbf{k}^{(2)})\cdot\mathbf{X}-2\omega_0\tau_0)} + B_2^{(1,-2)}(\tau_2)e^{2i(\mathbf{k}^{(1)}-\mathbf{k}^{(2)})\cdot\mathbf{X}} + \text{c.c.}
\end{aligned}
$$

(as before, we omit the general solution of the homogeneous system). We obtain the following expressions for the coefficients:

$$
\begin{aligned}
A_2^{(1,1)} &= \frac{2\gamma A^{(1)2}}{3(i+2k^2)},\ B_2^{(1,1)} = -2iA_2^{(1,1)}, \\
A_2^{(2,2)} &= \frac{2\gamma A^{(2)2}}{3(i+2k^2)},\ B_2^{(2,2)} = -2iA_2^{(2,2)}, \\
A_2^{(1,2)} &= \frac{2\gamma(1+\zeta)A^{(1)}A^{(2)}}{i\zeta(2+\zeta)+(1+\zeta)(1+2\zeta)k^2},\ B_2^{(1,2)} = -i(1+\zeta)A_2^{(1,2)}, \\
A_2^{(1,-2)} &= 0,\ B^{(1,-2)} = -\gamma A^{(1)}A^{(2)*},
\end{aligned}
$$

where $\zeta = \mathbf{k}^{(1)}\cdot\mathbf{k}^{(2)}/k^2$.

The solvability condition of the third-order equations determines a pair of Landau equations for the amplitudes evolution:

$$
\begin{aligned}
\frac{dA^{(1)}}{d\tau_2} &= \frac{k^2}{2}(\mu_2 A^{(1)} - \lambda_0|A^{(1)}|^2A^{(1)} - \lambda(\zeta)|A^{(2)}|^2A^{(1)}), \\
\frac{dA^{(2)}}{d\tau_2} &= \frac{k^2}{2}(\mu_2 A^{(2)} - \lambda_0|A^{(2)}|^2A^{(2)} - \lambda(\zeta)|A^{(1)}|^2A^{(2)}),
\end{aligned} \tag{4.87}
$$

where

$$
\lambda_0 = 3 - \frac{4\gamma^2(2k^2-i)}{3(1+4k^2)}, \tag{4.88}
$$

$$
\lambda(\zeta) = 6 - \frac{4\gamma^2(1+\zeta)[k^2(1+\zeta)(1+2\zeta)-i\zeta(2+\zeta)]}{k^4(1+\zeta)^2(1+2\zeta)^2+\zeta^2(2+\zeta)^2}. \tag{4.89}
$$

The system of equations (4.87) describes the nonlinear evolution of two interacting harmonic waves. It can be easily shown that in the case where $0 < \text{Re}\lambda_0 < \text{Re}\lambda(\zeta)$ for any ζ (Re denotes the real part), the one-dimensional traveling wave solutions $|A^{(1)}|^2 = \mu_2/\text{Re}\lambda_0$, $|A^{(2)}|^2 = 0$, and $|A^{(2)}|^2 = \mu_2/\text{Re}\lambda_0$, $|A^{(1)}|^2 = 0$ are stable as $\mu_2 > 0$ in frames of the system (4.87). In the case $\text{Re}\lambda(\zeta) < \text{Re}\lambda_0$, the nonlinear superposition of two waves $|A^{(1)}|^2 = |A^{(2)}|^2 = \mu_2/(\text{Rc}\lambda_0 + \text{Re}\lambda(\zeta))$ is stable if $\mu_2 > 0$, and $\text{Re}\lambda_0 + \text{Re}\lambda(\zeta) > 0$.

Using expressions (4.88) and (4.89), we find that the necessary condition of the stability of one-dimensional traveling wave solutions is

$$\mathrm{Re}(\lambda(\zeta)-\lambda_0)=3+k^2\gamma^2\left[\frac{8}{3(4k^4+1)}-\frac{4(1+\zeta)^2(1+2\zeta)}{k^4(1+\zeta)^2(1+2\zeta)^2+\zeta^2(2+\zeta)^2}\right]>0. \tag{4.90}$$

If $\gamma = 0$ (the quadratic term in equation (4.80) is absent), the one-dimensional traveling wave is stable with respect to disturbances with the same wave number moving in any direction.

If γ is small, expression (4.90) is negative only in the region of small wave numbers ($|k| < \sqrt{4/3}\gamma + O(\gamma^5)$), and positive otherwise. Expression (4.90) is minimal for small values of the parameter ζ: $\zeta \approx -k^4/4$. Thus, the transition to nearly square patterns is expected for sufficiently small k. Let us note that square patterns were predicted in the case of the dispersion relation $\omega \sim k^2$ by Pismen (1988). For finite values of γ, the traveling wave is unstable with respect to disturbances with $\zeta = 0$, generating square patterns, if its wave number satisfies the relation $|k| < k_*(\gamma)$, where

$$\gamma^2 = \frac{9k^2(4k^4+1)}{4(3+10k^4)}.$$

Finite-Amplitude Traveling Waves

In the previous paragraphs, we analyzed small-amplitude traveling waves near the neutral curve. Now, we shall consider finite-amplitude spatially periodic traveling wave solutions

$$H = H(\xi),\ H_* = H_*(\xi),\ \xi = X - c\tau, \tag{4.91}$$

$$H(\xi+L) = H(\xi),\ H_*(\xi+L) = H_*(\xi). \tag{4.92}$$

Substituting expressions (4.91) into equations (4.80) and (4.81), eliminating $H_*(\xi)$, and integrating once the obtained equation, we arrive at

$$H_{\xi\xi\xi} - \frac{H_{\xi\xi}}{c} + \mu H_\xi - cH + (\gamma H^2 - H^3)_\xi = -c\langle H\rangle, \tag{4.93}$$

where the subscript ξ denotes differentiation with respect to ξ, and the integration constant

$$\langle H\rangle = \frac{1}{L}\int_0^L H(\xi)d\xi$$

is the mean value of the function $H(\xi)$ and, therefore, should be put equal to zero, because $H(\xi)$ is proportional to the deviation of the fluid level from its mean value.

Spatially periodic waves (4.91) and (4.92) correspond to limit cycles of the dynamical system (4.93). Equation (4.93) with the periodicity condition $H(\xi + L) = H(\xi)$ is a nonlinear eigenvalue problem for the phase velocity $c(L)$. Because the dependence $c(L)$ is unknown from the beginning, in our calculations we actually change the parameter c arbitrarily and find the corresponding value of the limit cycle's period $L(c)$.

Equation (4.93) can be simplified in the limit case of small c. Multiplying equation (4.93) by H and integrating over the period, we find that $\langle H_\xi^2 \rangle = c^2 \langle H^2 \rangle$. Thus, the limit $c \to 0$ corresponds to the long-wave limit $L \to \infty$.

In this long-wave limit, it is convenient to use the variable $\zeta = c\xi$ and to rewrite equation (4.93) in the form

$$-c^2 H_{\zeta\zeta\zeta} + H_{\zeta\zeta} + (H^3 - \gamma H^2 - \mu H)_\zeta + H = 0. \tag{4.94}$$

Let us note that $H_* = -H_\zeta$. In the limit $c \to 0$, if the periodic solutions of equation (4.94) reveal no boundary layers, they tend to the periodic solutions of the second-order equation

$$H_{\zeta\zeta} + (H^3 - \gamma H^2 - \mu H)_\zeta + H = 0. \tag{4.95}$$

For the case $\gamma = 0$, where the traveling waves are expected to be stable with respect to two-dimensional disturbances, equation (4.95) turns into the familiar Van der Pol equation. The latter is known to yield limit cycles whose shape varies from harmonic oscillations (in case of small μ) to the strongly nonlinear relaxation oscillations for large values of μ. In case of non-zero γ, the invariance of the equation with respect to the change of sign of H is broken: The relaxation-like oscillations are still observed, but the characteristic Van der Pol symmetry between the humps and the valleys is absent.

Some additional simplification of equation (4.95) can be obtained in the limit $\gamma \gg 1$, $\mu \ll 1$. If we assume that $H = \gamma^{-1}\tilde{H}$, $\tilde{H} = O(1)$, and omit the term containing μ, we find in the leading order

$$\tilde{H}_{\zeta\zeta} - (\tilde{H}^2)_\zeta + \tilde{H} = 0. \tag{4.96}$$

The trajectory corresponding to the unbounded solution

$$\tilde{H} = \zeta/2 \tag{4.97}$$

separates the phase plane $(\tilde{H}, \tilde{H}_* = -\tilde{H}_\zeta)$ into two regions. In the region $\tilde{H}_* < -1/2$, the trajectories tend to infinity as $\zeta \to \pm\infty$. The half-plane $\tilde{H}_* > -1/2$ is filled by periodic orbits determined by the equation

$$\tilde{H}_* - \ln(1 + 2\tilde{H}_*) + \frac{1}{2}\tilde{H}^2 = E, \; E = const > 0.$$

If $E \gg 1$, the wave relief has a "saw-like" shape. On a long interval of the length $O(\sqrt{E})$, $\tilde{H}_* = -1/2 + O(\exp(-E))$, so that the relief of $\tilde{H}$ is exponentially close to the linear one (4.97). On a short interval of the length $O(1/\sqrt{E})$, $\tilde{H}_* > 0$, $\tilde{H}_* = O(E)$, which corresponds to a steep wave front of $\tilde{H}$.

The typical wave profiles $H(\zeta)$ and $H_*(\zeta)$ calculated by means of the symmetric ($\gamma = 0$) and "asymmetric" ($\gamma = 1$) (Eq.(4.94)) are plotted in Figs. 4.8(a) and Fig. 4.8(b), respectively. These two plots as well as the following ones present

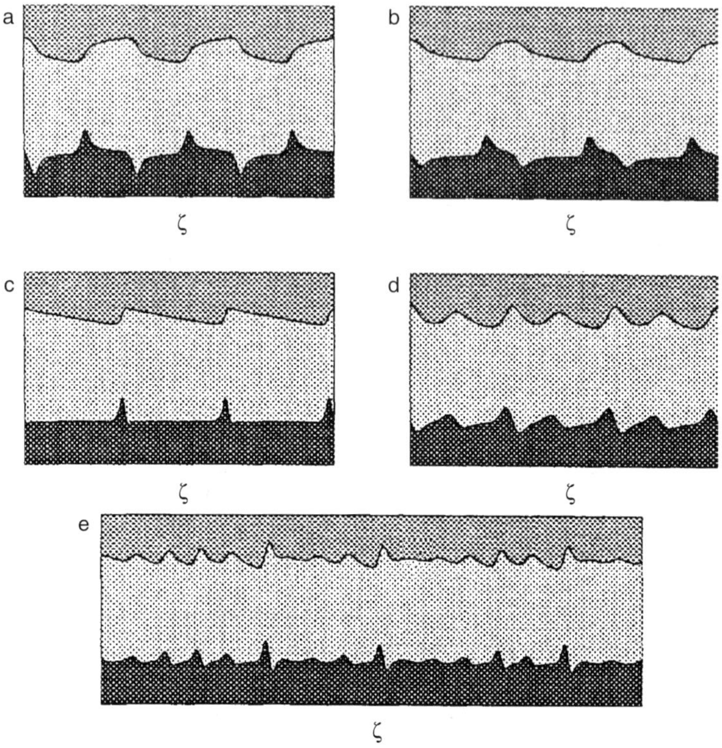

FIGURE 4.8. Examples of traveling wave trains according to equation (4.95): (a) $\gamma = 0$; $c = 0.222$; $\mu = 2$; (b) $\gamma = 1$; $c = 0.08$; $\mu = 0.368$; (c) $\gamma = 5$; $c = 0.36033$; $\mu = 0.03$; (d) $\gamma = 3$; $c = 1$; $\mu = 0.7$; (e) $\gamma = 3$; $c = 0.8$; $\mu = 0.368$.

the longitudinal shape of the waves in the correct way. On the contrary, the vertical displacements, as related to the thicknesses of the layers, cannot be recovered from the equations and, therefore, are represented only qualitatively.

In case of vanishing or small values of γ, this regular pattern with one hump on a period is the only nontrivial bounded solution of equation (4.93) and exists only for positive values of μ; the velocity of the traveling waves obeys the inequality $c^2 < \mu$, and for each value of c, there exists only one periodic solution. The situation changes for $\gamma > \sqrt{3}$, when the above inequality does not necessarily hold. In this case, the increase of μ can create the finite-amplitude periodic solutions through a saddle-node bifurcation. Under $\gamma > 2.8$, the traveling waves of this kind can be found even for negative values of μ. For sufficiently large γ, the upper interface displays the characteristic saw-like oscillations, whereas the lower one consists of the long, almost horizontal segments separated by short elevations (Fig. 4.8(c)).

Moderate and large values of γ allow for more complicated patterns of traveling waves: In this parameter domain, one encounters further bifurcations of periodic solutions that include cascades of period-doublings (in fact, these doublings with respect to the variable ξ are doublings of the spatial period) and onset of chaotic wave profiles. Respective deformations of the interfaces are plotted in Figs. 4.8(d) and 4.8(e). In the last three cases, the periodic pattern delivered by equation (4.94) under the fixed values of γ and μ for a given velocity c is not unique (in the very last case, there is apparently an infinity of different periodic solutions).

4.3.3 Results of Numerical Simulations

One-Dimensional Waves

The one-dimensional version of equations (4.80) and (4.81),

$$\frac{\partial H}{\partial \tau} + \frac{\partial^2}{\partial X^2}\left(\frac{\partial^2 H}{\partial X^2} + \mu H + \gamma H^2 - H^3 + H_*\right) = 0, \tag{4.98}$$

$$\frac{\partial H_*}{\partial \tau} - \frac{\partial^2 H}{\partial X^2} = 0, \tag{4.99}$$

was simulated numerically under periodic boundary conditions. The pseudo-spectral technique was employed for the spatial discretization and the Adams' scheme for the time advance. The standard routines CO6EAF and CO6EBF for the FFT, and DO2CBF for the Adams' scheme from the NAG routines library were used. The number of spatial discretization points was chosen in such a way that the typical wavelength $\lambda_c = 2\pi/k_c$ of the most unstable (in the linear approximation) mode with wave number k_c was covered by at least 10 points. The latter ensures fair resolution of the calculated solution. The time step was chosen automatically.

Both small-amplitude random fields and regular functions with amplitudes of $O(1)$ were used as initial data in all simulations. In a number of cases, the result of evolution depends on initial conditions. This means that equations (4.98) and (4.99) have several coexisting attractors. We considered the same values of parameters (γ, μ) as in Fig. 4.8.

For the parameters (γ, μ) from Fig. 4.8(a), two different interfacial configurations were obtained. Random small-amplitude initial conditions evolved to the profile shown in Fig. 4.9, whereas the regular $O(1)$ initial conditions settled to the wave are plotted in Fig. 4.10. Interfaces form the traveling waves moving with

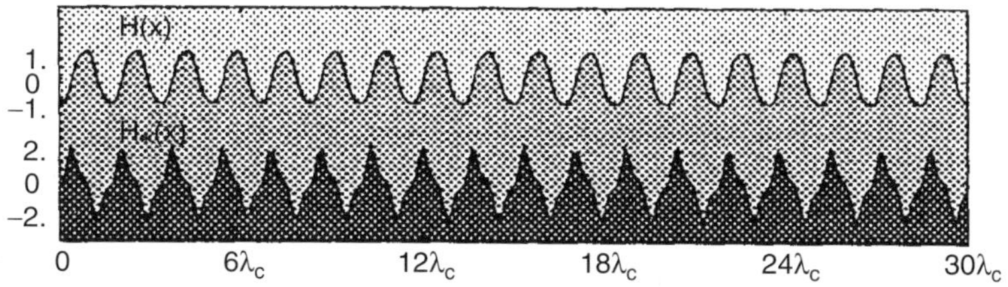

FIGURE 4.9. Traveling wave for $\mu = 2, \gamma = 0$. Initial conditions: small-amplitude random field.

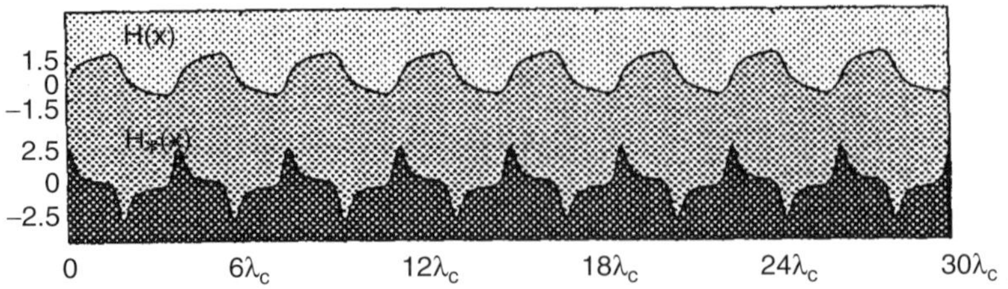

FIGURE 4.10. Traveling wave for $\mu = 2, \gamma = 0$. Initial conditions: regular function of $O(1)$.

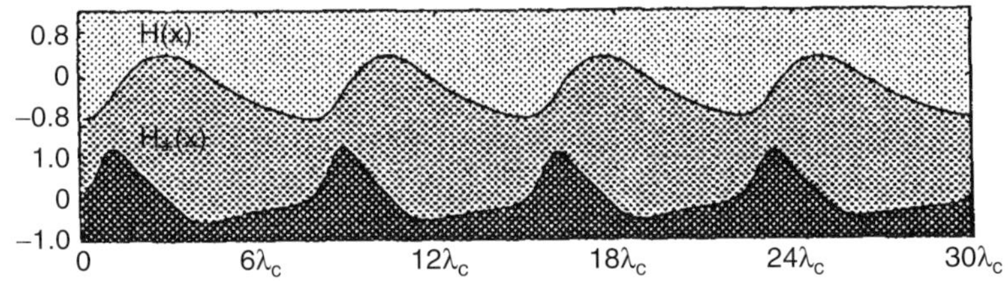

FIGURE 4.11. Traveling wave for $\mu = 0.368, \gamma = 1$.

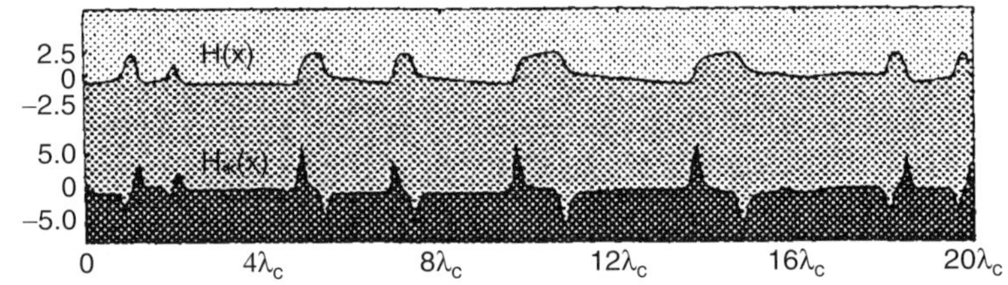

FIGURE 4.12. Typical interfacial profiles for $\mu = 0.7, \gamma = 3$.

velocities $c = -0.0015$ and $c = -0.01$, respectively. In both cases the typical wavescale is larger than that predicted by the linear theory, λ_c.

A few different interfacial configurations were observed for the pair $\mu = 0.368, \gamma = 1$ (Fig. 4.8(b)). The number of settled waves varies from one to six; that is, there is significant enlargement of the wavelength compared with results of linear stability analysis. Figure 4.11 shows the interfaces with four typical humps that form the traveling wave with velocity $c = -0.00197$.

Only the blow-up for all initial conditions was observed in the final stage of the evolution for the pair $\mu = 0.03, \gamma = 1$.

Typical interfacial profiles observed for the pairs $\mu = 0.7, \gamma = 3$ (Fig. 4.8(d)) and $\mu = 0.368, \gamma = 3$ (Fig. 4.8(e)), are shown in Figs. 4.12 and 4.13, respectively. The dynamics, observed for all initial conditions are unsteady. The typical wavescale in both cases is larger than λ_c.

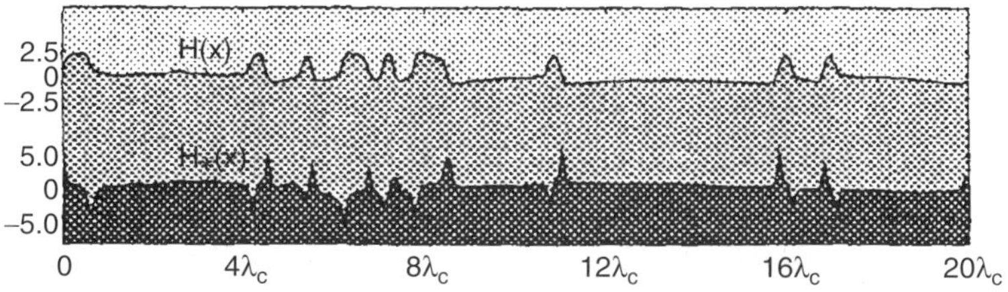

FIGURE 4.13. Typical interfacial profiles for $\mu = 0.368$, $\gamma = 3$.

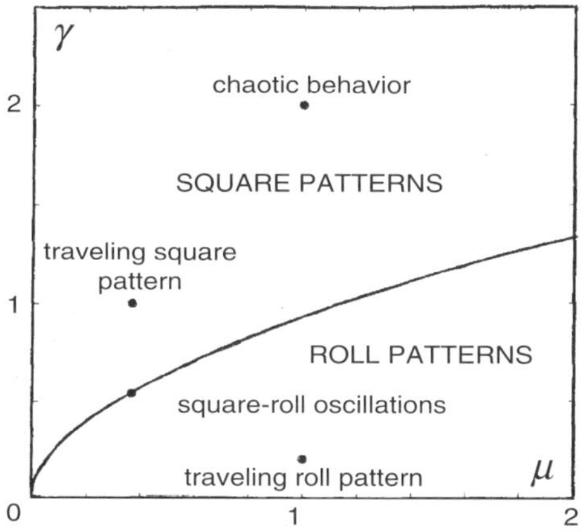

FIGURE 4.14. Parameter regions corresponding to wave patterns with different symmetries. Circles correspond to various patterns found in numerical simulations.

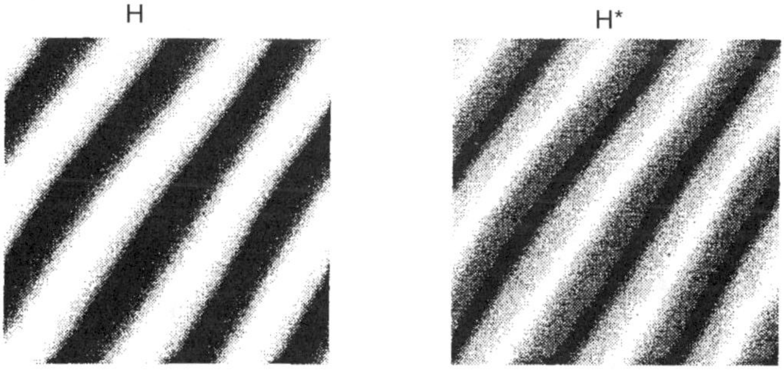

FIGURE 4.15. Snapshot of a traveling roll pattern, $\mu = 1.0$, $\gamma = 0.2$.

Two-Dimensional Simulations.

The results of direct numerical two-dimensional simulations of the system (4.80), (4.81) have been presented by Nepomnyashchy *et al.* (2003). The results of simulations are summarized in Fig. 4.14. The most typical wavy patterns are traveling rolls (Fig. 4.15) and traveling squares (Fig. 4.16). Near the boundary between the regions of above-mentioned patterns, alternating rolls were observed (Fig. 4.17).

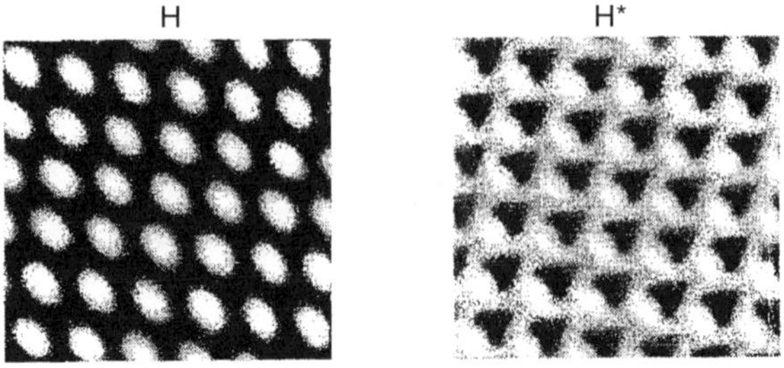

FIGURE 4.16. Snapshot of a traveling square pattern, $\mu = 0.368$, $\gamma = 1.0$.

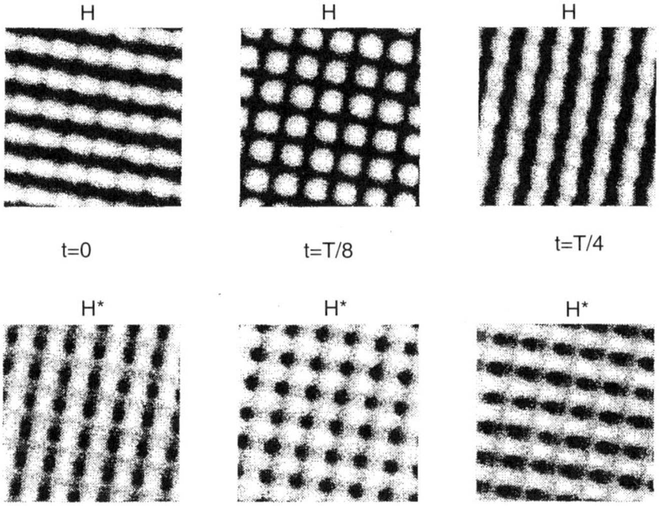

FIGURE 4.17. Alternating rolls, different oscillation phases; $\mu = 0.368$, $\gamma = 0.5415$.

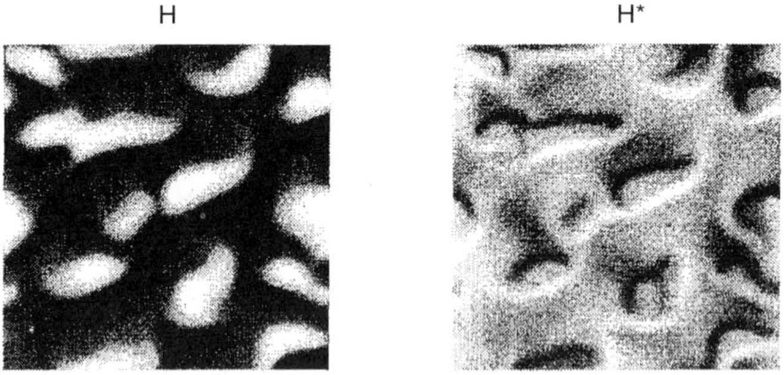

FIGURE 4.18. Snapshot of a chaotic pattern, $\mu = 1.0$, $\gamma = 2.0$.

This pattern is a nonlinear superposition of two systems of standing waves with orthogonal wave vectors. The temporal phase shift between standing waves of different spatial orientations is equal to $T/4$, where T is the full period of oscillations. Thus, one observes some kind of roll pattern that change their orientation with the time interval $T/4$. Note that the alternating roll pattern is one of the generic wave patterns that appear in a rotationally invariant systems due to a primary oscillatory instability of the spatially homogeneous state (Pismen, 1986b; Swift, 1988; Silber and Knobloch, 1991). Also, some spatially chaotic patterns were found. In the latter flow regime, the deformation of the upper interface displays irregular "spots" of a nearly flat interface that split and merge in a chaotic manner (Fig. 4.18).

5

Stability of Flows

In the present chapter, the stability of convective flows in multilayer systems is considered. First, we will discuss general mechanisms of instabilities for flows generated by a surface-tension gradient applied along the surface (Section 5.1). Then, we present a linear analysis of thermocapillary flows in two-layer systems subject to an inclined temperature gradient (Section 5.2). Results of the investigation of buoyancy-thermocapillary convection in two-layer and three-layer systems are given in Sections 5.3 and 5.4, respectively. Section 5.5 contains an analysis of deformational instabilities of thermocapillary flows in three-layer systems.

5.1 Mechanisms of Instabilities for Flows Generated by a Surface-Tension Gradient Applied Along the Surface

Consider a very long liquid layer (with the non-dimensional length $L \gg 1$) on a horizontal rigid plane with a free surface subject to a horizontal temperature gradient A in the direction of axis x. Except relatively small region near the lateral boundaries, the flow can be considered as a parallel one. We will distinguish between two typical situations:
(i) free *open flow* with a zero pressure gradient and nonzero fluid flux ("unidirectional flow"); in the case of a purely thermocapillary convection ($R = 0$), this flow has a linear velocity profile, and it is called "linear flow";
(ii) flow in a very long but *closed* vessel with counteracting surface tangential stress and volume pressure gradient, which generate together a parabolic velocity profile with zero fluid flux ("return flow").

5.1.1 Purely Thermocapillary Flows

First, we will discuss the main types of instability in the case of purely thermocapillary flows which are appropriate mainly under microgravity conditions.

For purely thermocapillary flows ($R = 0$), the velocity profile is

$$u = -Mz \tag{5.1}$$

in the case of the unidirectional (linear) flow, and

$$u = M\left(-\frac{3z^2}{4} + \frac{z}{2}\right) \tag{5.2}$$

in the case of the return flow. Assume that the rigid boundary is heat insulated. The dimensionless temperature field generated by the parallel thermocapillary flow is described by the formula

$$T(x, z) = x + \tau(z), \tag{5.3}$$

where

$$\tau = M\left(-\frac{z^3}{6} + \frac{z}{2}\right) \tag{5.4}$$

in the case of the linear flow, and

$$\tau = -M\left(\frac{z^4}{16} - \frac{z^3}{12} + \frac{z}{48}\right) \tag{5.5}$$

in the case of the return flow. According to the definition of the function $\tau(z)$, $\tau(1) = 0$ in both cases. It would be noted, that in the former case $\tau(z) > 0$, $\tau'(z) \leq 0$ as $0 \leq z < 1$, thus the layer is effectively "heated from below" by the parallel thermocapillary flow. In the latter case $\tau(z) < 0$, $\tau'(z) \geq 0$ as $0 \leq z < 1$; e.g., the layer is "cooled from below".

According to Smith and Davis, 1983a, 1983b; Davis 1987, the instabilities can be separated into several classes.

1. In the case of the linear flow, which produces some heating from below, the usual Marangoni instability may appear. It prevails for large Prandtl numbers and generates stationary longitudinal rolls, e.g., the disturbances of all physical fields depend periodically on the transverse horizontal coordinate y and the vertical coordinate z but do not depend on the longitudinal coordinate x.

2. The basic shear flows with a free boundary are subject to instabilities of *isothermal nature*. The prevailing type of the instability may be short-wave instability (for the linear flow) or long-wave instability (for the return flow). These instabilities appear if the capillary number C is not too small.

3. There are also two types of instabilities which are essentially connected with both the thermocapillary effect and the existence of the flow. These instabilities are called *hydrothermal* instabilities (Smith, 1986).

The first type of hydrothermal instabilities appears mainly for relatively small Prandtl numbers and generates waves propagating *across* the basic flow. As we discussed formerly, the origin of the oscillatory instability is the existence of compensating mechanisms tending to diminish the disturbances but acting with some time delay. Let us consider, for instance, the linear flow (generating heating from below) and discuss the evolution of a surface temperature disturbance in the form of a hot line parallel to the flow direction. As usually, the surface flow directed away from this line generates *an upflow* supporting the temperature disturbance.

However, it is necessary to take into account the fact that the fluid rises from the region with smaller velocity of the basic flow into a region with larger velocity of the basic flow. That is why the upflow generates an *upstream* velocity disturbance on the surface which *cools* the surface. The time delay between two processes is the origin of oscillations.

The second type of hydrothermal instabilities is characteristic for large Prandtl numbers. Let us consider the return flow (generating heating from above) and a surface hot line oriented across the stream. The surface temperature disturbance creates an upflow under the line itself and downflows on a certain distance from the line. The downflow arising upstream from the line moves upstream, generating an upstream traveling wave. Actually, because of the interaction of both types of instability, an oblique motion of rolls is observed (Smith and Davis, 1983a).

5.1.2 Flows Under Combined Action of Thermocapillarity and Buoyancy

Let us discuss now the main results of the instability analysis for a return flow between heat insulated boundaries which is generated by a longitudinal temperature gradient under a simultaneous action of thermocapillarity and buoyancy obtained by Parmentier *et al.* (1993). The deformation of the surface is disregarded.

Pure Buoyancy Instability

It turns out that there are two different types of hydrothermal waves, one for small values of the Prandtl number, $0 < P < P_1 \approx 0.41$, and another one for intermediate values of the Prandtl number, $P_1 < P > P_2 \approx 2.6$. The small-Prandtl-number instability is characterized by a nearly longitudinal orientation of rolls and relatively small values of the wave number. Quite opposite, the moderate-Prandtl-number waves rapidly tend to a nearly transverse orientation of rolls, as the Prandtl number grows, and their wave numbers are never small.

Coupled Buoyancy and Thermocapillarity

The joint action of the buoyancy and thermocapillarity leads to an instability for any values of P, including $P > P_2$. In the latter region, there is a remarkable transition from an inclined orientation of hydrothermal waves at small $B_d = R/M$ to a transverse orientation at larger values of B_d.

Thus, one could expect that by growing B_d one should observe a transition from inclined, three-dimensional, hydrothermal waves to transverse, two-dimensional, waves.

Surprisingly, the experiments (Schwabe *et al.*, 1992; De Saedeleer *et al.*, 1996; Riley and Neitzel, 1998) showed the appearance of *steady* transverse periodic patterns, which demonstrated an apparent disagreement between the theory and the experiment. The explanation of this paradox, given by Priede and Gerberth (1997), is discussed in Outlook.

5.2 Thermocapillary Flows in Two-Layer Systems

5.2.1 Basic Equations and Boundary Conditions

In the following section, we will consider thermocapillary flows in the system air-water. Because it is more natural to define the dimensionless parameters of the system using the physical parameters of water rather than those of air, we will denote the variables of the bottom layer by subscript 1, and the variables of the top layer by subscript 2. Let the space between two parallel rigid plates $z = -a_1$ and $z = a_2$ be filled by two immiscible viscous fluids (see Fig. 5.1). The temperature on these plates is fixed in the following way: $T(x, y, -a_1) = Ax + \Theta$, $T(x, y, a_2) = Ax$. Thus, a constant temperature gradient A is imposed in the direction of the axis x. For fixed values of x and y, the difference between the temperatures of the lower plate and the upper plate is equal to Θ. It is assumed that the interfacial tension coefficient σ decreases linearly with temperature: $\sigma = \sigma_0 - \alpha T$. The buoyancy force is neglected.

In the present section, the interface is assumed to be a plane: $z = 0$. Strictly speaking, the interface can be perfectly flat only if horizontal pressure gradients appearing in both fluids are equal. Generally, these gradients are not equal, so that the interface is deformed in such a way that the pressure difference generated by the thermocapillary motion is balanced by the hydrostatic pressure and the interfacial tension. However, in some cases, the deformation is negligible. The relevant parameters characterizing the interface deformation are $\epsilon = Al_x\alpha/\sigma_0$ (Pshenichnikov and Tokmenina, 1983) and $R = Al_x\alpha/(\rho_2 - \rho_1)ga_2$ (Tan *et al.*, 1990); g is the acceleration of gravity, and l_x is the characteristic scale of the region in the x-direction. In the present section, we assume that these parameters

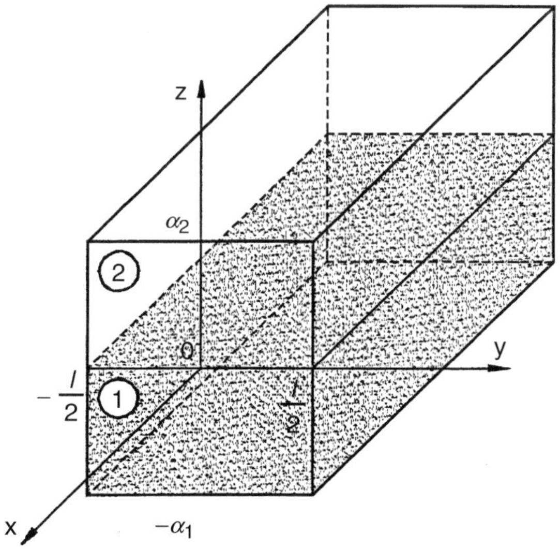

FIGURE 5.1. Geometric configuration of the region and coordinate axes.

are small; hence, the interface deformation caused by the difference of horizontal pressure gradients can be neglected.

Obviously, under the assumption $z = 0$, we disregard the deformational instabilities discovered by Smith and Davis (1983b). One can expect these instabilities to be unimportant if the crispation numbers $C_j = \eta_j \chi_j / a_j \sigma_0 \ll 1$ and the Galileo numbers $G_j = g a_j^3 / \nu_j \chi_j \gg 1$, $j = 1, 2$.

The units of length, time, velocity, and pressure are chosen as in Chapter 1. As the unit of temperature, we use Aa_1.

The complete system of nonlinear equations can be written in the following dimensionless form:

$$\frac{\partial \mathbf{v}_1}{\partial t} + (\mathbf{v}_1 \cdot \nabla)\mathbf{v}_1 = -\nabla p_1 + \Delta \mathbf{v}_1, \tag{5.6}$$

$$\begin{aligned} &\frac{\partial T_1}{\partial t} + \mathbf{v}_1 \cdot \nabla T_1 = \frac{1}{P} \Delta T_1, \ \nabla \cdot \mathbf{v}_1 = 0; \\ &\frac{\partial \mathbf{v}_2}{\partial t} + (\mathbf{v}_2 \cdot \nabla)\mathbf{v}_2 = -\rho \nabla p_2 + \frac{1}{\nu} \Delta \mathbf{v}_2, \\ &\frac{\partial T_2}{\partial t} + \mathbf{v}_2 \cdot \nabla T_2 = \frac{1}{\chi P} \Delta T_2, \ \nabla \cdot \mathbf{v}_2 = 0, \end{aligned} \tag{5.7}$$

where $P = \nu_1 / \chi_1$ is the Prandtl number of the bottom fluid.

On the rigid horizontal plates, the following boundary conditions are used:

$$z = -1: \ \mathbf{v}_1 = 0, \ T_1 = x, \tag{5.8}$$

$$z = a: \ \mathbf{v}_2 = 0, \ T_2 = x - b, \tag{5.9}$$

where parameter $b = \Theta / Aa_1$ describes the relation between the characteristic vertical and horizontal temperature differences. At the interface, the normal components of the velocity vanish:

$$z = 0: \ v_{z1} = v_{z2} = 0; \tag{5.10}$$

and the continuity conditions for the tangential components of the velocity

$$z = 0: \ v_{x1} = v_{x2}, \ v_{y1} = v_{y2}, \tag{5.11}$$

for the tangential stresses

$$z = 0: \ \eta \frac{\partial v_{x1}}{\partial z} = \frac{\partial v_{x2}}{\partial z} - \frac{M\eta}{P} \frac{\partial T_1}{\partial x}, \ \eta \frac{\partial v_{y1}}{\partial z} = \frac{\partial v_{y2}}{\partial z} - \frac{M\eta}{P} \frac{\partial T_1}{\partial y}, \tag{5.12}$$

for the temperature

$$T_1 = T_2, \tag{5.13}$$

and for the heat fluxes

$$\kappa \frac{\partial T_1}{\partial z} = \frac{\partial T_2}{\partial z} \tag{5.14}$$

are fulfilled. Here, $M = \alpha A a_1^2 / \eta_1 \chi_1$ is the Marangoni number.

In the limit of an infinite layer, it is necessary to impose some additional conditions determining the pressure gradients in the system. If the flow occurs in a channel that connects two vessels kept under the same pressure, the mean longitudinal pressure gradient in the system is zero. The corresponding thermocapillary flow is usually called "linear flow" (Davis 1987). In the case of a closed cavity, the mean longitudinal flux of fluid is zero, so that the "return flow" occurs which is characterized by a nonzero longitudinal pressure gradient. In the latter case,

$$\int_{-1}^{0} dz\, U_1^{(0)}(z) = 0, \quad \int_{0}^{a} dz\, U_2^{(0)}(z) = 0. \tag{5.15}$$

5.2.2 Stationary Flow Profiles

The boundary value problem (5.6) – (5.14) has an exact solution corresponding to a parallel flow in the direction opposite to the direction of the temperature gradient:

$$\mathbf{v}_i = U_i^{(0)}(z)\mathbf{e}_x,\; p_i = B_i^{(0)}x,\; T_i = x + \Theta_i^{(0)}(z),\; i = 1, 2,$$

where $\mathbf{e}_x$ is the unit vector of the axis x (Doi and Koster, 1993).

In the case of the zero longitudinal pressure gradient (linear flow),

$$U_1^{(0)}(z) = -\frac{a\eta M}{P(1+a\eta)}(z+1); \tag{5.16}$$

$$U_2^{(0)}(z) = \frac{\eta M}{P(1+a\eta)}(z-a); \tag{5.17}$$

$$\Theta_1^{(0)}(z) = -\frac{a\eta M}{6(1+a\eta)}\left[(z+1)^3 - \frac{1+2\chi a^2+3\kappa a}{1+\kappa a}(z+1)\right] - \frac{b(z+1)}{1+a\kappa}; \tag{5.18}$$

$$\Theta_2^{(0)}(z) = \frac{\eta M}{6(1+a\eta)}\left[-\chi(-z+a)^3 + \frac{a(\chi\kappa a^2+2\kappa+3\chi a)}{1+\kappa a}(-z+a)\right] - \frac{b(\kappa z+1)}{1+a\kappa}; \tag{5.19}$$

$$B_1^{(0)} = B_2^{(0)} = 0. \tag{5.20}$$

Note that in the case $b \geq 0$, the deviations of the temperature $\Theta_i(z) \leq 0$.

In the case of the zero longitudinal fluxes of fluids (return flow),

$$U_1^{(0)}(z) = -\frac{a\eta M\left(1+4z+3z^2\right)}{4P(1+a\eta)}; \tag{5.21}$$

$$U_2^{(0)}(z) = -\frac{\eta M\left(a^2-4az+3z^2\right)}{4aP(1+a\eta)}; \tag{5.22}$$

$$\Theta_1^{(0)}(z) = \frac{a\eta M(1+z)\left[a(a\chi+\kappa)-(1+a\kappa)\left(z+5z^2+3z^3\right)\right]}{48(1+a\eta)(1+a\kappa)} - \frac{b(z+1)}{1+a\kappa}; \tag{5.23}$$

$$\Theta_2^{(0)}(z) = \frac{\eta M(a-z)\left[a^2(a\chi+\kappa)+\chi(1+a\kappa)\left(a^2z-5az^2+3z^3\right)\right]}{48a(1+a\eta)(1+a\kappa)} - \frac{b(\kappa z+1)}{1+a\kappa}; \tag{5.24}$$

$$B_1^{(0)} = -\frac{3a\eta M}{2(1+a\eta)P}; \tag{5.25}$$

$$B_2^{(0)} = -\frac{3M}{2aP(1+a\eta)}. \tag{5.26}$$

The velocity is negative near the interface and positive near the rigid walls. On the interface, the absolute value of velocity is equal to

$$v_s = |U_1^{(0)}(0)| = a\eta M/4P(1+a\eta). \tag{5.27}$$

This parallel flow is an analogue of the return thermocapillary flow considered in the one-layer approach (Davis 1987).

The temperature profiles (5.23) and (5.24) are generated by the combined action of the heat advection by the parallel flow (the terms proportional to M) and the external heating from below (the terms proportional to b). The functions $\Theta_1^{(0)}(z)$ and $\Theta_2^{(0)}(z)$ can be nonmonotonic. For instance, the vertical component of the temperature gradient in the lower fluid has equal minima in the points $z=-1$ and $z=0$,

$$\Theta_1^{(0)\prime}(-1) = \Theta_1^{(0)\prime}(0) = -\frac{a\eta M(1-\chi a^2)}{48(1+a\eta)(1+a\kappa)} - \frac{b}{1+a\kappa}, \tag{5.28}$$

and a maximum in the point $z=-1/3$,

$$\Theta_1^{(0)\prime}(-1/3) = \frac{a\eta M[a(a\chi+\kappa)+(7/9)(1+a\kappa)]}{48(1+a\eta)(1+a\kappa)} - \frac{b}{1+a\kappa}. \tag{5.29}$$

Thus, if $1-\chi a^2 > 0$, the vertical component of the temperature gradient is negative near the boundaries $z=-1$ and $z=0$, but it is positive in the middle of the layer if b/M is less than a certain value. The sign of the interfacial temperature

$$\Theta_s = \Theta_1^{(0)} = \frac{\eta M a^2(a\chi+\kappa)}{48(1+a\eta)(1+a\kappa)} - \frac{b}{1+a\kappa} \tag{5.30}$$

is also determined by the ratio b/M. If $1-\chi a^2 < 0$, the temperature profile is nonmonotonic in the upper fluid.

5.2.3 Linear Stability Theory

Let us investigate now the linear stability of the parallel thermocapillary flow (5.21) – (5.26). We will follow a recent study by Nepomnyashchy *et al.* (2001).

We linearize the fields of all the variables that appear in the boundary value problem (5.6) – (5.14) around the stationary solution (5.21) – (5.26):

$$\mathbf{v}_i(\mathbf{x}, z, t) = U_i^{(0)}(z)\mathbf{e}_x + \tilde{\mathbf{v}}_i(z)\exp(i\mathbf{k}\cdot\mathbf{x} + \lambda t),$$

$$T_i(\mathbf{x}, z, t) = \Theta_i^{(0)}(z) + \tilde{T}_i(z)\exp(i\mathbf{k}\cdot\mathbf{x} + \lambda t),$$

$$p_i(\mathbf{x}, z, t) = B_i^{(0)}x + \tilde{p}_i(z)\exp(i\mathbf{k}\cdot\mathbf{x} + \lambda t),$$

where $\mathbf{x} = (x, y)$, and $\mathbf{k} = (k_x, k_y)$ are horizontal two-dimensional vectors.

Let $k_x = k\sin\alpha$, and $k_y = k\cos\alpha$. It is convenient to perform a transformation of horizontal coordinates $x = X\cos\alpha + Y\sin\alpha$, and $y = -X\sin\alpha + Y\cos\alpha$. After this transformation, the wave vector $\mathbf{k}$ is directed along axis Y, and the disturbances do not depend on X.

We obtain the following equations for disturbances of the parallel flow:

$$\lambda\tilde{v}_{Xi} + ik\sin\alpha\cdot U_i^{(0)}\tilde{v}_{Xi} + \cos\alpha\cdot U_i^{(0)\prime}\tilde{v}_{zi} = c_i(\tilde{v}_{Xi}'' - k^2\tilde{v}_{Xi}); \tag{5.31}$$

$$\lambda\tilde{v}_{Yi} + ik\sin\alpha\cdot U_i^{(0)}\tilde{v}_{Yi} + \sin\alpha\cdot U_i^{(0)\prime}\tilde{v}_{zi} = -ike_i\tilde{p}_i + c_i(\tilde{v}_{Yi}'' - k^2\tilde{v}_{Yi}); \tag{5.32}$$

$$\lambda\tilde{v}_{zi} + ik\sin\alpha\cdot U_i^{(0)}\tilde{v}_{zi} = -e_i\tilde{p}_i' + c_i(\tilde{v}_{zi}'' - k^2\tilde{v}_{zi}); \tag{5.33}$$

$$\lambda\tilde{T}_i + ik\sin\alpha\cdot U_i^{(0)}\tilde{T} + \Theta_i^{(0)\prime}\tilde{v}_{zi} + \cos\alpha\cdot\tilde{v}_{Xi} + \sin\alpha\cdot\tilde{v}_{Yi}$$
$$= (d_i/P)(\tilde{T}'' - k^2\tilde{T}); \tag{5.34}$$

$$\tilde{v}_{zi}' + ik\tilde{v}_{Yi} = 0, \quad i = 1, 2, \tag{5.35}$$

where $c_1 = d_1 = e_1 = 1$, $c_2 = 1/\nu$, $d_2 = 1/\chi$, and $e_2 = \rho$; $'$ denotes differentiation with respect to z.

Introducing the stream-function disturbance $\tilde{\psi}_i$, $\tilde{v}_{zi} = -ik\tilde{\psi}_i$, $\tilde{v}_{Yi} = \tilde{\psi}'$ and eliminating the pressure disturbance, we obtain the following eigenvalue problem describing the stability of the parallel flow:

$$c_i(\tilde{\psi}_i'''' - 2k^2\tilde{\psi}_i'' + k^4\tilde{\psi}_i'') - ik\sin\alpha[U_i^{(0)}(\tilde{\psi}_i'' - k^2\tilde{\psi}_i) - U_i^{(0)\prime\prime}\tilde{\psi}_i]$$
$$-\lambda(\tilde{\psi}_i'' - k^2\tilde{\psi}_i) = 0; \tag{5.36}$$

$$\lambda\tilde{v}_{Xi} + ik\sin\alpha U_i^{(0)}\tilde{v}_{Xi} - ik\cos\alpha U_i^{(0)\prime}\tilde{\psi}_i = c_i(\tilde{v}_{Xi}'' - k^2\tilde{v}_{Xi}); \tag{5.37}$$

$$\lambda\tilde{T}_i + ik\sin\alpha U_i^{(0)}\tilde{T}_i - ik\Theta_i^{(0)\prime}\tilde{\psi}_i + \cos\alpha\tilde{v}_{Xi} + \sin\alpha\tilde{\psi}_i'$$
$$= (d_i/P)(\tilde{T}_i'' - k^2\tilde{T}_i); \tag{5.38}$$

$$z = -1: \ \tilde{\psi}_1 = 0;\ \tilde{\psi}_1' = 0;\ \tilde{v}_{X1} = 0;\ \tilde{T}_i = 0; \tag{5.39}$$

$$z = a: \ \tilde{\psi}_2 = 0; \ \tilde{\psi}_2' = 0; \ \tilde{v}_{X2} = 0; \ \tilde{T}_i = 0; \tag{5.40}$$

$$z = 0: \ \tilde{v}_{X1}' - \eta^{-1}\tilde{v}_{X2}' = 0; \ -\tilde{\psi}_1'' + \eta^{-1}\tilde{\psi}_2'' = (ikM/P)\tilde{T}_1; \tag{5.41}$$

$$\tilde{v}_{X1} = \tilde{v}_{X2}; \ \tilde{\psi}_1 = \tilde{\psi}_2 = 0; \ \tilde{\psi}_1' = \tilde{\psi}_2'; \tag{5.42}$$

$$\tilde{T}_1 = \tilde{T}_2; \ \kappa\tilde{T}_1' = \tilde{T}_2'. \tag{5.43}$$

Let us present the results of calculations performed for the real air/water system (at 20° and 1 bar) with the following parameters: $\eta = 55.3$, $\nu = 0.0659$, $\kappa = 23.3$, $\chi = 0.00667$, $P = 6.96$, and $a = 1$ (Perry, 1997; Vargaftik, 1975).

Horizontal temperature gradient

In the case $b = 0$, which corresponds to the thermocapillary flow generated by a horizontal temperature gradient, the instability appears with respect to *oblique* hydrothermal waves; i.e., the direction of the critical wave vector is characterized by a certain value of α, $0 < |\alpha| < 90°$. For any orientation of the disturbance wave vector characterized by the angle α, the minimum of the neutral stability curve $M = M_o(k, \alpha)$ was found, and the corresponding values M_c, $\omega_c = -\text{Im}\lambda_c$, and k_c are presented as functions of α in Fig. 5.2 (these results have been obtained by L. Braverman, unpublished). Only one of two curves that correspond to opposite values of ω_c at opposite values of α are shown.

In the case of the return flow, the sign of ω_c does not depend on α. The minimum of the curve $M_c(\alpha)$ is achieved as $\alpha = -57.6°$. The positive value of ω_c at the negative value of α (as well as the negative value of ω_c at the positive value of α, which characterizes the branch not shown in Fig. 5.2) corresponds to the propagation of the oblique hydrothermal wave toward the hot end (i.e., opposite to the direction of the flow at the interface (see Davis, 1987)).

In the case of the linear flow, the sign of ω_c depends on α. Near the minimum of the curve, $M_c(\alpha)$, ω_c is negative. That means that the hydrothermal wave propagates toward the cold end. That difference may be caused by the difference in the transverse temperature profiles for the linear flow and the return flow. Note that the dependence of M_c on α is extremely weak in the region $-90° < \alpha < -40°$.

Inclined temperature gradient

Let us investigate now the case $b \neq 0$ (inclined temperature gradient). We will consider only the case of the return flow. Note that for the chosen set of parameters $1 - \chi a^2 > 0$; thus, in the basic flow the vertical component of the temperature gradient in the upper layer is always negative, whereas in the lower layer, there exists a region with a positive value of the temperature gradient if $b/M < 0.864$. The temperature Θ_s on the interface is positive if $b/M < 0.477$ and negative in the opposite case.

The instability with respect to oblique hydrothermal waves is important in the interval $0 < b < b_F$, $b_F \approx 162.4$. A typical neutral curve calculated for the critical inclination angle is shown in Fig. 5.3 (line 2). The dependence of the frequency $\omega = |\text{Im}\lambda|$ on the wave number k for the critical inclination angle is

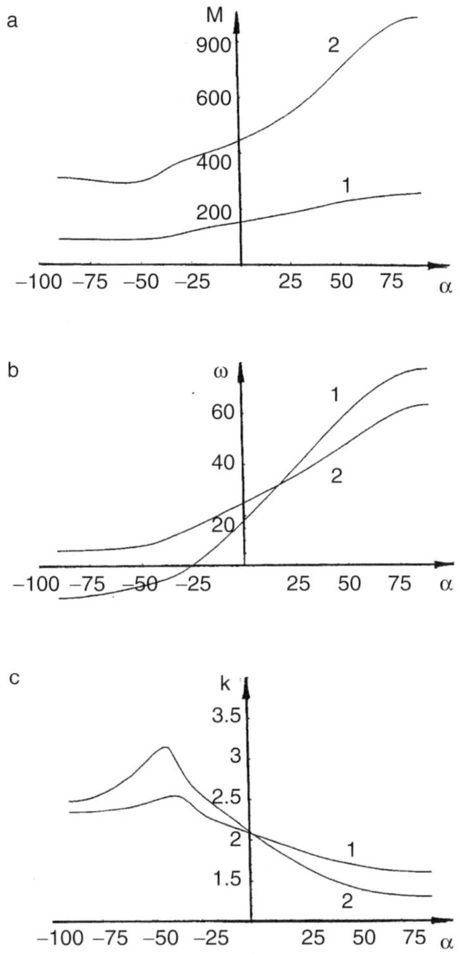

FIGURE 5.2. (a) The critical Marangoni number, (b) the critical frequency, and (c) the critical wave number as the functions of the angle for the linear flow (lines 1) and for the return flow (lines 2).

shown in Fig. 5.4 (line 2). The critical value of the Marangoni number M_c, which corresponds to the minimum of the neutral curve, grows from $M = M_H \approx 263.9$ as $b = 0$ to $M = M_F \approx 314.9$ as $b = b_F$ (see line 3 in Fig. 5.5), whereas the angle $|\alpha|$ decreases from $\alpha_H \approx 57.6°$ to $\alpha_F \approx 30°$. The critical wave number k_c and the critical frequency ω_c decrease slowly when b grows ($k_H \approx 2.67$, $\omega_H \approx 5.91$; $k_F \approx 2.58$, $\omega_F \approx 4.95$). Note that the phase velocity of the hydrothermal wave $v_{ph} = \omega_c/k_c$ is essentially smaller than the fluid velocity at the interface v_s (see Eq. (5.27)). For instance, in the point H, $v_{ph} \approx 2.21$, $v_s \approx 9.31$, and in the point F, $v_{ph} \approx 1.92$, $v_s \approx 11.1$. The group velocity $v_{gr} = d\omega/dk$ calculated in the point $k = k_c$, $M = M_c$ is rather small for hydrothermal waves (in the point H, $v_{gr} \approx 0.507$; in the point F, $v_{gr} \approx 0.0885$).

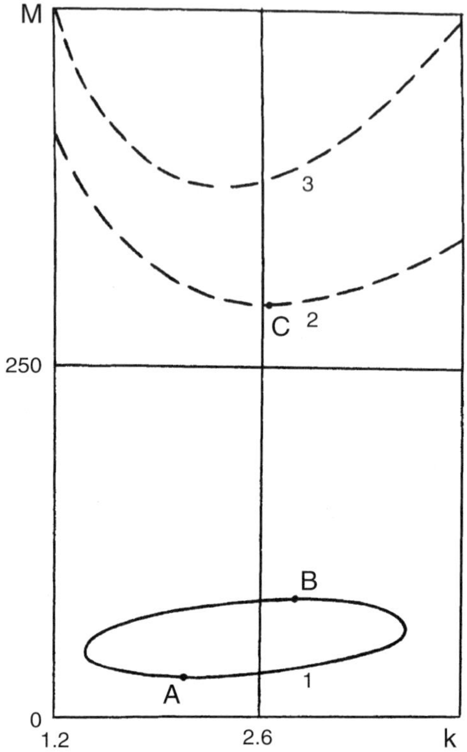

FIGURE 5.3. Neutral curves for $b = 100$: the closed neutral curve for stationary longitudinal rolls, $\alpha = 0$ (line 1); the neutral curve for hydrothermal waves at the critical angle $\alpha = \alpha_c$ (line 2); the neutral curve for hydrothermal waves, $\alpha = 0$ (line 3). Points A and B determine the existence interval of stationary longitudinal rolls; point C determines the onset threshold of oblique hydrothermal waves.

The opposite case $b \rightarrow \infty$ corresponds to the problem of the Pearson's Marangoni convection (Pearson 1958) by heating from below in the absence of a horizontal component of the temperature gradient. In this limit, the critical Marangoni number $M_c \sim M'_c/b$, where M' is the Marangoni number defined by means of the transverse temperature difference, which remains finite and does not depend on the direction of the wavevector. The instability is monotonic, and the stability boundary can be calculated analytically (see Smith 1966). For the air/water system with $a = 1$, $M'_c \approx 21$, and $k_c \approx 2$ (see Gilev et al., (1987b)). At arbitrary large but finite b, the degeneracy of $\lambda(\alpha)$ is broken. Only longitudinal rolls with the axis parallel to the horizontal component of the temperature gradient ($\alpha = 0$) grow monotonically, and any other disturbances grow in an oscillatory way because of the drift by the main flow. We have found that at large values of b, the disturbances with $\alpha = 0$ are the most unstable ones. The corresponding stability boundary is shown in Fig. 5.5 (line 1).

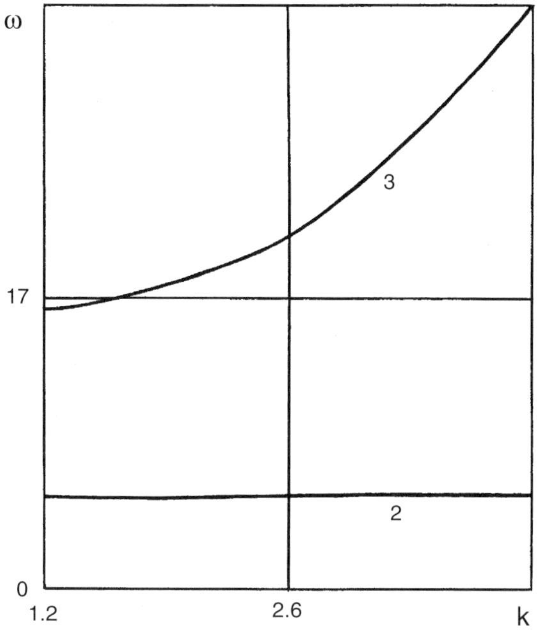

FIGURE 5.4. Dependence $\omega(k)$ for hydrothermal waves at $\alpha = \alpha_c$ (line 2) and at $\alpha = 0$ (line 3); $b = 100$.

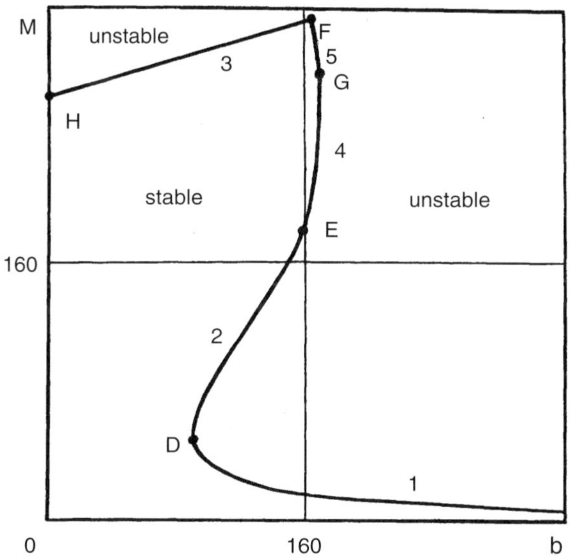

FIGURE 5.5. Boundaries of instability with respect to stationary longitudinal rolls ($\alpha = 0$; lines 1 and 2), oblique hydrothermal waves ($\alpha = \alpha_c$; line 3), and transverse traveling rolls ($\alpha = 90°$; lines 4 and 5). Point D corresponds to the disappearance of longitudinal rolls; point E determines the transition between longitudinal rolls and transverse traveling rolls; point F determines the transition between oblique hydrothermal waves and transverse traveling rolls; point G corresponds to the disappearance of transverse traveling rolls.

Surprisingly, the Pearson's Marangoni instability disappears at rather large values of parameter b, $b = b_D \approx 88.5$ (see point D in Fig. 5.5). Note that for $b = b_D$, the horizontal temperature gradient is less than the *mean* vertical one in two orders of magnitude. Nevertheless, such a small horizontal temperature gradient turns out to be sufficient for a complete suppression of Pearson's instability. The latter kind of instability appears only at $b > b_D$ in a small closed region in the space of parameters (α, k, M) around the point $\alpha = 0$, $k = k_D \approx 2.31$, $M = M_D \approx 47.9$ (see line 1 in Fig. 5.3). The explanation of this paradox is as follows. Because the heat conductivity of the water is much larger than that of the air ($\kappa = 23.3$), the vertical gradient in the water is actually much less than b. Using formula (5.30), which determines the temperature on the interface Θ_s (note that $\Theta_s \approx -b/\kappa + M/48$ since $\kappa \gg 1 \gg \chi, \eta \gg 1$), and evaluating expression (5.30) in the point ($b = b_D$, $M = M_D$), we find that Pearson's instability is damped when the mean vertical temperature gradient *in the water layer* is only 2.7 times larger than the horizontal temperature gradient.

For the fixed value of $\alpha = 0$, the stability boundary is a closed curve in the plane (k, M) (line 1 in Fig. 5.3), which is located much lower than the stability boundary for the hydrothermal waves (line 2 in Fig. 5.3). The minimum and maximum points of the closed neutral curve (points A and B in Fig. 5.3) determine the interval of stationary instability $M_A < M < M_B$. If the heating is enhanced while maintaining constant the ratio of characteristic vertical and horizontal temperature differences b, the parallel flow first becomes unstable with respect to longitudinal rolls at $M = M_A$, but then it is *restabilized* as $M > M_B$. The parallel flow becomes unstable again (this time with respect to inclined hydrothermal waves), only for much larger values of the Marangoni number $M > M_C$, where point C corresponds to the minimum of the neutral curve for hydrothermal waves at the critical angle $|\alpha| = \alpha_c$ (line 2 in Fig. 5.3). The dependences of M_A, M_B, and M_C on b are presented in Fig. 5.5 (lines 1, 2, and 3, respectively). Lines 1 and 2 merge in the point (b_D, M_D) (point D in Fig. 5.5).

When b increases, the interval of angles $|\alpha| < \alpha_m$ where Pearson's instability takes place, grows. At $b = b_1$, $b_1 \approx 118$, α_m reaches 90°; that is, at $b > b_1$ for any direction of the wave vector, there exists an interval of instability. However, the critical Marangoni number M_A corresponds to $\alpha = 0$ in the whole region $b > b_D$. The maximum of the neutral surface $M = M(k, \alpha)$, M_B, corresponds to the value $\alpha = 0$ only in the interval $b_D < b < b_E$, $b_E \approx 154.6$.

For values of b slightly smaller than b_E, a new maximum appears at $|\alpha| = 90°$. It is caused by a new instability mode which corresponds to two-dimensional rolls directed perpendicularly to the direction of the basic flow. For $b = b_E$, both maxima have the same height $M = M_E \approx 181.5$ (see point E in Fig. 5.5). As $b > b_E$, the maximum at $|\alpha| = 90°$ becomes higher than the maximum at $\alpha = 0$. Thus, the lower boundary of the stability gap between the regions of Pearson's instability and the instability with respect to hydrothermal waves is connected with two-dimensional rolls. The dependence of the corresponding critical Marangoni number on b is shown in Fig. 5.5 as line 4. Similarly, when b increases, an additional minimum appears at $|\alpha| = 90°$ on the neutral surface

$M = M(k, \alpha)$ for hydrothermal waves, which competes with the minimum at $|\alpha| = \alpha_c$, $\alpha_c \neq 90°$. The former minimum corresponds to waves moving in the direction of the flow at the interface (i.e., to drifted rolls). Both minima provide the same critical Marangoni number $M = M_F$ as $b = b_F$ (codimension-2 point; see point F in Fig. 5.5). As $b > b_F$, the upper boundary of the stability gap is connected with two-dimensional disturbances (line 5 in Fig. 5.5) (as well as its lower boundary). At $b = b_G$, $b_G \approx 164.3$, lines 4 and 5 merge with $M = M_5 \approx 276.1$ (point G in Fig. 5.5), and the stability gap disappears.

Note that for the two-dimensional instability mode described above, $\omega \neq 0$, because the rolls are driven by the basic flow. Unlike hydrothermal waves, the drifted convective rolls move in the *same* direction as the flow at the interface. Moreover, the phase velocity of disturbances can be even *larger* than the fluid velocity on the interface v_s (see Eq. (5.27)). At the same time, the group velocity of waves is always smaller than v_s. For instance, in the point F, $v_s \approx 11.1$, $v_{ph} \approx 13.4$, and $v_{gr} \approx 10.1$.

Similarly, when b increases, an additional minimum appears at $|\alpha| = 90°$ on the neutral surface $M = M(k, \alpha)$ for hydrothermal waves, which competes with the minimum at $|\alpha| = \alpha_c$, $\alpha_c \neq 90°$. The former minimum corresponds to waves moving in the direction of the flow at the interface (i.e., to drifted rolls). Both minima provide the same critical Marangoni number $M = M_F$ as $b = b_F$, $b_F \approx 162.4$ (codimension-2 point; see point F in Fig. 5.5). As $b > b_F$, the upper boundary of the stability gap is connected with two-dimensional disturbances (line 5 in Fig. 5.5) (as well as its lower boundary).

At $b = b_G$, $b_G \approx 164.3$, lines 4 and 5 merge with $M = M_5 \approx 276.1$ (point G in Fig. 5.5), and the stability gap disappears.

The transition between the inclined hydrothermal waves moving upstream and the two-dimensional waves moving downstream takes place at the value $b_F/M_F \approx 0.515$. Note that this value is close to the value $b/M \approx 0.477$, where the quantity Θ_s, which characterizes the mean vertical temperature gradient in water, changes its sign. Indeed, the explanation of the upstream motion of hydrothermal waves in a one-layer system (Davis, 1987) is based on the fact that a downward flow generates a subsurface heating, which takes place because of the positive temperature gradient. In our case, because the heat diffusivity of air is much larger than that of water, the heat advection by a flow disturbance in air can be ignored; thus, the direction of the temperature gradient in water is crucial. For relatively small b/M, the mean vertical temperature gradient in water is mainly positive, and the direction of the wave propagation can be explained as in the case of a one-layer system. For larger values of b/M, the downward flow causes a subsurface cooling, which influences the propagation direction of a temperature disturbance in an opposite way.

Let us summarize the main predictions of the linear theory. For relatively small values of b ($0 < b < b_F$) and large values of M, the excitation of inclined hydrothermal waves is expected (line 3 in Fig. 5.5). These waves move in the opposite direction to that of the flow at the interface. For relatively large values of b ($b > b_D$) and small values of M, the theory predicts the appearance of stationary convective rolls due to Pearson's instability (lines 1 and 2 in Fig. 5.5). The axes of rolls are ordered by the thermocapillary flow along the direction of the imposed horizontal temperature gradient. For intermediate values of M, the convective rolls are ordered *across* the direction of the horizontal temperature gradient, and they are drifted by the thermocapillary flow. Unlike the hydrothermal waves, the drifted rolls move in the same direction as the flow at the interface.

5.2.4 Nonlinear Patterns

We present now the results of numerical simulations of the flow regimes predicted by the linear stability theory.

Description of the method

We will describe in more detail the numerical approach for nonlinear simulations of the longitudinal rolls (spiral flows) that appear due to the instability of the basic thermocapillary flow with respect to monotonically growing disturbances with $\alpha = 0$. One can expect that the corresponding solutions of the nonlinear boundary value problem (5.6) – (5.14) have the following structure:

$$\mathbf{v}_i = \mathbf{v}_i(y, z), \; p_i = p_i(y, z) + B_i x, \; T_i = x + \Theta_i(y, z), \; i = 1, 2. \qquad (5.44)$$

We will assume that the motion is spatially periodic in y with a certain period $L = l/a_1$. Constants B_i are unknown and should be determined from the conditions of vanishing mean horizontal fluxes of fluids:

$$\int_{-L/2}^{L/2} dy \int_{-1}^{0} dz\, v_{x1}(y, z) = 0, \quad \int_{-L/2}^{L/2} dy \int_{0}^{a} dz\, v_{x2}(y, z) = 0. \qquad (5.45)$$

For spiral flows (5.44), the continuity equations

$$\frac{\partial v_{yi}}{\partial y} + \frac{\partial v_{zi}}{\partial z} = 0, \quad i = 1, 2,$$

do not include v_{xi}. That is why it is possible to define the stream functions ψ_i of the transverse flow in the following way:

$$v_{yi} = \frac{\partial \psi_i}{\partial z}, \quad v_{zi} = -\frac{\partial \psi_i}{\partial y}. \qquad (5.46)$$

After elimination of the pressure fields $p_i(y, z)$ in the usual manner, we obtain the following nonlinear boundary value problem:

$$\frac{\partial}{\partial t}\Delta_\perp \psi_i + \frac{\partial \psi_i}{\partial z}\frac{\partial}{\partial y}\Delta_\perp \psi_i - \frac{\partial \psi_i}{\partial y}\frac{\partial}{\partial z}\Delta_\perp \psi_i = c_i \Delta_\perp^2 \psi_i; \qquad (5.47)$$

$$\frac{\partial}{\partial t}U_i + \frac{\partial \psi_i}{\partial z}\frac{\partial}{\partial y}U_i - \frac{\partial \psi_i}{\partial y}\frac{\partial}{\partial z}U_i = c_i \Delta_\perp^2 U_i - e_i B_i; \qquad (5.48)$$

$$\frac{\partial}{\partial t}\Theta_i + \frac{\partial \psi_i}{\partial z}\frac{\partial}{\partial y}\Theta_i - \frac{\partial \psi_i}{\partial y}\frac{\partial}{\partial z}\Theta_i + U_i = \frac{d_i}{P}\Delta_\perp \Theta_i; \tag{5.49}$$

$$z = -1: \ \psi_1 = 0, \ \frac{\partial \psi_1}{\partial z} = 0, \ U_1 = 0, \ \Theta_1 = 0; \tag{5.50}$$

$$z = a: \ \psi_2 = 0, \ \frac{\partial \psi_2}{\partial z} = 0, \ U_2 = 0, \ \Theta_2 = -b; \tag{5.51}$$

$$z = 0: \ \psi_1 = \psi_2 = 0, \ \frac{\partial \psi_1}{\partial z} = \frac{\partial \psi_2}{\partial z}, \ U_1 = U_2; \tag{5.52}$$

$$\eta \frac{\partial^2 \psi_1}{\partial z^2} = \frac{\partial^2 \psi_2}{\partial z^2} - \frac{M\eta}{P}\frac{\partial \Theta_1}{\partial y}; \tag{5.53}$$

$$\eta \frac{\partial U_1}{\partial z} = \frac{\partial U_2}{\partial z} - \frac{M\eta}{P}; \tag{5.54}$$

$$\Theta_1 = \Theta_2; \tag{5.55}$$

$$\kappa \frac{\partial \Theta_1}{\partial z} = \frac{\partial \Theta_2}{\partial z}; \tag{5.56}$$

$$\begin{aligned} &\psi_i(y+L, z) = \psi_i(y, z); \ U_i(y+L, z) = U_i(y, z); \\ &\Theta_i(y+L, z) = \Theta_i(y, z), \ i = 1, 2, \end{aligned} \tag{5.57}$$

where

$$\Delta_\perp = \frac{\partial^2}{\partial y^2} + \frac{\partial^2}{\partial z^2}, \ U_i = v_{xi}, \ c_1 = d_1 = e_1 = 1, \ c_2 = 1/\nu, \ d_2 = 1/\chi, \ e_2 = \rho.$$

Constants B_i $(i = 1, 2)$ should be found from the conditions

$$\int_{-L/2}^{L/2} dy \int_{-1}^{0} dz\, U_1 = 0, \ \int_{-L/2}^{L/2} dy \int_{0}^{a} dz\, U_2 = 0. \tag{5.58}$$

For the calculation of B_i and the fields U_i that satisfy conditions (5.58), the following procedure is applied. Functions U_i are presented in the form

$$U_i = \tilde{U}_i + \sum_{j=1}^{2} e_j B_j V_{ij}, \ i = 1, 2, \ j = 1, 2, \tag{5.59}$$

where $\tilde{U}_i$ and V_{ij} satisfy the following equations and boundary conditions:

$$\frac{\partial}{\partial t}\tilde{U}_i + \frac{\partial \psi_i}{\partial z}\frac{\partial}{\partial y}\tilde{U}_i - \frac{\partial \psi_i}{\partial y}\frac{\partial}{\partial z}\tilde{U}_i = c_i \Delta_\perp U_i; \tag{5.60}$$

$$z = -1: \ \tilde{U}_1 = 0; \ z = a: \ \tilde{U}_2 = 0; \tag{5.61}$$

$$z = 0: \ \tilde{U}_1 = \tilde{U}_2; \ \eta \frac{\partial \tilde{U}_1}{\partial z} = \frac{\partial \tilde{U}_2}{\partial z} - \frac{M\eta}{P}; \tag{5.62}$$

$$\tilde{U}_i(y+L,z)=\tilde{U}_i(y,z); \tag{5.63}$$

$$\frac{\partial}{\partial t}V_{ij}+\frac{\partial \psi_i}{\partial z}\frac{\partial}{\partial y}V_{ij}-\frac{\partial \psi_i}{\partial y}\frac{\partial}{\partial z}V_{ij}=c_i\Delta_\perp V_{ij}-\delta_{ij}-\frac{\partial \ln B_j}{\partial t}V_{ij}; \tag{5.64}$$

$$z=-1:\ V_{1j}=0;\ z=a:\ V_{2j}=0; \tag{5.65}$$

$$z=0:\ V_{1j}=V_{2j};\ \eta\frac{\partial V_{1j}}{\partial z}=\frac{\partial V_{2j}}{\partial z}; \tag{5.66}$$

$$V_{ij}(y+L,z)=V_{ij}(y,z); \tag{5.67}$$

$i=1,2,\ j=1,2$, and δ_{ij} is the Kronecker symbol. After the calculation of $\tilde{U}_i$ and V_{ij} (the last term in equation (5.64) is evaluated from the previous time steps), expression (5.59) is substituted into conditions (5.58) and the system of two linear algebraic equations for B_j is solved.

Solutions of the boundary value problem (5.47) – (5.58) are obtained by the finite-difference method. The variables "stream function - vorticity" are used to solve the equation (5.47) with corresponding boundary conditions (for details, see Simanovskii and Nepomnyashchy (1993)). Equations and boundary conditions (5.47) – (5.57) are approximated on a uniform mesh using a second order approximation for the spatial coordinates. The integration of evolution equations is performed by means of an explicit scheme. The rectangular mesh 28×56 was used. The time step was chosen from the stability conditions.

For simulation of inclined hydrothermal waves, we performed the transformation of variables

$$x=X\cos\alpha+Y\sin\alpha,\ y=-X\sin\alpha+Y\cos\alpha \tag{5.68}$$

and presented the solution in the form:

$$\mathbf{v}_i=\mathbf{v}_i(Y,z),\ p_i=p_i(Y,z)+B_iX,\ T_i=X\cos\alpha+Y\sin\alpha+\Theta_i(Y,z),\ i=1,2. \tag{5.69}$$

In this case, equation (5.49) is replaced by

$$\frac{\partial}{\partial t}\Theta_i+\frac{\partial \psi_i}{\partial z}\left(\frac{\partial}{\partial y}\Theta_i+\sin\alpha\right)-\frac{\partial \psi_i}{\partial y}\frac{\partial}{\partial z}\Theta_i+U_i\cos\alpha=\frac{d_i}{P}\Delta_\perp\Theta_i, \tag{5.70}$$

and boundary conditions (5.53) and (5.54) take the form

$$\eta\frac{\partial^2\psi_1}{\partial z^2}=\frac{\partial^2\psi_2}{\partial z^2}-\frac{M\eta}{P}\left(\frac{\partial}{\partial y}\Theta_1+\sin\alpha\right); \tag{5.71}$$

$$\eta\frac{\partial U_1}{\partial z}=\frac{\partial U_2}{\partial z}-\frac{M\eta}{P}\cos\alpha. \tag{5.72}$$

The obtained boundary value problem was solved in the same way as in the case of longitudinal rolls.

In simulation of drifted transverse rolls α, was taken equal to 90° in equations (5.68) – (5.72).

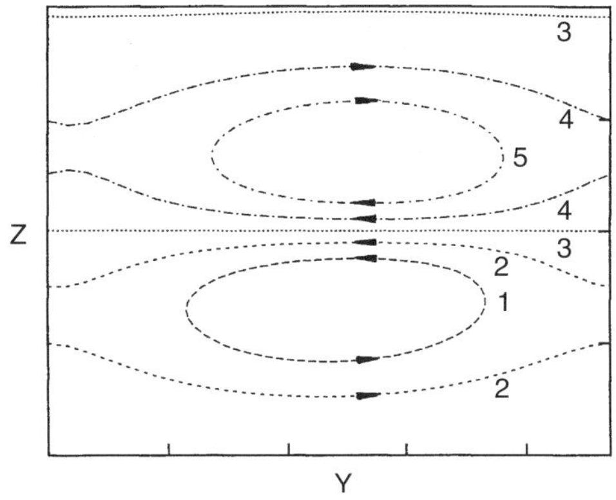

FIGURE 5.6. The stream function for inclined hydrothermal waves; $M = 328$, $b = 100$, $\alpha = 41°$, $L = 2.35$.

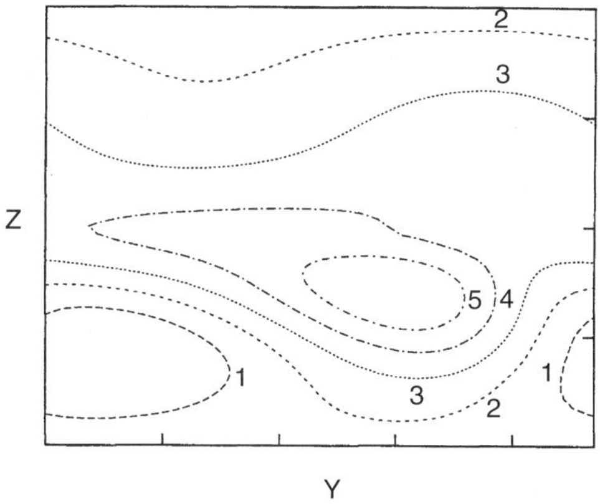

FIGURE 5.7. The temperature disturbance for inclined hydrothermal waves; $M = 328$, $b = 100$, $\alpha = 41°$, $L = 2.35$.

Nonlinear flow regimes

We have investigated the nonlinear flow regimes for the same air/water system as in Sec. 4.3.2.

First, let us consider the hydrothermal waves with an oblique wave vector ($0 < \alpha < 90°$). This type of wave is developed for relatively small values of b/M, when there exists a region with a positive vertical component of the temperature gradient in the lower layer, and the interfacial temperature Θ_s is positive (see equation (5.30)). A snapshot of the fields of the stream functions $\psi_i(Y, z)$, the *disturbances* of the temperature $\Theta_i(Y, z) - \Theta_i^{(0)}(z)$, and the longitudinal velocity $U_i(Y, z)$ ($i = 1, 2$) is shown in Figs. 5.6, 5.7, and 5.8, respectively. Because of some technical reasons, the nondimensional values of ψ_i, U_i, and $\Theta_i - \Theta_i^{(0)}$ are given in units ν_2, ν_2/a_2, and Θ, respectively. The motion is a traveling wave,

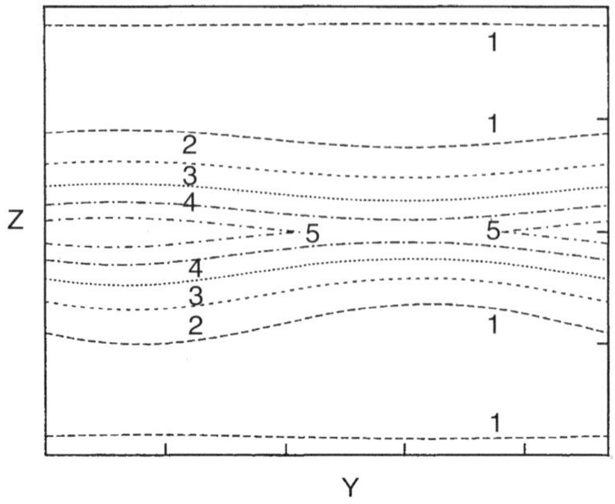

FIGURE 5.8. The longitudinal velocity for inclined hydrothermal waves; $M = 328$, $b = 100$, $\alpha = 41°$, $L = 2.35$.

$f_i(Y, z, t) = f_i(Y - ct, z)$, $i = 1, 2$, where $f_i = (\psi_i, U_i, \Theta_i - \Theta_i^{(0)})$, with a *positive* phase velocity c; that is, the waves move in the direction *opposite* to the Y-component of the flow velocity at the interface. The explanation of this phenomenon given by Davis (1987) is based on the fact that the vertical component of the temperature gradient in water is positive in a certain region. Let us consider the fields of variables shown in Figs. 5.6 and 5.7. A negative disturbance of the *interfacial* temperature is situated to the left of the computation region center (see Fig. 5.7), and it generates a corresponding descending motion in the water. Because of the incompressibility of the fluid, the ascending motion appears in the region to the right of the center (see Fig. 5.6). The latter flow leads to a *cooling* in the region *under the interface*, where the vertical component of the temperature gradient is positive. This effect tends to shift the minimum of the interfacial temperature to the right. That is why the wave moves to the right, despite the advection of the interfacial temperature field by the Y-component of the interfacial velocity. Because of the large thermal diffusivity of air, the influence of the motion in the upper fluid can be neglected.

Let us discuss now the results of nonlinear simulations of two-dimensional (transverse) rolls moving in the same direction as the interfacial flow. This kind of waves appears at larger values of b/M, where the region of a positive vertical temperature gradient is relatively small, and the sign of the interfacial temperature (5.30) in water is negative (i.e., the external heating from below prevails the "heating from above" caused by the basic flow). The snapshots of the fields of variables during one-half of the period are shown in Figs. 5.9, and 5.10. Let us emphasize that the temperature in the region of an ascending flow is now higher than that in the region of a descending flow. Thus, the mechanism driving the wave in the direction opposite to the interfacial flow is switched off. The minimum of the interfacial temperature moves now in the same direction as the interfacial flow; that is, the wave moves to the left ($c < 0$).

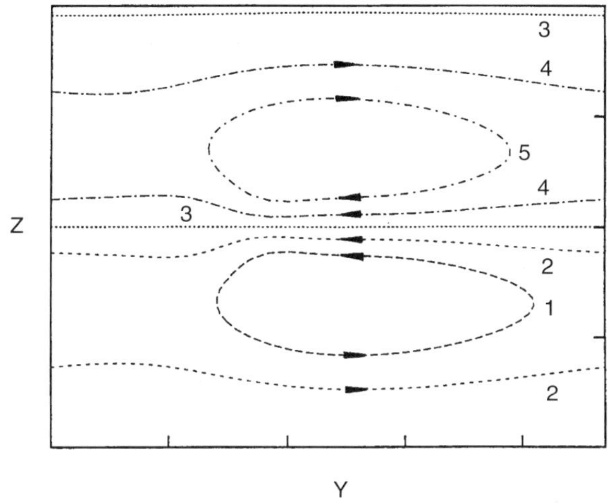

FIGURE 5.9. The stream function for transverse traveling rolls; $M = 200$, $b = 200$, $\alpha = 90°$, $L = 2.35$.

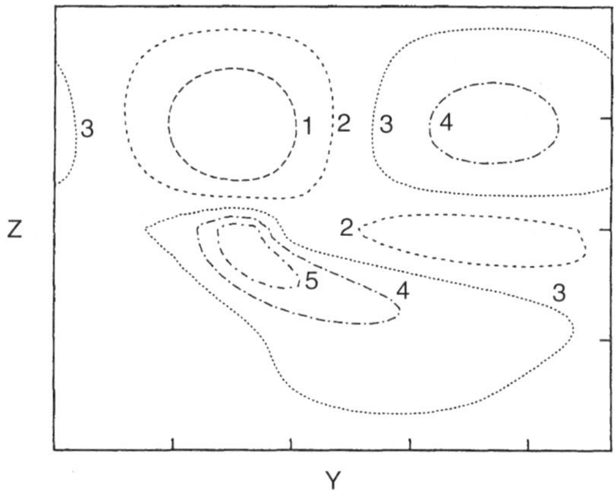

FIGURE 5.10. The temperature disturbance for transverse traveling rolls; $M = 200$, $b = 200$, $\alpha = 90°$, $L = 2.35$.

Finally, let us consider longitudinal rolls (spiral flows). The simulations have been fulfilled for $b = 100$ and $L = 2.75$ (i.e., $k = 2\pi/L \approx 2.285$). The linear theory predicts an instability in the interval between $M = M_A \approx 28.47$ (with the critical wave number $k = 2.11$) and $M = M_B \approx 82.59$ (with the critical wave number $k = 2.825$). The numerical simulations reveal the existence of stable spiral flows with the chosen period L in the interval $M_1 < M < M_2$, $M_1 \approx 30$, and $M_2 \approx 80$. No subcritical flows have been found either as $M > M_A$ or as $M < M_B$. Thus, the prediction of the linear theory concerning the restabilization of the parallel flow above line 2 of Fig. 5.5 has been justified by our numerical

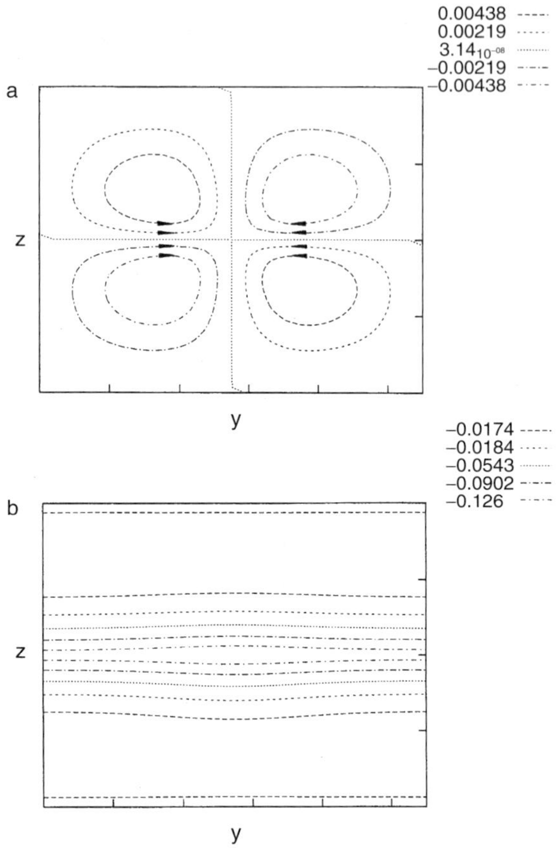

FIGURE 5.11. Isolines of (a) the stream-function fields and (b) longitudinal velocity fields for periodic stationary spiral flow at $M = 61.2$, $b = 100$, $\alpha = 0$, $L = 2.75$.

simulations. The isolines of the stream function fields ψ_i and the longitudinal velocity fields U_i calculated for $M = 61.2$ are shown in Fig. 5.11. The secondary motion $\psi_i(y, z)$ has a four-vortex structure. The fields of variables satisfy the following symmetry conditions:

$$\psi_i(y, z) = -\psi_i(-y, z),\ U_i(y, z) = U_i(-y, z),\ \Theta_i(y, z) = \Theta_i(-y, z). \quad (5.73)$$

5.3 Buoyancy-thermocapillary Convection in Two-layer Systems

In the present section, we include the action of buoyancy. We return to the standard notation: the variables referring to the top layer are marked by subscript 1, and the variables referring to the bottom layer are marked by subscript 2;

i.e., the system is bounded from above and from below by two rigid plates, $z = a_1$ and $z = -a_2$. We consider only the case of a horizontal temperature gradient. Due to technical reasons, we assume in this section that the constant temperature gradient is imposed in the direction opposite to that of the axis x: $T_1(x, y, a_1, t) = T_2(x, y, -a_2, t) = -Ax + const,\ A > 0$. Recently, Madruga et al. (2003, 2004) studied the linear stability of two superposed horizontal liquid layers bounded by two solid planes and subjected to a horizontal temperature gradient. The analysis has revealed a variety of instability modes. Specifically, for the system 5cS silicone oil - HT70, the analysis predicts a change in the direction of the wave propagation with the growth of the ratio of the layers thicknesses.

In the present section, we describe results of nonlinear simulations of the wavy convective regimes for the above-mentioned system of liquids. We perform nonlinear simulations of two-dimensional flows ($v_{my} = 0$ $(m = 1, 2)$; the fields of physical variables do not depend on y). Introducing the stream functions ψ_m

$$v_{mx} = \frac{\partial \psi_m}{\partial z}, \quad v_{mz} = -\frac{\partial \psi_m}{\partial x}, \quad (m = 1, 2).$$

and vorticities

$$\phi_m = \frac{\partial v_{mz}}{\partial x} - \frac{\partial v_{mx}}{\partial z},$$

we obtain the following boundary-value problem:

$$\frac{\partial \phi_m}{\partial t} + \frac{\partial \psi_m}{\partial z} \cdot \frac{\partial \phi_m}{\partial x} - \frac{\partial \psi_m}{\partial x} \cdot \frac{\partial \phi_m}{\partial z} = c_m \nabla^2 \phi_m + b_m G \frac{\partial T_m}{\partial x}, \tag{5.74}$$

$$\nabla^2 \psi_m = -\phi_m, \tag{5.75}$$

$$\frac{\partial T_m}{\partial t} + \frac{\partial \psi_m}{\partial z} \cdot \frac{\partial T_m}{\partial x} - \frac{\partial \psi_m}{\partial x} \cdot \frac{\partial T_m}{\partial z} = \frac{d_m}{P} \nabla^2 T_m \tag{5.76}$$

$$(m = 1, 2).$$

$$z = 1: \quad \psi_1 = \frac{\partial \psi_1}{\partial z} = 0; \quad T_1 = T_0 - x; \tag{5.77}$$

$$z = -a: \quad \psi_2 = \frac{\partial \psi_2}{\partial z} = 0; \quad T_2 = T_0 - x; \tag{5.78}$$

$$z = 0: \quad \psi_1 = \psi_2 = 0, \quad \frac{\partial \psi_1}{\partial z} = \frac{\partial \psi_2}{\partial z}, \quad \phi_2 = \eta \phi_1 + \frac{\eta M}{P} \frac{\partial T_1}{\partial x}; \tag{5.79}$$

$$T_1 = T_2, \quad \kappa \frac{\partial T_1}{\partial z} = \frac{\partial T_2}{\partial z}. \tag{5.80}$$

The coefficients b_m, c_m, d_m have been defined in Sec. 2.1.

For studying spatially-periodic flows, it is convenient to use the transformation of variables $T_m = \Theta_m - x$, $m = 1, 2$. The following boundary value problem is solved:

$$\frac{\partial \phi_m}{\partial t} + \frac{\partial \psi_m}{\partial z} \cdot \frac{\partial \phi_m}{\partial x} - \frac{\partial \psi_m}{\partial x} \cdot \frac{\partial \phi_m}{\partial z} = c_m \nabla^2 \phi_m + b_m G(\frac{\partial \Theta_m}{\partial x} - 1), \tag{5.81}$$

$$\nabla^2 \psi_m = -\phi_m, \tag{5.82}$$

$$\frac{\partial \Theta_m}{\partial t} + \frac{\partial \psi_m}{\partial z} \cdot (\frac{\partial \Theta_m}{\partial x} - 1) - \frac{\partial \psi_m}{\partial x} \cdot \frac{\partial \Theta_m}{\partial z} = \frac{d_m}{P} \nabla^2 \Theta_m \tag{5.83}$$

$$(m = 1, 2).$$

$$z = 1: \quad \psi_1 = \frac{\partial \psi_1}{\partial z} = 0; \quad \Theta_1 = 0; \tag{5.84}$$

$$z = -a: \quad \psi_2 = \frac{\partial \psi_2}{\partial z} = 0; \quad \Theta_2 = 0; \tag{5.85}$$

$$z = 0: \quad \psi_1 = \psi_2 = 0, \quad \frac{\partial \psi_1}{\partial z} = \frac{\partial \psi_2}{\partial z}, \quad \phi_2 = \eta \phi_1 + \frac{\eta M}{P}(\frac{\partial \Theta_1}{\partial x} - 1); \tag{5.86}$$

$$\Theta_1 = \Theta_2, \quad \kappa \frac{\partial \Theta_1}{\partial z} = \frac{\partial \Theta_2}{\partial z}; \tag{5.87}$$

on the lateral boundaries of the computational region we apply periodic boundary conditions:

$$\psi_m(x + L, z) = \psi_m(x, z); \quad \phi_m(x + L, z) = \phi_m(x, z);$$
$$\Theta_m(x + L, z) = \Theta_m(x, z); \tag{5.88}$$

The computations have been done for the system 5cs silicone oil - HT70 with the following set of parameters: $\eta = 5.476$, $\nu = 10$, $\kappa = 1.67$, $\chi = 1.847$, $\beta = 0.955$; $P = 62.512$ (Madruga et al., 2003). For any values of M, G, the boundary value problem (5.81) – (5.87) has a solution $\psi_m = \psi_m^0(z)$, $\Theta_m = \Theta_m^0(z)$, $m = 1, 2$ corresponding to a parallel flow. Exact expressions for $\psi_m^0(z)$ and $\Theta_m^0(z)$ are as follows:

$$\psi_1^0(z) = -\frac{G}{24}\left[z^4 - z^3 \frac{4\beta + 3\beta\eta a + \nu a^3}{2\beta(1 + \eta a)} + z^2 \frac{\beta + \nu a^3}{\beta(1 + \eta a)} + z \frac{(\beta\eta - \nu a^2)a}{2\beta(1 + \eta a)}\right]$$
$$+ \frac{M\eta a}{4(1 + \eta a)P}(z^3 - 2z^2 + z), \tag{5.89}$$

$$\psi_2^0(z) = -\frac{G}{24\beta}\left[\nu z^4 + z^3 \frac{3a^2\nu + 4a^3\nu\eta + \beta\eta}{2a(1 + \eta a)} + z^2 \frac{(\nu a^3 + \beta)\eta}{1 + \eta a} + z \frac{(\beta\eta - \nu a^2)a}{2(1 + \eta a)}\right]$$
$$+ \frac{M\eta}{4a(1 + \eta a)P}(z^3 + 2az^2 + a^2 z); \tag{5.90}$$

$$\Theta_1^0(z) = P(z-1)\left\{\frac{G}{24}\left[\frac{z^4}{5} - z^3\frac{12\beta + 7\beta\eta a + 5\nu a^3}{40\beta(1+\eta a)}\right.\right.$$
$$+z^2\frac{4\beta + 25\nu a^3 - 21\beta\eta a}{120\beta(1+\eta a)} + z\frac{4\beta - 5\nu a^3 + 9\beta\eta a}{120\beta(1+\eta a)}$$
$$\left. + a\frac{\kappa(4\beta - 5\nu a^3 + 9\beta\eta a) - \chi a^2(4\nu a^3 - 5\beta\eta + 9\nu a^2)}{120\beta(1+\eta a)(1+\kappa a)}\right]$$
$$\left. - \frac{M\eta a}{4P(1+\eta a)}\left[\frac{z^3}{4} - \frac{5z^2}{12} + \frac{z}{12} + \frac{a(\kappa + \chi a)}{12(1+a\kappa)}\right]\right\}, \tag{5.91}$$

$$\Theta_2^0(z) = P(z+a)\left\{\frac{G\chi}{24\beta}\left[\frac{\nu z^4}{5} + \frac{7\nu a^2 + 12a^3\nu\eta + 5\beta\eta}{40a(1+\eta a)}z^3\right.\right.$$
$$+ \frac{4\nu\eta a^3 + 25\beta\eta - 21\nu a^2}{120(1+\eta a)}z^2 + \frac{a(5\beta\eta - 9\nu a^2 - 4\nu\eta a^3)}{120(1+\eta a)}z$$
$$\left. - \frac{\kappa(4\beta - 5\nu a^3 + 9\beta\eta a) - \chi a^2(9\nu a^2 + 4\nu\eta a^3 - 5\beta\eta)}{120\chi(1+\eta a)(1+\kappa a)}\right]$$
$$\left. - \frac{M\eta\chi}{4Pa(1+\eta a)}\left[\frac{z^3}{4} + \frac{5az^2}{12} + \frac{a^2 z}{12} - \frac{a^2(\kappa + \chi a)}{12\chi(1+\kappa a)}\right]\right\}. \tag{5.92}$$

One can see that the stream function and the temperature profiles are composed in additive way from two parts. The first part is caused by buoyancy and proportional to G, the second part is caused by the thermocapillary effect and proportional to M.

With the growth of M and G, the parallel flow becomes unstable. The type of the instability mode depends on the dynamic Bond number $B = G/M$ and the ratio of layers thicknesses a.

Madruga et al. (2003) have investigated the transitions between different instability modes for the fixed value of the total depth (i.e. $B^{1/2}(1+a) = const$) with the change of the relative thickness of the bottom layer $a/(1+a)$. Specifically, the authors have predicted a transition from waves moving in the direction of the applied temperature gradient to another type of waves, moving in the opposite direction, with larger wave numbers and frequencies, at $a = 1.7$ for a layer with the total depth 6 mm (which corresponds to $B = 0.010$).

Below we describe results of computations which are carried out for the system of equations (5.81) – (5.83) with boundary conditions (5.84) – (5.87) on the horizontal boundaries and with periodic boundary conditions (5.88) on the lateral boundaries. The ratios of layers thicknesses have been chosen $a = 1$, or $a = 2$.

The Case $a = 1$

In that case, two kinds of parallel flow structures are observed. When the thermocapillary component (proportional to M) in expressions (5.89) – (5.92) prevails, the fluid moves to the right near the interface and to the left near the rigid planes. When the buoyancy component (proportional to G) prevails, a "two-store" structure is developed in the top layer: the fluid moves to the right both near the interface and the top rigid plane and to the left in the middle of the top layer. The temperature deviation $\Theta_m^0(z)$ is positive in the top layer and in the upper part of the bottom layer, and it is negative in the lower part of the bottom layer.

For sufficiently large G and M the parallel flow becomes unstable, and a travelling wave,

$$\psi_m(x,z,t) = \psi_m(x+ct,z), \quad \Theta_m(x,z,t) = \Theta_m(x+ct,z) \tag{5.93}$$

is developed. For any B, the wave moves to the *left* near the instability threshold ($c > 0$), in accordance with the predictions of Madruga et al., 2003. However, the structure of the wavy flow strongly depends on B.

In the case of a purely buoyant flow ($G \neq 0$, $M = 0$), convective cells of both signs which occupy a large part of the area, are developed in the top layer (see Fig. 5.12).

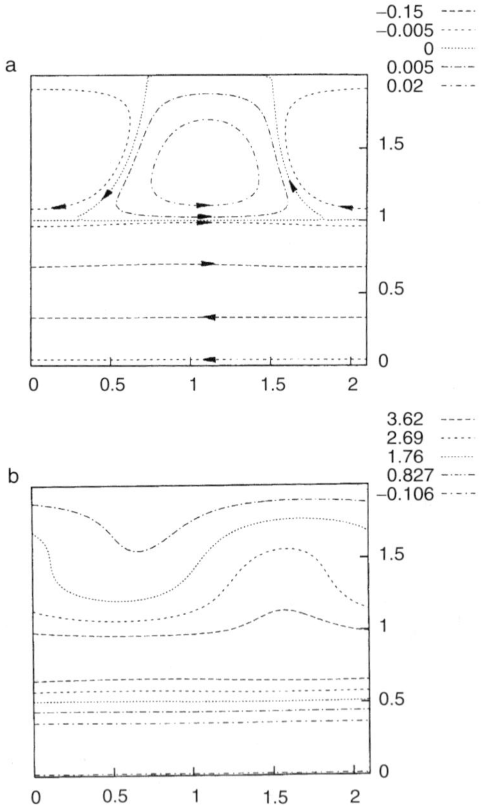

FIGURE 5.12. Snapshot of (a) streamlines and (b) isolines of the temperature deviation for a buoyant traveling wave ($a = 1$, $L = 2.1$, $G = 7$, $M = 0$).

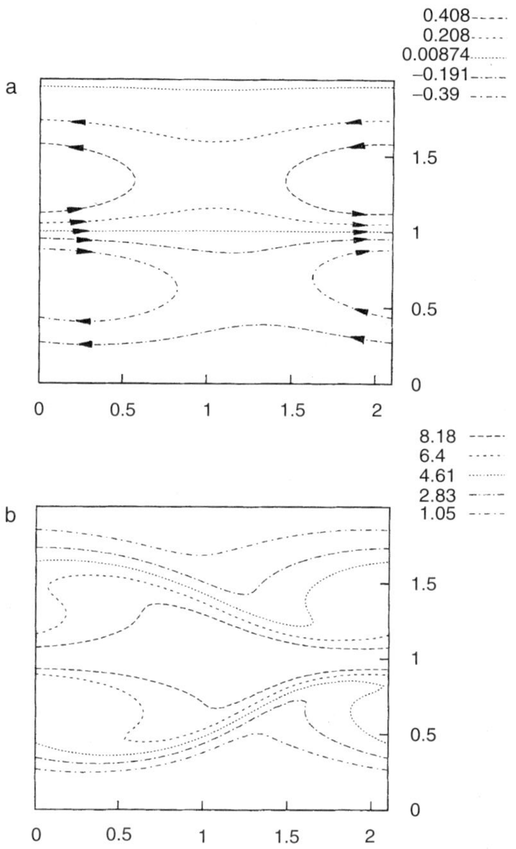

FIGURE 5.13. Snapshot of (a) streamlines and (b) isolines of the temperature deviation for a thermocapillary traveling wave with $c > 0$ ($a = 1$, $L = 2.1$, $G = 0$, $M = 913$).

In the case of a purely thermocapillary flow ($G = 0$, $M \neq 0$), the waves are developed simultaneously in both layers (see Fig. 5.13). All the vortices are positive in the top layer and negative in the bottom layer. It is interesting to note that the direction of the wave propagation is changed with the growth of M (see Fig. 5.14, line 1).

The flow patterns under the simultaneous action of the buoyancy and the thermocapillary effect ($G \neq 0$, $M \neq 0$), are shown in Fig. 5.15. The wavy motion takes place mainly in the top layer. One observes an essential asymmetry between the positive vortices, which occupy a large area in the top layer, and rather compact negative vortices localized near the upper rigid plane. The direction of the wave propagation is changed with the growth of M (see Fig. 5.14, lines 2 and 3). Note that for sufficiently large G, the wave velocity c is changed in a non-monotonic way.

In a long computational region with $L = 16$ all the types of traveling waves described above keep their perfect periodicity (see Figs. 5.16 – 5.18). Thus, there

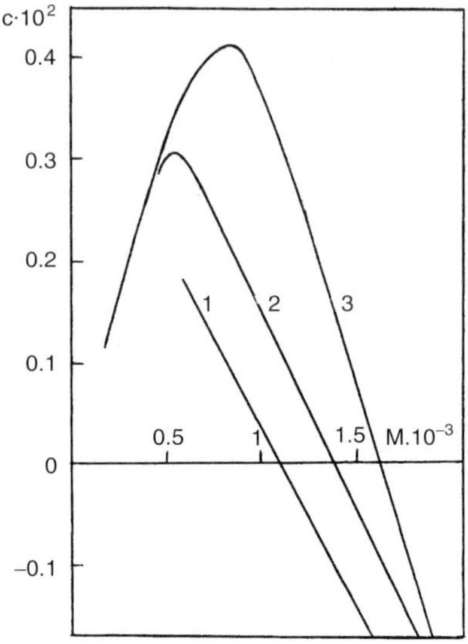

FIGURE 5.14. Dependence of the wave velocity c on the Marangoni number M for $G = 0$ (line 1), 1.5 (line 2), 4 (line 3).

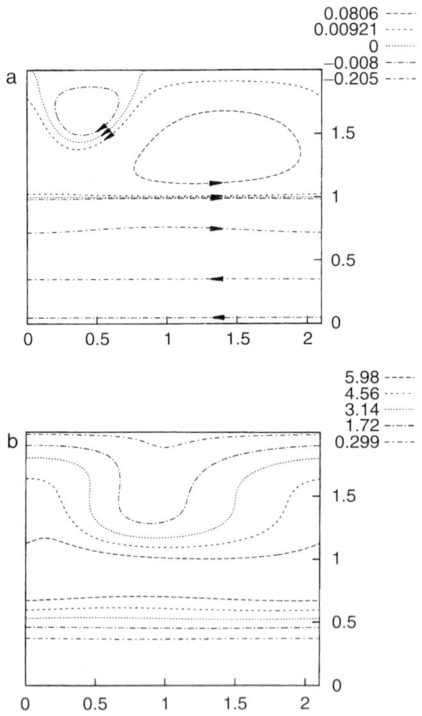

FIGURE 5.15. Snapshot of (a) streamlines and (b) isolines of the temperature deviation for a buoyant - thermocapillary traveling wave ($a = 1$, $L = 2.1$, $G = 7$, $M = 171.2$).

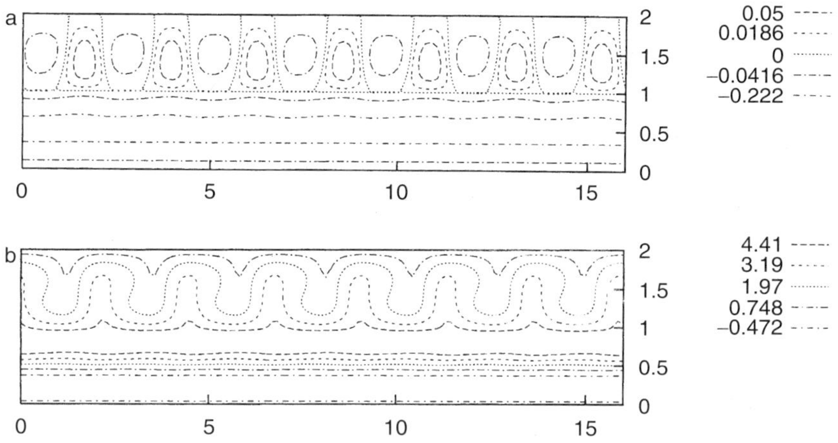

FIGURE 5.16. Snapshot of (a) streamlines and (b) isolines of the temperature deviation for a buoyant traveling wave in a long computational region ($a = 1$, $L = 16$, $G = 10$, $M = 0$).

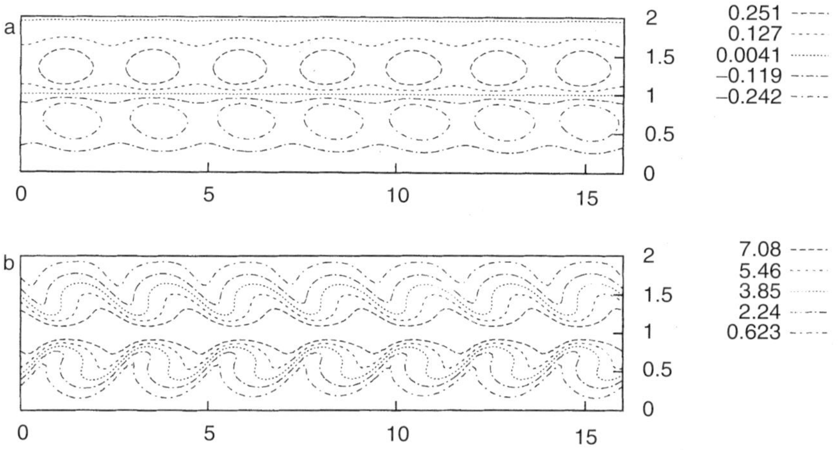

FIGURE 5.17. Snapshot of (a) streamlines and (b) isolines of the temperature deviation for a thermocapillary traveling wave in a long computational region ($a = 1$, $L = 16$, $G = 0$, $M = 571$).

is no long - wave modulational instability for buoyant - thermocapillary traveling waves at the moderate values of G and M.

The Case $a = 2$.

In accordance with the predictions of the linear stability theory (Madruga et al., 2003), the instability in the case $a = 2$ is generated by another mode of disturbances than in the case $a = 1$. Near the threshold, the traveling wave moves to the *right*. The characteristic wave length is shorter, than that in the previous case,

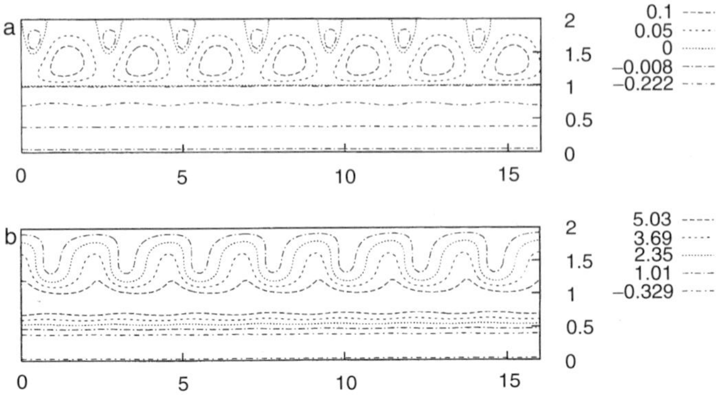

FIGURE 5.18. Snapshot of (a) streamlines and (b) isolines of the temperature deviation for a buoyant - thermocapillary traveling wave in a long computational region ($a = 1$, $L = 16$, $G = 7$, $M = 171.2$).

so one observes two spatial periods of the wave in the computational region with $L = 2.1$ (see Fig. 5.19). The phase velocity of the threshold waves moving to the right (for $a = 2$) is larger than that of moving to the left (for $a = 1$). It is interesting to note that the waves are excited *in the subcritical way*. Depending on the initial conditions, the system evolves either to the traveling wave or to the parallel flow (Nepomnyashchy and Simanovskii, 2006).

5.4 Buoyancy-thermocapillary convection in three-layer systems

In the present section, we describe results of the numerical simulation of convective flows in a closed cavity filled by three immiscible viscous fluids.

5.4.1 Formulation of the Problem

Let the rectangular cavity with rigid boundaries be filled by three immiscible viscous fluids (see Fig. 3.18). The lateral plates $x = -l/2$ and $x = l/2$ are kept at different constant temperatures $T = \Theta$ and $T = 0$, respectively. The temperature on the horizontal plates $z = a_3$ and $z = -(a_1 + a_2)$ is fixed: $T(x, z) = \Theta(1/2 - x/l)$. It is assumed that surface tension coefficients on the upper and lower interfaces σ and σ_* decrease linearly with temperature: $\sigma = \sigma_0 - \alpha T$, and $\sigma_* = \sigma_{0*} - \alpha_* T$. We neglect the deformations of interfaces.

Let us use the following notations:

$$\nu = \nu_3/\nu_2, \ \ \nu_* = \nu_3/\nu_1, \ \ \eta = \eta_3/\eta_2, \ \ \eta_* = \eta_3/\eta_1, \ \ \kappa = \kappa_3/\kappa_2, \ \ \kappa_* = \kappa_3/\kappa_1,$$

$$\chi = \chi_3/\chi_2, \ \ \chi_* = \chi_3/\chi_1, \ \ \beta = \beta_3/\beta_2, \ \ \beta_* = \beta_3/\beta_1, \ \ \bar{\alpha} = \alpha_*/\alpha,$$

$$a_* = a_1/a_3, \ \ a = a_2/a_3, \ \ L = l/a_3.$$

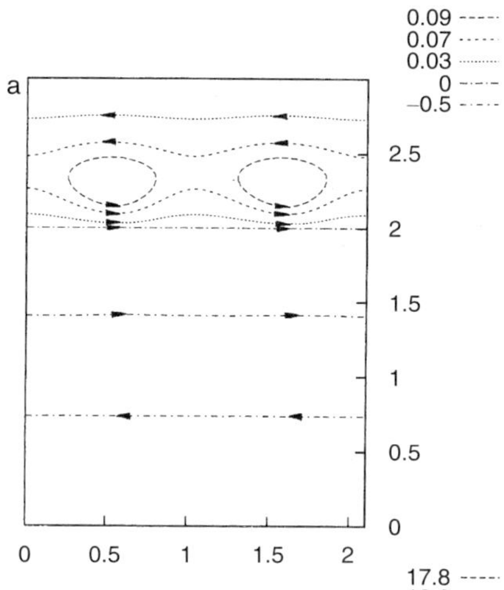

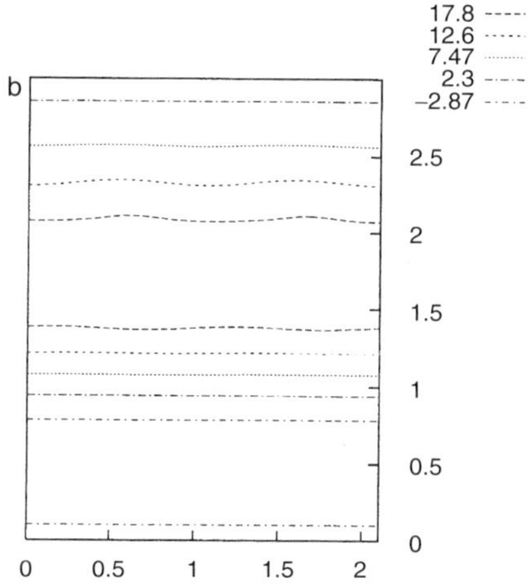

FIGURE 5.19. Snapshot of (a) streamlines and (b) isolines of the temperature deviation for the traveling wave with $c < 0$ ($a = 2$, $L = 2.1$, $G = 1.08$, $M = 137$).

Here, ν_m, η_m, κ_m, χ_m, β_m, and a_m are respectively kinematic and dynamic viscosities, heat conductivity, heat diffusivity, heat expansion coefficient, and thickness of the mth layer ($m = 1, 2, 3$). As the units of length, time, velocity, pressure, and temperature, we use a_3, a_3^2/ν_3, ν_3/a_3, $\rho_3\nu_3^2/a_3^2$, and Θ.

Introducing the stream function ψ and the vorticity ϕ, we can write the dimensionless equations in the following form:

$$\frac{\partial \phi_m}{\partial t} + \frac{\partial \psi_m}{\partial z}\frac{\partial \phi_m}{\partial x} - \frac{\partial \psi_m}{\partial x}\frac{\partial \phi_m}{\partial z} = d_m \Delta \phi_m + b_m G \frac{\partial T_m}{\partial x},$$

$$\Delta \psi_m = -\phi_m, \tag{5.94}$$

$$\frac{\partial T_m}{\partial t} + \frac{\partial \psi_m}{\partial z}\frac{\partial T_m}{\partial x} - \frac{\partial \psi_m}{\partial x}\frac{\partial T_m}{\partial z} = \frac{c_m}{P}\Delta T_m \quad (m = 1, 2, 3).$$

Here, $d_3 = b_3 = c_3 = 1$; $d_1 = 1/\nu_*$, $b_1 = 1/\beta_*$, $c_1 = 1/\chi_*$; $d_2 = 1/\nu$, $b_2 = 1/\beta$, $c_2 = 1/\chi$; $G = g\beta_3\Theta a_3^3/\nu_3^2$ is the Grashof number, and $P = \nu_3/\chi_3$ is the Prandtl number for the liquid in layer 3.

At the interfaces, normal components of velocity vanish. We will take into account both viscous coupling (the continuity conditions for tangential components of velocity and viscous stresses) and thermal coupling (the continuity conditions for temperatures and heat fluxes) between the layers:

$$z = 0: \ \psi_2 = \psi_3 = 0; \ \frac{\partial \psi_2}{\partial z} = \frac{\partial \psi_3}{\partial z}; \ T_2 = T_3;$$

$$\frac{1}{\kappa}\frac{\partial T_2}{\partial z} = \frac{\partial T_3}{\partial z}; \ \eta\frac{\partial^2 \psi_3}{\partial z^2} = \frac{\partial^2 \psi_2}{\partial z^2} + Mr\frac{\partial T_3}{\partial x}; \tag{5.95}$$

$$z = -a: \ \psi_1 = \psi_2 = 0; \ \frac{\partial \psi_1}{\partial z} = \frac{\partial \psi_2}{\partial z}; \ T_1 = T_2;$$

$$\frac{1}{\kappa_*}\frac{\partial T_1}{\partial z} = \frac{1}{\kappa}\frac{\partial T_2}{\partial z}; \ \eta_*\eta^{-1}\frac{\partial^2 \psi_2}{\partial z^2} = \frac{\partial^2 \psi_1}{\partial z^2} + Mr_*\frac{\partial T_2}{\partial x}. \tag{5.96}$$

Here, $Mr = \eta M/P$, $Mr_* = \eta_*\bar{\alpha}M/P$; $M = \alpha\Theta a_3/\eta_3\chi_3$ is the Marangoni number. On the horizontal rigid plates, the boundary conditions have the following form:

$$z = 1: \ \psi_3 = \frac{\partial \psi_3}{\partial z} = 0; \ T_3 = 1/2 - x/L; \tag{5.97}$$

$$z = -a - a_*: \ \psi_1 = \frac{\partial \psi_1}{\partial z} = 0; \ T_1 = 1/2 - x/L. \tag{5.98}$$

We consider the following conditions on vertical boundaries:

$$x = -L/2: \ \psi_m = \frac{\partial \psi_m}{\partial x} = 0; \ T_m = 1; \ m = 1, 2, 3. \tag{5.99}$$

$$x = L/2: \ \psi_m = \frac{\partial \psi_m}{\partial x} = 0; \ T_m = 0; \ m = 1, 2, 3. \tag{5.100}$$

Prakash and Koster (1994a, 1994b) presented an asymptotic analysis of the problem separately for the buoyancy convection ($M = 0$) and for the thermocapillary convection ($G = 0$) by means of matched asymptotic expansions.

The boundary value problem (5.94) – (5.100) contains 17 independent nondimensional parameters. The parametric investigation of this problem is thus impossible. We will concentrate on some particular systems of fluids demonstrating various characteristic phenomena.

The boundary value problem (5.94) – (5.100) is solved by the finite-difference method. Equations (5.94) are approximated on a uniform mesh using a second-order approximation for the spatial coordinates. The nonlinear equations are solved using the explicit scheme on a rectangular 120 × 144 mesh (see Géoris *et al.*, 1997).

5.4.2 Results of Numerical Simulations

Buoyancy-Thermocapillary Flows

First, we take the system air / ethylene glycol / Fluorinert FC75 (system 3 in Table 2). Let us consider a purely thermogravitational convection flow ($M/G \to 0$), which corresponds to the case of relatively thick layers. In an infinite layer of a liquid with kinematic viscosity $\bar{\nu}$, heat diffusivity $\bar{\chi}$, and heat expansion coefficient $\bar{\beta}$ subject to a horizontal temperature gradient A, the maximal value of velocity $||v||$ is proportional to $g\bar{\beta}A\bar{a}^3/\bar{\nu}$, whereas the inhomogeneity of the temperature across the layer $||T||$ is proportional to $g\bar{\beta}A^2\bar{a}^5/\bar{\nu}\bar{\chi}$, where $\bar{a}$ is the thickness of the layer (Birikh, 1966). Because of this, we can expect that in the case of a three-layer system, the following estimates can be used in the limit $L \gg 1$:

$$\delta_{32} = \frac{||v_3||}{||v_2||} = \frac{\beta}{a^3\nu}, \quad \delta_{31} = \frac{||v_3||}{||v_1||} = \frac{\beta_*}{a_*^3\nu_*},$$

$$\epsilon_{32} = \frac{||T_3||}{||T_2||} = \frac{\beta}{a^5\nu\chi}, \quad \epsilon_{31} = \frac{||T_3||}{||T_1||} = \frac{\beta_*}{a_*^5\nu_*\chi_*}.$$

For the system under consideration, $\delta_{32} = 6.1$, $\delta_{31} = 0.14$, $\epsilon_{32} = 0.0282$, and $\epsilon_{31} = 0.000230$. Thus, we find

$$||v_2|| < ||v_3|| < ||v_1||, \quad ||T_3|| < ||T_2|| < ||T_1||.$$

This prediction coincides with the results of numerical simulations for long layers (see Fig. 5.20).

For shorter layers, the obtained prediction is valid only for relatively small values of G (see Fig. 5.21). With increasing G, the intensity of the motion in the upper layer becomes higher than that in the lower one (see Fig. 5.22). The dependence of the maximal value of the modulus of the stream function on G is shown in Fig. 5.23.

Since the coefficient $\bar{\alpha}$ is less than unity, the pure thermocapillary convection is more intensive in the upper layer and in the middle layer (see Fig. 5.24). Let us note that the direction of the thermocapillary motion in the upper and lower layers is opposite to that of the thermogravitational motion. The inclusion of G leads to the appearance of an intensive thermogravitational motion in the lower layer and a thermogravitational motion of less intensity in the upper layer (see Fig. 5.25).

Thermocapillary Flows

The results of this paragraph have been obtained by Shevtsova *et al.* (2005). Let us consider a pure thermocapillary flow ($G = 0$) in the symmetric system silicone oil 47v2/water/silicone oil 47v2 with the following set of parameters: $\eta = 1.7375$, $\nu = 2$, $\kappa = 0.184$, $\chi = 0.778$, $\eta_* = \nu_* = \kappa_* = \chi_* = 1$, and $P = 25.7$. We take $a = a_* = 1$. It means that the exterior layers have the same

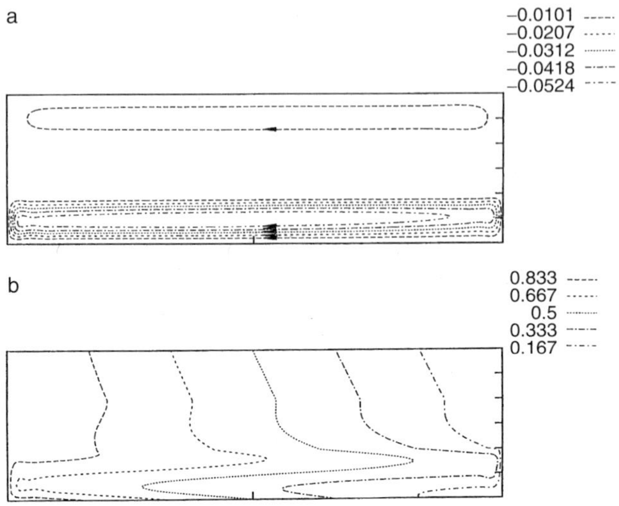

FIGURE 5.20. (a) Stream lines and (b) isotherms for $G = 75,\ M = 0,\ L = 10$.

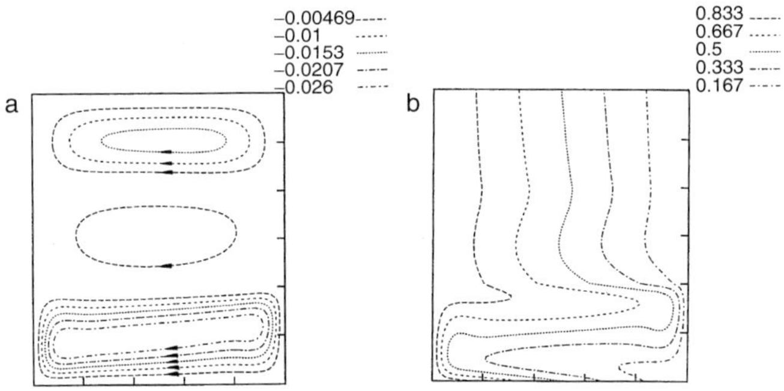

FIGURE 5.21. (a) Stream lines and (b) isotherms for $G = 20,\ M = 0,\ L = 2.5$.

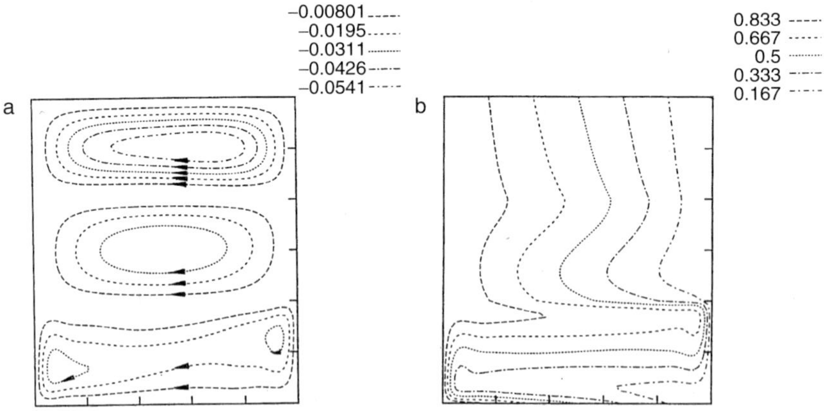

FIGURE 5.22. (a) Stream lines and (b) isotherms for $G = 75,\ M = 0,\ L = 2.5$.

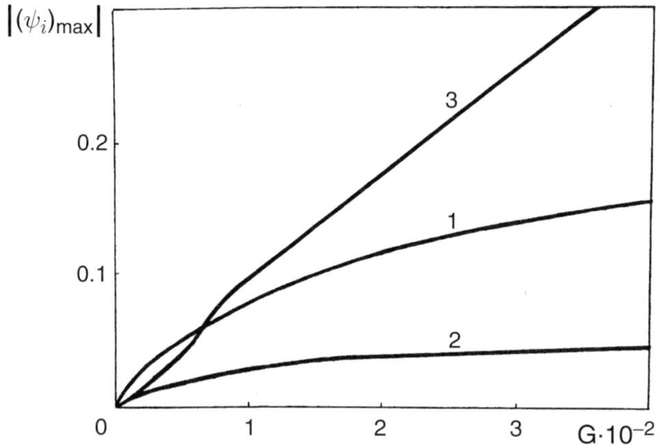

FIGURE 5.23. Dependence of $|\Psi|_{i,max}$ $(i = 1,\ 2,\ 3)$ on G.

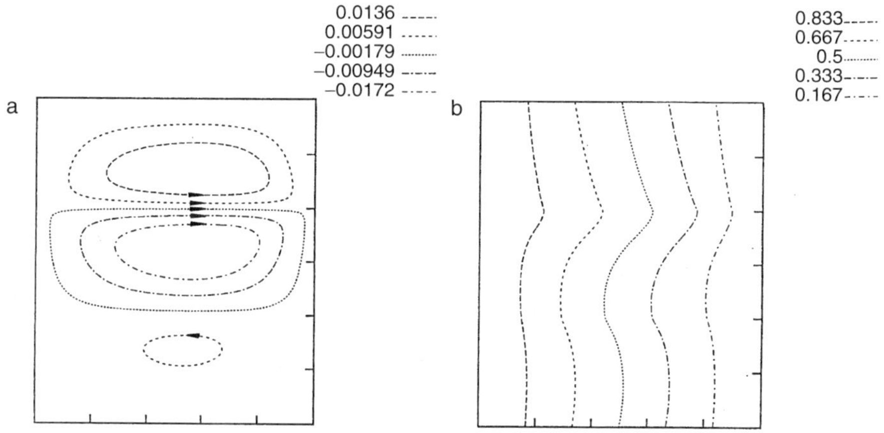

FIGURE 5.24. (a) Stream lines and (b) isotherms for $G = 0,\ M = 10^3,\ L = 2.5$.

thermophysical properties. First, let us consider the cavity with the small aspect ratio ($L = 3.2$). Even for small values of the Marangoni number ($M \neq 0$), the mechanical equilibrium state is impossible and a steady motion takes place in the system (see Fig. 5.26). One can see that in the central part of the cavity the flow is nearly parallel. Along the interfaces, the fluids move from the hot wall to the cold wall. In the middle layer, the motion consists of two vortices of different signs and has the two-storied structure. The flow fields in different layers are coupled by viscous stresses. Near the lateral walls, the fluid may move both upward and downward. For relatively small values of M, the flows are quite symmetric with respect to the vertical axis $x = 0$.

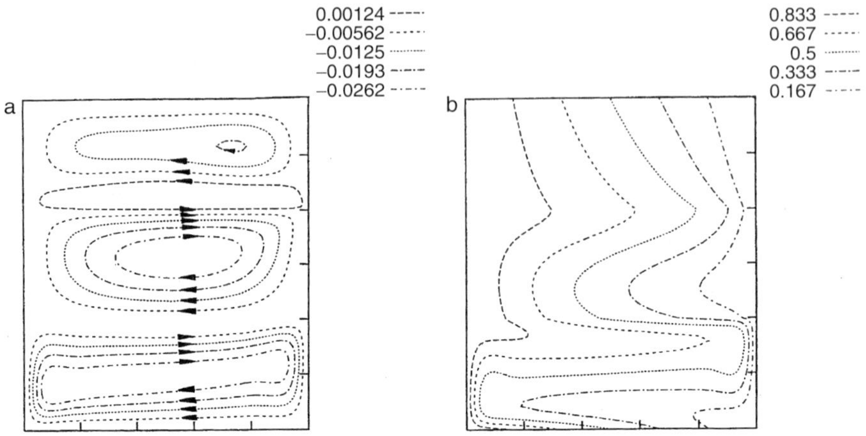

FIGURE 5.25. (a) Streamlines and (b) isotherms for $G = 35$, $M = 10^3$, $L = 2.5$.

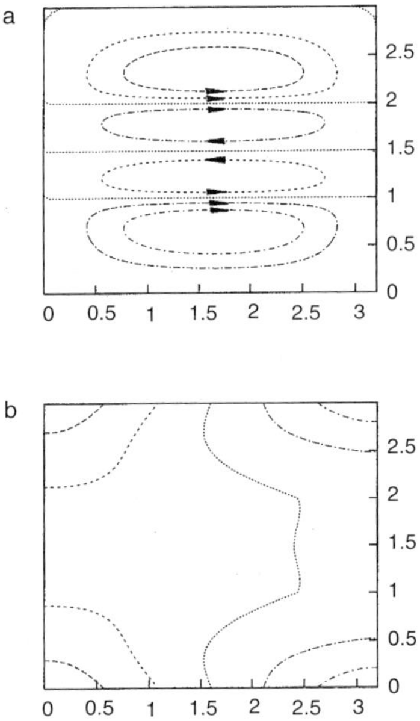

FIGURE 5.26. (a) Stream lines and (b) isotherms for the steady motion at $M = 10^3$, $L = 3.2$.

With the increase of the Marangoni number, the intensity of the flow near the hot wall becomes higher than that near the cold wall. The steady motion becomes unstable and the regular oscillations develop in the system. The snapshots of the stream lines for one period of oscillations are presented in Fig. 5.27. During the oscillatory process, the vortices become longer in the horizontal direction (Fig. 5.27(b)) and then the additional vortices appear in the middle layer (Fig. 5.27(c)). These new vortices move to the right; then they are reflected by the cold lateral wall and start to move in the opposite direction (Figs. 5.27(e) – 5.27(g)). Finally, the new vortices couple with the main vortices in the middle layer and the oscillatory process is repeated.

The time evolution of the quantities $S_{l,m}(t)$ $(m = 1, 3)$ and $S_{r1}(t)$ are shown in Fig. 5.28. Let us remember that these variables are the sums of stream functions in all the nodes of the calculating mesh in the left or in the right halves of the corresponding layer. One can see that during the oscillatory process, the integral variables do not change their signs.

Let us now consider the longer cavity with $L = 16$. As in the case of the short cavity, for relatively small values of the Marangoni number ($M \neq 0$) the steady motion takes place in the system (see Fig. 5.29). At the larger values of M ($M_* > 130,000$), the steady state becomes unstable and the oscillations appear in the system. The snapshots of stream lines for half of the period of oscillations are presented in Fig. 5.30; the vortices in the middle layer have the chess-order configuration. During the oscillatory process, the number of vortices is changed in the layers. The wave moves in the direction of the temperature gradient (i.e., from the cold end to the hot end). This direction of motion is characteristic for hydrothermal waves. For the cavities with $L = 32$, the transition from the steady state to the oscillatory flow (see Fig. 5.31) takes place at ($M_* \approx 80,000$).

Let us note that presented type of oscillations in the symmetric system may be fulfilled in experiments under microgravity conditions.

5.5 Deformational Instabilities of Thermocapillary Flows in Three-Layer Systems

Recall that Smith and Davis (1982, 1983a, 1983b) investigated the main types of instability of thermocapillary parallel flows in one-layer systems. These instabilities can be divided into two classes. The instabilities that belong to the first class are produced by the interaction of velocity and temperature disturbances and are relatively insensitive to the interface deformations. The second class is a surface-wave instability, where free-surface deformations are crucial. The latter kind of instability has been identified for different forms of basic velocity field. In particular, a long-wavelength instability of "negative viscosity type" ($\lambda_r \sim \Lambda k^2$) was revealed (Smith and Davis, 1982, 1983b).

The instabilities of the first kind for parallel flows in *two-layer* systems have been studied in Sec. 4.3. In the present section, we consider the surface kind of instability in *three-layer* thermocapillary creeping flows. It is shown that in this system, new types of instability arise that do not appear in one- and two-layer systems. Nonlinear equations governing interfacial evolution are derived and simulated numerically.

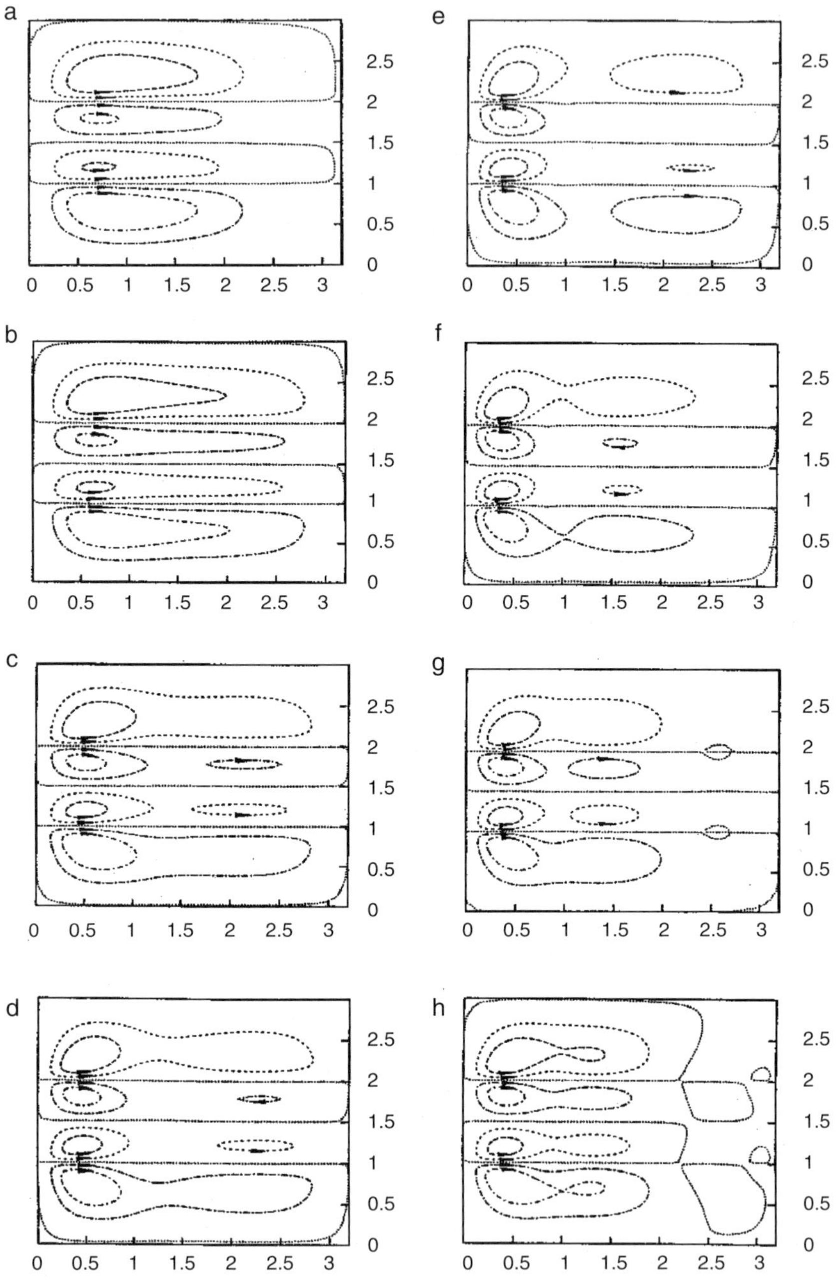

FIGURE 5.27. Snapshots of stream lines for the oscillatory motion during one period at $M = 350,000$, $L = 3.2$.

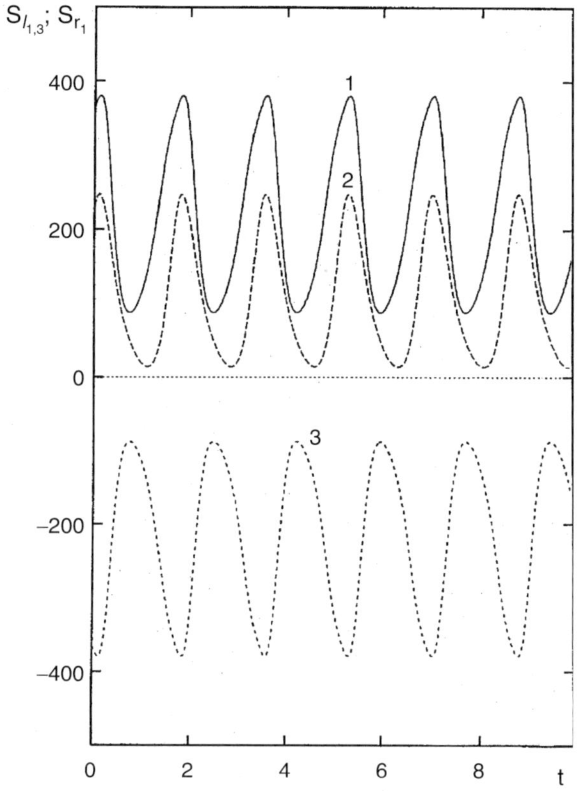

FIGURE 5.28. Time evolution of integral characteristics $S_{l,1}$ (line 1), $S_{r,1}$ (line 2), and $S_{l,3}$ (line 3) for $M = 350,000$, $L = 3.2$.

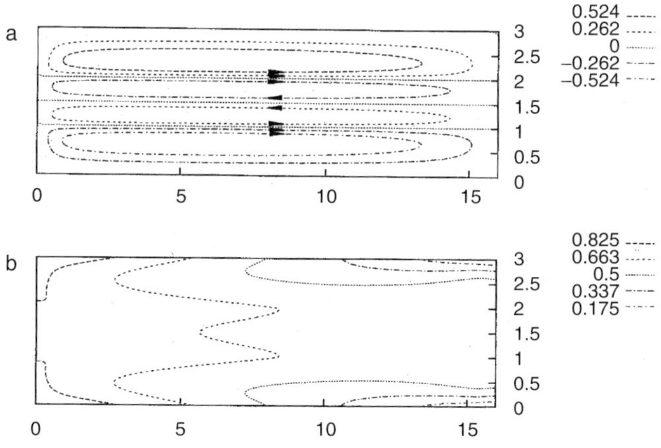

FIGURE 5.29. (a) Stream lines and (b) isotherms for the steady motion at $M = 110,000$, $L = 16$.

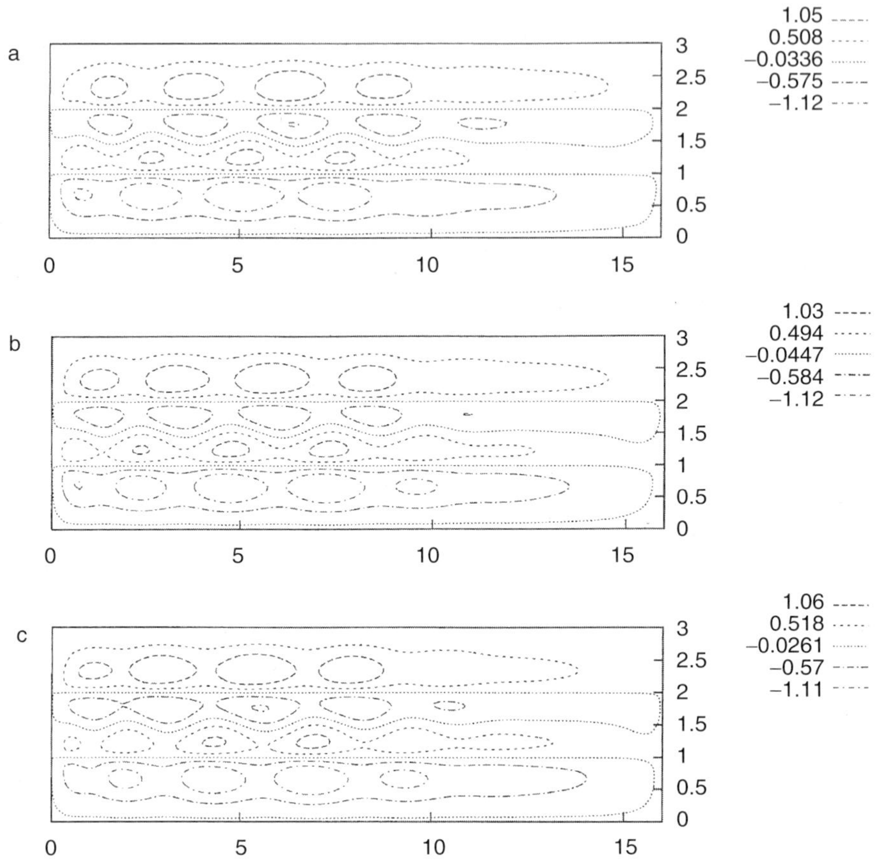

FIGURE 5.30. Snapshots of stream lines for the oscillatory motion during half of the period at $M = 250,000,\ L = 16$.

5.5.1 Formulation of the Model

We consider slow *creeping* flows (i.e., flows in the limit of small Reynolds numbers, when the inertial effects are not important). In order to capture the essence of the pertinent dynamics, one may well consider a quasisteady and quasiparallel three-layer creeping flow between rigid well-conducting plates (Fig. 5.32). For simplicity, the temperature is fixed as $T = T_0 - Ax$, $A > 0$; that is, the temperature disturbances are disregarded. This corresponds to the case of fast decay of temperature fluctuations (i.e., to small Prandtl numbers). The surface tension coefficients, which are equal at both interfaces, decrease linearly with the temperature and increases in longitudinal direction:

$$\sigma = \sigma_0 - \alpha(T - T_0) = \sigma_0 + \gamma x, \gamma = \alpha A > 0. \tag{5.101}$$

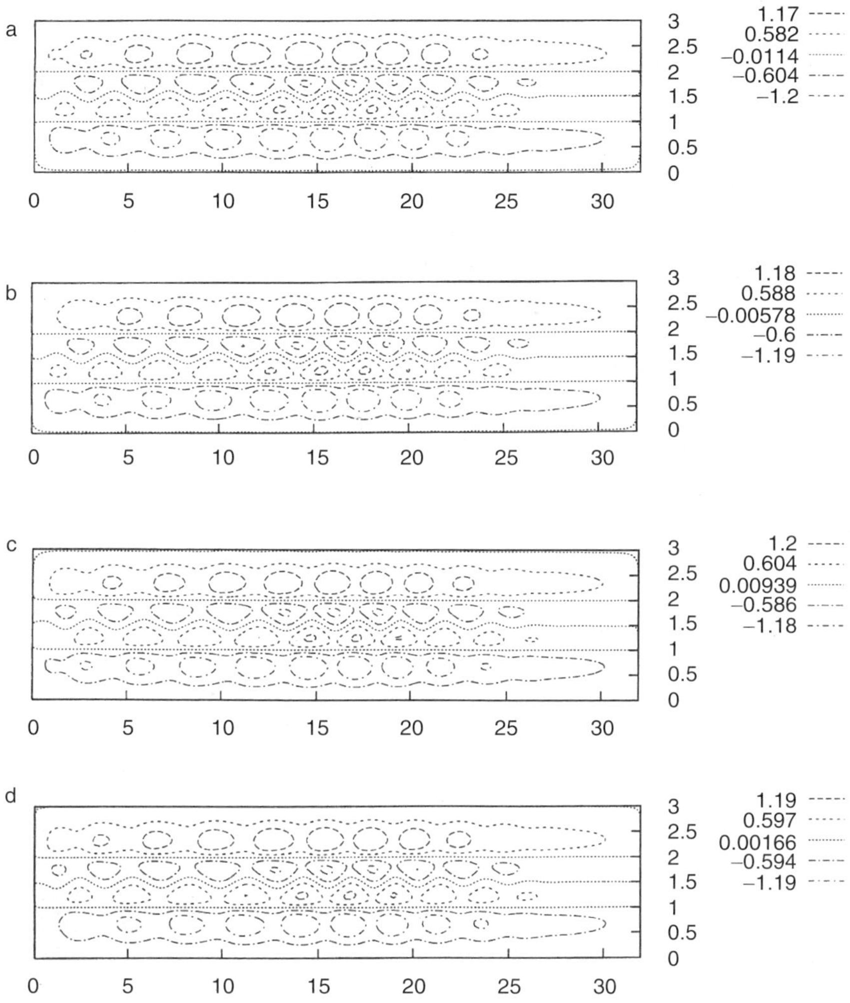

FIGURE 5.31. Snapshots of stream lines for the oscillatory motion during one period at $M = 110,000$, $L = 32$.

We consider the simplest case of the flow in an open infinite channel without external pressure gradient; as a result, the basic parallel flow is linear (Fig. 5.32).

As a further simplification, the liquids in upper and lower layers are assumed to be identical. The densities ρ of all liquids are assumed to be equal.

The surface tension σ is further assumed to be large, and the flow is therefore considered as quasiplane and quasisteady; that is, the resulting interfacial dynamics is expected to be long wave and slowly varying. In the long-wave approximation, the corresponding boundary value problem is essentially simplified.

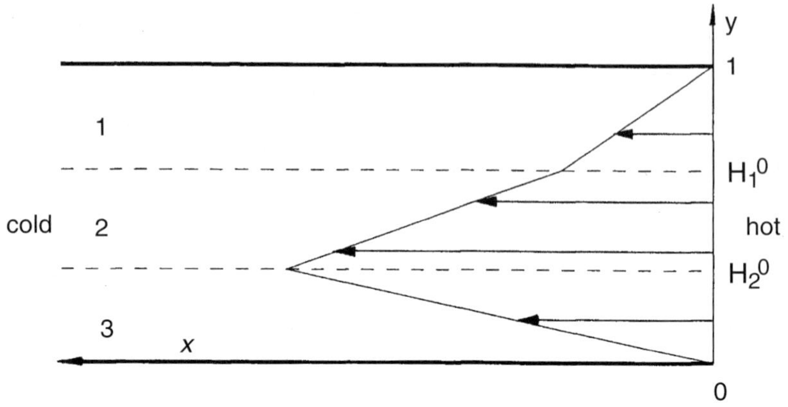

FIGURE 5.32. Undisturbed unidirectional velocity profiles in a three-layer system.

In suitably chosen nondimensional variables, the pertinent system of Stokes equations may be written as follows:

$$m_i \frac{\partial^2 u_i}{\partial y^2} = \frac{\partial p_i}{\partial x}, \tag{5.102}$$

$$\frac{\partial p_i}{\partial y} = 0, \tag{5.103}$$

$$\frac{\partial u_i}{\partial x} + \frac{\partial v_i}{\partial y} = 0. \tag{5.104}$$

At the bottom and top planes and the interfaces, the following conditions are imposed:

(i) nonslip conditions at the bottom and top planes:

$$u_3 = v_3 = 0 \text{ at } y = 0, \tag{5.105}$$

$$u_1 = v_1 = 0 \text{ at } y = 1; \tag{5.106}$$

(ii) continuity of the velocity at the inner interfaces:

$$u_i = u_{i+1}, \; v_i = v_{i+1} \text{ at } y = H_i, \; i = 1, 2; \tag{5.107}$$

(iii) conditions relating the shear jump to variations of the surface tension:

$$m_i \frac{\partial u_i}{\partial y} = m_{i+1} \frac{\partial u_{i+1}}{\partial y} + 1 \text{ at } y = H_i, \; i = 1, 2; \tag{5.108}$$

(iv) conditions relating the pressure jump to the surface tension:

$$p_i - p_{i+1} = \Sigma \frac{\partial^2 H_i}{\partial x^2} \text{ at } y = H_i, \; i = 1, 2; \tag{5.109}$$

(v) impermeability of the interfaces (kinematic conditions):

$$\frac{\partial H_i}{\partial t} + u_i \frac{\partial H_i}{\partial x} - v_i = 0 \text{ at } y = H_i,\ i = 1, 2. \tag{5.110}$$

The index $i = 1, 2, 3$ labels the layers $1 > y > H_1$, $H_1 > y > H_2$, and $H_2 > y > 0$; $H_1(x, t)$, and $H_2(x, t)$ are the flow interfaces in units of d, the total width of the three-layer system; u and v are the flow velocity components, referred to $U = \gamma \rho d/\mu_1$; μ_i are dynamic viscosity of ith layer; $m_i = \mu_i/\mu_1$ are the viscosity ratios ($m_1 = m_3 = 1$); x and y are the spatial coordinates in units of d; t is time in units of $\mu_1/\gamma \rho$, and p is the pressure in units of $\gamma \rho$. We disregard the dependence of σ on x in the boundary condition (5.109): $\Sigma = \sigma_0/\gamma \rho d$.

Through the combination of the continuity equation (5.104), and boundary conditions (5.105) – (5.107), kinematic relations (5.110) may be expressed in the conservative form:

$$\frac{\partial H_1}{\partial t} + \frac{\partial}{\partial x}\left(\int_1^{H_1} u_1\, dy\right) = 0, \tag{5.111}$$

$$\frac{\partial H_2}{\partial t} + \frac{\partial}{\partial x}\left(\int_1^{H_2} u_3\, dy\right) = 0. \tag{5.112}$$

Equations (5.102) – (5.104) considered jointly with the conditions (5.105) – (5.109), (5.111), and (5.112) define a free boundary problem for the evolving interfaces $H_1(x, t)$ and $H_2(x, t)$.

5.5.2 Derivation of the Interface Evolution Equations

Combination of boundary conditions (5.105) – (5.107) and the continuity equation (5.104) leads to the following equivalent boundary condition for u_i:

$$\frac{\partial}{\partial x}\left[\int_0^{H_2} u_3 dy + \int_{H_2}^{H_1} u_2 dy + \int_{H_1}^{1} u_1\, dy\right] = 0, \tag{5.113}$$

or

$$\int_0^{H_2} u_3 dy + \int_{H_2}^{H_1} u_2 dy + \int_{H_1}^{1} u_1\, dy = C(t), \tag{5.114}$$

where $C(t)$ is a function of time found below.

The velocity components u_i, defined by equation (5.102), may be written as

$$u_i = \frac{1}{2m_i}\frac{\partial p_i}{\partial x} y^2 + A_i y + B_i. \tag{5.115}$$

From condition (5.109) follows that

$$p_1 = p_2 + \Sigma \frac{\partial^2 H_1}{\partial x^2},\quad p_3 = p_2 - \Sigma \frac{\partial^2 H_2}{\partial x^2}. \tag{5.116}$$

Substitution of equation (5.116) in equation (5.115) reduces the problem for u_i to searching of seven y - independent functions p_2, A_i, and B_i, $i = 1, 2, 3$, from seven linear boundary conditions (5.105) – (5.108), and (5.114). As a result, p_2 is expressed through H_i, $\partial^3 H_i/\partial x^3$, and $C(t)$:

$$\frac{\partial p_2}{\partial x} = F_1(H_i)\frac{\partial^3 H_1}{\partial x^3} + F_2(H_i)\frac{\partial^3 H_2}{\partial x^3} + G(H_i)\, C(t). \tag{5.117}$$

Here, F_1, F_2, and G are the rational functions of H_i. Coefficients A_i, and B_i are the linear functions of $\partial^3 H_i/\partial x^3$, and $C(t)$, with coefficients depending on H_i. Finally, the x, t dependence of velocity components u_i involves three groups of terms: (i) corresponding to the mean motion because of the Marangoni effect, (ii) corresponding to the influence of the surface tension at the interfaces, and (iii) corresponding to the "global control" function $C(t)$, appearing due to the infinite velocity of disturbances propagation.

Integration of expression (5.117) over the length of the system together with the condition of the pressure absence on the channel edges $(-L, L)$ leads to

$$p_2(L) - p_2(-L) = 0 = \int_{-L}^{L} \left[F_1\frac{\partial^3 H_1}{\partial x^3} + F_2\frac{\partial^3 H_2}{\partial x^3} + GC(t) \right] dy. \tag{5.118}$$

Substituting the resulting expressions for u_i into the kinematic conditions (5.111) and (5.112) and expressing $C(t)$ from equation (5.118), one ends up with the sought for system for the interacting interfaces:

$$\frac{\partial H_i}{\partial t} + \frac{\partial}{\partial x}(Q_i + C(t)S_i) + \sigma\frac{\partial}{\partial x}\left(G_{ij}\frac{\partial^3 H_j}{\partial x^3} \right) = 0, \tag{5.119}$$

$$C(t) = \frac{\int_{-L}^{L} \left[F_1\frac{\partial^3 H_1}{\partial x^3} + F_2\frac{\partial^3 H_2}{\partial x^3} \right] dy}{\int_{-L}^{L} G dy}. \tag{5.120}$$

Here, Q_i, S_i, and G_{ij} are rational functions of H_1, H_2; $m = m_2$. The evaluation of Q_i, S_i, and G_{ij} involves straightforward algebra and was effectively conducted with the assistance of the MATHEMATICA software; the explicit expressions of these functions being awkward are not presented here.

Equations (5.119) and (5.120) describe the interfacial dynamics of the three-layer thermocapillary flow. The complete investigation of equations seems to be rather complicated. We consider a number of tractable cases when equations (5.119) and (5.120) can be studied analytically and numerically.

5.5.3 Linear Stability Analysis

The linearized version of equation (5.119) is

$$\frac{\partial h_i}{\partial t} + \alpha_{ij}\frac{\partial h_j}{\partial x} + \Sigma\chi_{ij}\frac{\partial^4 h_j}{\partial x^4} = 0, \tag{5.121}$$

$$h_i(x,t) = H_i - H_i^0, \tag{5.122}$$

$$\alpha_{ij} = \frac{\partial Q_i}{\partial H_j}(H_1^0, H_2^0), \ \ \chi_{ij} = G_{ij}(H_1^0, H_2^0), \ i, j = 1, 2. \tag{5.123}$$

Here, h_i are the perturbations of the planar interfaces, and H_i^0 correspond to the interfaces of the undisturbed flow (Fig. 5.32). The constant tensors α_{ij} and χ_{ij} are determined by the basic distribution of viscosities μ_i and the interfaces positions H_i^0. The matrix χ_{ij} ensures dissipation of the short-wavelength disturbances as k^4.

Linearization of equation (5.120) together with the assumption that small perturbations are periodic over the channel, $h(-L) = h(L)$, leads to

$$C(t) = 0. \tag{5.124}$$

Substitution of

$$h_i(x,t) = a_i \exp(\lambda t + ikx), \tag{5.125}$$

in equation (5.121) yields the following eigenvalue problem:

$$\lambda a_i + (ik\alpha_{ij} + \Sigma k^4 \chi_{ij}) a_j = 0. \tag{5.126}$$

We consider the temporal formulation of the stability problem with real k and complex λ in the long-wavelength limit.

As the first step, we analyze the shortened long-wavelength version of equation (5.126):

$$\omega a_i = -ik\alpha_{ij} a_j. \tag{5.127}$$

Since, in general, the matrix α_{ij} is non-Hermitian, one may well end up with its complex eigenvalues $c_j^{(\alpha)}$. In this case,

$$\lambda_r \sim |k|, \ k \ll 1. \tag{5.128}$$

The direct numerical inspection shows that in certain $m = m_2$, H_1^0, and H_2^0, the parameter range complex λ indeed arise. A similar purely kinetic mode of the instability may occur in viscosity stratified Couette, Poiseuille, and film flows (Li, 1969; Kliakhandler and Sivashinsky, 1995, 1997) and is also known to emerge in magneto-hydrodynamics or in anisotropic three-dimensional flows of homogeneous fluids (so-called α-effect) (Zeldovich *et al.*, 1983; Moiseev *et al.*, 1983; Frisch *et al.*, 1987; Khomenko *et al.*, 1991). Further, this kind of instability is termed the α-effect.

The marginal stability curves for the shortened equation (5.127) are presented in Fig. 5.33. The corresponding eigenvalue problem (5.127) yields either unstable ($\lambda_r > 0$) or neutrally stable ($\lambda_r = 0$) modes. For a fixed value of the viscosity ratio m, the instability region has a plume-like form. As it is readily seen, the instability sets in when the system geometry is not symmetric. The numerical checks suggest that the instability is likely to emerge only in the case if the viscosity of the intermediate layer is lower than viscosities of the adjacent layers.

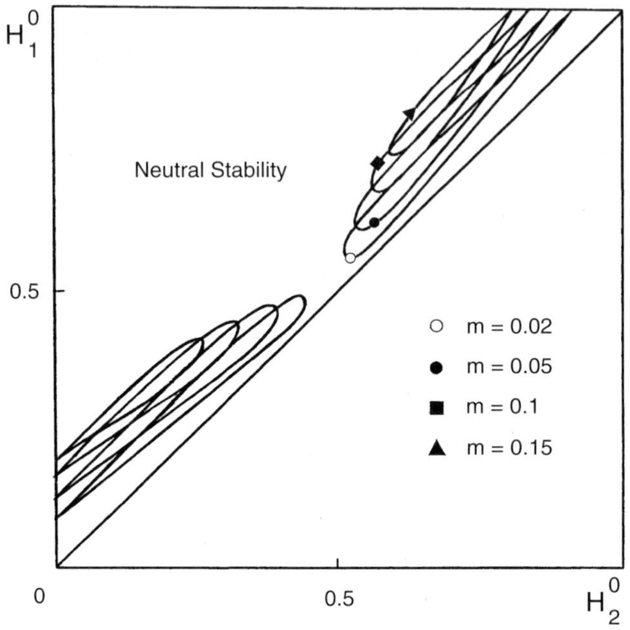

FIGURE 5.33. Marginal stability curves for the α-effect in the (H_1^0, H_2^0) parameter plane.

A similar feature of the α-effect was also identified in three-layer Poiseuille flow (Kliakhandler and Sivashinsky, 1995).

In the absence of the α-effect, the eigenvalues $c_j^{(\alpha)}$ of matrix α_{ij} are real and λ is pure imaginary. This fact, however, does not imply that the same statement is true for the complete system (5.126). The long-wavelength expansion of λ may be written as

$$\lambda_j \simeq -ikc_j^{(\alpha)} + k^4 \Lambda_j, \; k \ll 1, \; j = 1, 2, \tag{5.129}$$

where the second term stems from the fourth-derivative surface-tension term. Here,

$$\Lambda_j = -\frac{\alpha_{12}\chi_{21} - \alpha_{22}\chi_{11} + \alpha_{21}\chi_{12} - \alpha_{11}\chi_{22} + c_j^{(\alpha)}(\chi_{11} + \chi_{22})}{\alpha_{11} + \alpha_{22} - 2c_j^{(\alpha)}}. \tag{5.130}$$

Numerical inspection shows that in wide parameter ranges, some Λ_j become positive, implying the long-wavelength instability of the parallel shear flow. In this case, equation(5.121) yields a long-wavelength instability of the "surface-tension-induced" type:

$$\omega_r \sim k^4, \; k \ll 1, \; \text{and } \lambda_r \sim -k^4, \; k \gg 1. \tag{5.131}$$

The marginal stability curves for the surface-tension-induced instability in the (H_2^0, H_1^0) parameter plane are presented in Fig. 5.34. The domains of the

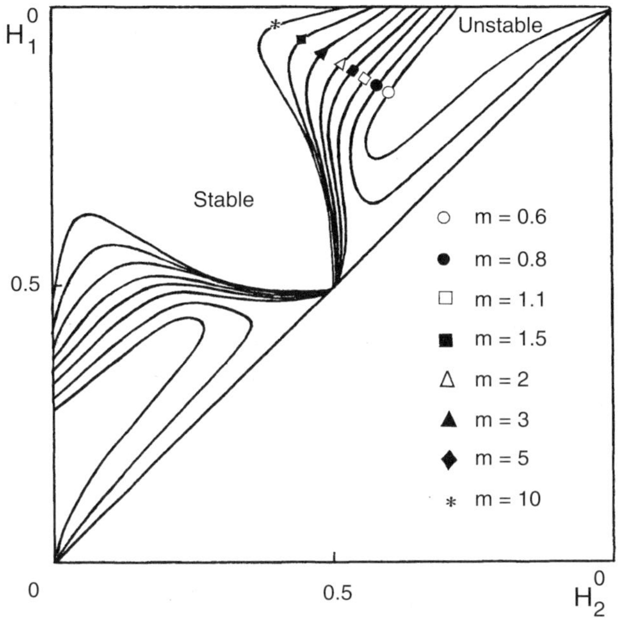

FIGURE 5.34. Marginal stability curves for the surface-tension-induced instability in the (H_1^0, H_2^0) parameter plane.

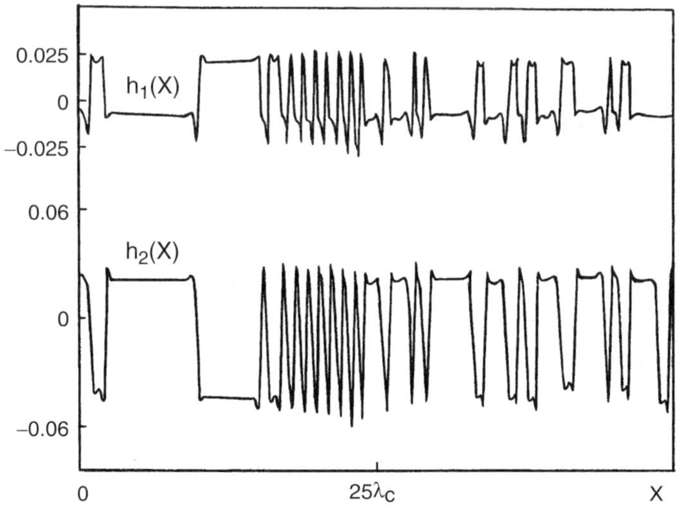

FIGURE 5.35. Typical solutions of the system (5.134) for the α-effect-type instability at $H_1^0 = 0.4$, $H_2^0 = 0.25$, $m = 0.15$, $\sigma = 100$.

instability adjoin to the corners and to the diagonal $H_1^0 = H_2^0$; along the marginal stability lines $H_1^0 = 1$, $H_2^0 = 0$, and $H_1^0 = H_2^0$, the system becomes to be two-layer or homogeneous where the instability disappears.

A conclusion from the above-mentioned statement is the following. The surface-tension-dissipation term, χ_{ij}, while suppressing the short-wavelength disturbances, may play a destabilizing role in the long-wavelength range. The same situation takes place in the case of three-layer Poiseuille flows (Kliakhandler and Sivashinsky, 1995). The mathematical feasibility of such a counterintuitive outcome for the second-order viscous-like dissipative terms has been ascertained by Majda and Pego (1985).

5.5.4 Weakly Nonlinear Model

Further, we derive weakly nonlinear equations describing the interfacial evolution. We consider the interfacial dynamics of the system, both for the α-effect and for the surface-tension-driven instability, near the instabilities thresholds. As was found in the previous subsection, the resulting interfacial dynamics is expected to be longwavelength (due to the large surface tension) and smallamplitude (due to the weak instability). The nonlinear terms restraining the growth of unstable modes are expected to be quadratic.

The dynamics is considered up to moderately large times, and for periodic boundary conditions on the channel ends,

$$h_i^{(n)}(-L) = h^{(n)}(L), \quad n = 0, 1, 2, 3. \tag{5.132}$$

Let us consider the term $C(t)\partial S_i/\partial x$. Expansion of S_i provides terms of the first order (recall that $C(t) = 0$ in the zeroth order, and therefore quadratic terms in the expansion for S_i will be canceled). However, for the functions h_i, periodic on the channel edges, the linear expansion of $C(t)$ yields

$$C(t) = 0, \tag{5.133}$$

similar to the same result for linear analysis.

Approximating, therefore, Q_i by their Taylor expansions around H_1^0 and H_2^0 up to the quadratic terms, one obtains

$$\frac{\partial h_i}{\partial t} + \alpha_{ij}\frac{\partial h_j}{\partial x} + \beta_{ijk}\frac{\partial h_j h_k}{\partial x} + \chi_{ij}\frac{\partial^4 h_j}{\partial x^4} = 0, \tag{5.134}$$

$$\beta_{ijk} = \frac{1}{2}\frac{\partial^2 Q_i}{\partial H_j \partial H_k}(H_1^0, H_2^0). \tag{5.135}$$

The constant tensor β_{ijk} is determined by the basic distribution of viscosities μ_i and the interfaces positions H_i^0. Figure 5.35 shows the pertinent wavy structure developing in the associated nonlinear system (5.134) for the α-effect type instability. The well-settled interfaces assume a form of irregular intermittent wavy

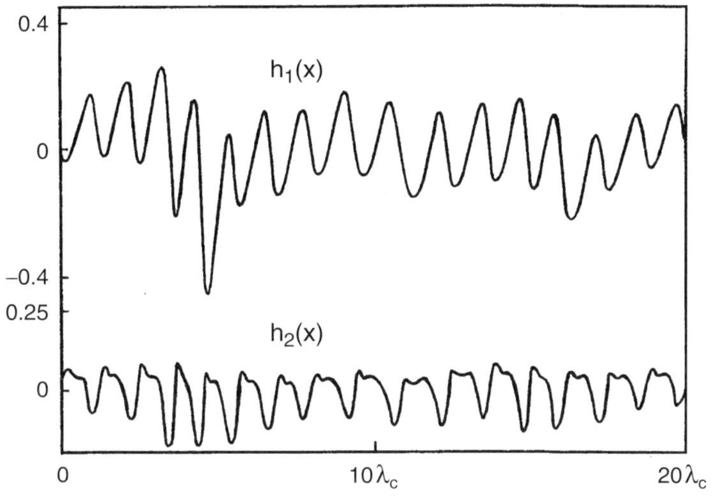

FIGURE 5.36. Typical solutions of the system (5.134) for the surface-tension-induced instability (5.131) at $H_1^0 = 0.4$, $H_2^0 = 0.2$, $m_2 = 4$, $\sigma = 100$.

structure whose typical scale considerably exceeds $2\pi / k_c$ given by the linear stability analysis. Here, k_c is the wave number of the most growing mode. A similar enlargement of typical wavescale was observed in interfacial dynamics of plane three-layer Poiseuille flow (Kliakhandler and Sivashinsky, 1995).

Some typical plots from the numerical simulations of the nonlinear systems (5.134) for the surface-tension-driven instability are presented in Fig. 5.36. The interfaces settle to a nearly periodic, varicose-like structures.

Restrictions of the presented approach relate to a small-amplitude approximation near the stability threshold and to the consideration over moderately large times when mean interfaces values are not changed markedly. At extremely large times, the thermocapillary flow in a three-layer system may essentially change the mean interfaces locus, so that one of the interfaces may just disappear: The liquid in the corresponding layer will be replaced by the liquid from another layer. Large-amplitude waves may provide the same result; the consideration of these complicated issues is remained for further investigations.

6

Flows in Ultra-Thin Films

In the present chapter, we will consider a specific class of interfacial phenomena related to the dynamics of very thin films, with the thickness less than 100 nm. Such kind of films has numerous technological applications (coating, flotation, biological membranes, adhesives etc.). The instabilities in thin films are of potential use in the formation of regular nanostructures and ordered porous membranes, in soft-lithographic techniques and in other areas of nanotechnology.

The description of the ultra-thin film dynamics needs a special approach. On one side, the small-scale flows are relatively simple because of the small values of the Reynolds number. On another side, as we noted in Sec. 2.4, the dynamics of very thin films and film rupture cannot be described without taking into account some additional physical phenomena. In the case where the film thickness is less than 100 nm, its dynamics is strongly influenced by effective molecular interactions between the film surface and the substrate. Typically, a film rupture due to molecular interactions leads to the formation of a network of liquid rims that may subsequently decay into small drops (Reiter, 1992).

Below, we will present basic approaches used for the description of the ultra-thin film dynamics.

6.1 Lubrication Approximation

In the case of *film flows,* when the fluid system is thin in a certain ('transverse') direction and extended in other ('longitudinal') directions, the nonlinear models governing three-dimensional flows with a deformable interface can be drastically simplified. A film flow is strongly affected by interfacial phenomena and *slaved* to the interface deformation. The idea of the approach has been explained in Sec. 2.4 (see Eq. (2.48)). By means of long-wave asymptotic expansions, one derives an evolution equation for the interface shape, $z = h(x, y, t)$, which due to the conservation of the fluid volume has the form:

$$\frac{\partial h}{\partial t} + \nabla \cdot \mathbf{q} = 0. \qquad (6.1)$$

The flow in a thin film is locally a superposition of a Poiseuille flow generated by the longitudinal pressure gradient ∇p (the pressure does not depend on the transverse coordinate z in the leading order), and a Couette flow created by the tangential stress $\tau(x, y, t)$ at the interface. Hence,

$$\mathbf{q} = -\frac{1}{3\eta} h^3 \nabla p + \frac{1}{2\eta} h^2 \tau. \tag{6.2}$$

In the formulas (6.1) and (6.2), ∇ means gradient with respect to longitudinal coordinates (x, y). Different physical factors (capillarity, gravity, thermocapillary effect etc.) can contribute into p and τ. For instance, for a liquid film on a solid horizontal plane, the pressure is the sum of the hydrostatic and capillary pressure:

$$p = \rho g h - \sigma_0 \nabla^2 h. \tag{6.3}$$

If the thermocapillary effect is taken into account,

$$\tau = \nabla \sigma = -\alpha \nabla T_s, \tag{6.4}$$

where T_s is the interfacial temperature. Its dependence on the local thickness h is obtained by solving the heat transfer equation with appropriate boundary condition in the long-wave limit. For instance, for a film with a fixed temperature T_0 on the bottom and the condition (1.13) on the free surface, one obtains the relation:

$$T_s(h) = T_g + \frac{\kappa(T_0 - T_g)}{\kappa + K h}. \tag{6.5}$$

Using dimensionless variables defined above, one obtains the following closed nonlinear equation for the evolution of the film thickness:

$$\frac{\partial h}{\partial t} + \nabla \cdot \left\{ \left[\frac{1}{3} W h^3 \nabla^2 - \frac{1}{3} G a h^3 + \frac{M B i h^2}{2P(1 + B i h)^2} \right] \nabla h \right\} = 0. \tag{6.6}$$

Various applications of this long-wave approximation described above (which is called also *lubrication approximation*) to different physical problems are described by Oron *et al.*, 1997.

6.2 Intermolecular Forces

As mentioned above, in the case of very thin (but still macroscopic) films, when the film thickness is less than about 100 nm, a new physical phenomenon has to be incorporated into the model. It is necessary to take into account the *long-range intermolecular forces* acting between molecules of the liquid and substrate (Israelashvili, 1985; Derjaguin *et al.*, 1987). It is essential that these forces act on distances *large* relative to interatomic distances. Hence, despite their microscopic (quantum) origin, they can be incorporated into a macroscopic theory (Lifshitz, 1956; Dzyaloshinskii and Pitaevskii, 1959; Dzyaloshinskii *et al.*, 1960). It can be shown that in the case where a pair potential between molecules is $U(r) \sim 1/r^n$, the effective energy of interaction between two planar surfaces on a distance h is $f(h) \sim 1/h^{n-4}$. In the framework of the continuum approach, the intermolecular forces manifest themselves as 'surface forces' or *'disjoining pressure'* $\Pi(h) = df(h)/dh$ (see, e.g., Williams and Davis, 1982; Sharma and Ruckenstein, 1986), which can be considered as a certain external normal stress imposed on the free surface (that is, p is replaced by $p + \Pi$ in the boundary condition (1.10)). The sign of the disjoining pressure can be either positive or negative.

If the film is formed from an apolar fluid, the only relevant kind of long-range intermolecular interaction is the van-der-Waals interaction $U(r) \sim 1/r^6$. In that case, the disjoining pressure can be taken as

$$\Pi(h) = A/6\pi h^3, \tag{6.7}$$

where A is the dimensionless *Hamaker constant* (Hamaker, 1937; Israelashvili, 1985; de Gennes *et al.*, 2004). Note that the Hamaker constant can be positive or negative. The law (6.7) is supported by many experimental data performed at $h < 30$ nm (see Derjaguin *et al.*, 1987). For larger distances between molecules, another law of the intermolecular interaction is expected due to the effect of electromagnetic retardation, $U(r) \sim 1/r^7$. That leads to the prediction

$$\Pi(h) = \frac{B}{10\pi h^4}, \tag{6.8}$$

which is supported by experiments done in the interval 40 nm$< h <$80 nm (Derjaguin *et al.*, 1987).

In the case of an electrolyte solution, the disjoining pressure includes an electrostatic component caused by a double electric layer on the liquid-gas surface characterized by the Debye screening length $1/\kappa$. In the simplest case, when the surface charge of the surface equals zero, the electrostatic contribution to the disjoining pressure in the region $\kappa h \ll 1$ is given by the Langmuir formula

$$\Pi_e(h) = (\pi\epsilon_0/8h^2)(kT/ez)^2, \tag{6.9}$$

where ϵ_0 is the dielectric constant, k is the Boltzmann constant, T is the temperature, e is the electron charge, and z is the ion valency. For $\kappa h \sim 1$, an approximate formula

$$\Pi_e(h) = Ce^{-\kappa h} \tag{6.10}$$

can be used.

Another possible component of the disjoining pressure, caused by a structural modification of polar liquids in thin films, has the form

$$\Pi_s(h) = K \exp(-h/\lambda), \tag{6.11}$$

where in the case of water the force magnitude K is of the order of 10^7 N/m^2, and the length λ varies from 2.3 to 3.3 nm (for more detail, see Churaev, 2003).

On very short (atomic) distances, the *steric repulsion* becomes relevant, which is due to overlapping electron shells and varies as $U \sim 1/r^{12}$, which yields $f(h) \sim 1/h^8$, $\Pi(h) \sim 1/h^9$.

In the literature, one can find some combinations of the expressions presented above. For instance, a combination of (6.7) and (6.10) leads to the (dimensionless) expression

$$\Pi(h) = \frac{b}{h^3} - e^{-h} \tag{6.12}$$

used by Sharma (1993) and Thiele *et al.* (2001a).

Disjoining pressure dominates at small film thicknesses and therefore governs the stability and wettability of ultra-thin films. Replacing p by $p + \Pi$ in the boundary condition for normal stresses, one obtains the relation,

$$p = \rho g h - \sigma_0 \nabla^2 h - \Pi(h), \tag{6.13}$$

to be substituted into (6.1) and (6.2). Typically, the gravity term can be neglected for an ultra-thin film.

Another example of the situation where the disjoining pressure has to be taken into account is the motion near the *contact line*. In the latter case, one has to take into account the dependence of the disjoining pressure not only on the film thickness h but also on the film slope, h_x (Hocking, 1993; Wu and Wong, 2004).

The latter problem demands one more modification of the standard model of the viscous fluid: the no-slip condition $\mathbf{v} = 0$ is violated near the contact line (Oron *et al.*, 1997; Shikhmurzaev, 1997). Indeed, the standard no-slip condition leads to singularities (Dussan V and Davis, 1974; Dussan V, 1979). Note that a real molecular slippage on the solid surface is justified by molecular-dynamic simulations (Koplik *et al.*, 1989; Thompson and Robbins, 1989; Barrat and Bocquet, 1999). The violation of the slip condition changes the velocity profile of the film flow and hence the relation (6.2) between $\mathbf{Q}$, ∇p and τ. For instance, if the slippage condition for the tangential velocity component $u = v_x$ is

$$u - \beta \frac{\partial u}{\partial z} = 0,$$

then

$$\mathbf{q} = -\frac{1}{\eta}\left(\frac{h^3}{3} + \beta h^2\right)\nabla p + \frac{1}{\eta}\left(\frac{h^2}{2} + \beta h\right)\tau \tag{6.14}$$

(see Oron *et al.*, 1997).

6.3 Generalized Cahn-Hilliard Equation

6.3.1 Model Formulation

In the framework of the lubrication approximation, the basic equation which describes the dynamics of a heated ultra-thin film on a horizontal solid substrate can be written as

$$\frac{\partial h}{\partial t} + \nabla \cdot \left\{ \frac{h^3}{3}[W\nabla^2\nabla h - G\nabla h + \nabla\Pi(h)] + \frac{MBih^2}{(1+Bih)^2}\nabla h \right\} = 0, \quad (6.15)$$

where $\Pi(h)$ is a (dimensionless) disjoining pressure. It is important that Eq. (6.15) can be presented as a *generalized Cahn-Hilliard equation*

$$\frac{\partial h}{\partial t} + \nabla \cdot \left\{ Q(h)\nabla \left[W\nabla^2 h - \frac{df(h)}{dh} \right] \right\} = 0, \quad (6.16)$$

where

$$Q(h) = \frac{h^3}{3},$$

$$\frac{df(h)}{dh} = Gh - \frac{3}{2}MBi\left[\ln\left(\frac{h}{1+Bih}\right) + \frac{1}{1+Bih}\right] - \Pi(h).$$

Hence, the evolution equation of a film in the horisontal substrate can be written in a *variational form*

$$\frac{\partial h}{\partial t} = \nabla \left[Q(h)\nabla \frac{\delta F[h]}{\delta h} \right], \quad (6.17)$$

where

$$F[h] = \int d\mathbf{r} \left[\frac{W}{2}(\nabla h)^2 + f(h) \right] \quad (6.18)$$

is the *Lyapunov functional* of the problem. Because

$$\frac{dF}{dt} = \int d\mathbf{r} \frac{\partial h}{\partial t}\frac{\delta F[h]}{\delta h} = -\int Q(h)\left[\nabla \frac{\delta F[h]}{\delta h}\right]^2 d\mathbf{r} \le 0,$$

the time evolution of the system leads to a quiescent state corresponding to a minimum of the functional $F[h]$, while persistent oscillations and chaos are impossible.

6.3.2 Film Rupture

Any value $h = h_0$ =const is a steady solution of (6.16), with the Lyapunov functional density ("free energy") equal to $f(h_0)$. However, if the Lyapunov functional can be diminished by creation of a non-uniform distribution of h (with the mean value $\langle h \rangle = h_0$ kept), a flat film of thickness h_0 is *unstable*. Linearization of Eq. (6.16) around h_0,

$$h(\mathbf{r}) = h_0 + \tilde{h} \exp(i\mathbf{k} \cdot \mathbf{r} + \sigma t),$$

leads to the dispersion relation

$$\sigma = Q(h_0) \left[-\frac{d^2 f}{dh^2}(h_0)k - Wk^4 \right]. \tag{6.19}$$

On the initial stage, the disturbances with the wave numbers in the interval $0 < k < k_c$ grow exponentially (with the preferred wave number $k_m = k_c/\sqrt{2}$, corresponding to the maximum growth rate).

The nonlinear evolution of unstable ultra-thin film depends on the behavior of the free energy $f(h)$ at small h. If $f(h)$ is not bounded from below (e.g., for a destabilizing disjoining pressure $\Pi(h) \sim A/6\pi h^3$, $A < 0$), a true rupture of the film takes place. This phenomenon has been studied in the framework of a linear theory (Ruckenstein and Jain, 1974; Gumerman and Homsy, 1975), as well as by means of some nonlinear approaches (Williams and Davis, 1982; Prévost and Galles, 1986; Sharma and Ruckenstein, 1986b; Burelbach *et al.*, 1988; Erneux and Davis, 1993). The film rupture leads to the appearance of a three-phase *contact line*.

Otherwise, if the disjoining pressure is stabilizing, the development of instability is similar to the *phase separation* when the unstable "phase" with the intermediate value of h is destroyed, and the system is separated into two stable "phases", one with small h (thin precursor film) and another one with large h (thick macroscopic film). The phases can form holes ("bubbles") and droplets, depending of parameters (Bestehorn *et al.*, 2003). Similarly to the usual phase transition, the film instability can develop either due to the growth of small disturbances, as explained above (*"spinodal decomposition"*) or through *nucleation* of critical depressions (Sharma, 2003; Thiele, 2003). Another similarity is observed in the coarsening process. Like in the Ostwald ripening in phase transitions, big droplets (holes) grows at the expense of small ones. Bestehorn *et al.* (2003) performed simulations based on equation (6.16) with an appropriate expression for $f(h)$ corresponding to the case of a repelling (stabilizing) disjoining pressure. Like in the case of a standard Cahn-Hilliard equation with a constant mobility, the instability is saturable and leads to the formation of drops and holes, depending on the value of α. The coarsening of drops or holes is characterized by the scaling law $\langle k \rangle \sim t^{-\beta}$, where $\langle k \rangle$ is the mean modulus of the structure's wave vector. Let us emphasize that the exponent $\beta \approx 0.21$ turns out to be independent of any parameters of the problem and different from the vale $1/3$ known for the standard Cahn-Hilliard equation.

The nonlinear evolution of thin film has been studied in the framework of Eq. (6.16) for different forms of $f(h)$ also by Sharma and Khanna (1998), Bestehorn and Neuffer (2001), and Sharma (2003).

6.3.3 Film on an Inclined Plane

For a film flowing on an inclined plane with inclination angle α, Eq. (6.16) is rewritten as

$$\frac{\partial h}{\partial t} + \nabla\cdot \left\{ \frac{h^3}{3}[W\nabla^2\nabla h - G\cos\alpha\nabla h + \nabla\Pi(h) + G\sin\alpha\mathbf{e}_x] + \frac{MBih^2}{(1+Bih)^2}\nabla h \right\} = 0, \qquad (6.20)$$

where $\mathbf{e}_x$ is the unit vector corresponding to the projection of the gravity acceleration vector on the inclined plane. It can be written as a *convective Cahn-Hilliard equation*

$$\frac{\partial h}{\partial t} + \nabla\cdot \left\{ Q(h)\nabla\left[W\nabla^2 h - \frac{df(h)}{dh}\right] + \chi(h)\mathbf{e}_x \right\} = 0 \qquad (6.21)$$

with $\chi(h) = Gh^3\sin\alpha/3$. Let us emphasize that the convective Cahn-Hilliard equation is not a variational system, and its dynamics can be complex (see, e.g., Golovin *et al.*, 2001).

Simulations of Eq. (6.21) have been used for studying decomposition of unstable films (Bestehorn and Neuffer, 2001; Bestehorn *et al.*, 2003), as well as for the exploration of dynamics of drops sliding on an inclined plane and instabilities of a contact line (see Sec. 6.5.3)

6.3.4 Two-Layer Films

Recently, a series of experiments has been carried out on the stability of systems which consist of two immiscible thin films on a solid substrate (Higgins and Jones, 2000; Lin *et al.*, 2001, 2002a, 2002b). A theoretical description of such systems has been developed by Pototsky *et al.* (2004, 2005), Bandyopadhyay *et al.* (2005), and Fisher and Golovin (2005). The system involves two deformable interfaces $z = h_1(x, y, t)$ (liquid-gas) and $z = h_2(x, y, t)$ (liquid-liquid). One has to take into account *two disjoining pressures* $\Pi_1(h_1, h_2)$ and $\Pi_2(h_1, h_2)$, applied at each interface:

$$\text{at } z = h_1,\ p_1 - p_0 = -\sigma_1\nabla^2 h_1 + \Pi_1(h_1, h_2);$$

$$\text{at } z = h_2,\ p_2 - p_1 = -\sigma_2\nabla^2 h_2 + \Pi_2(h_1, h_2),$$

where σ_1 and σ_2 stand for interfacial tensions at the corresponding interfaces, and p_0 is the atmospheric pressure.

The total potential energy of the van der Waals interactions $\Psi = \Psi_1 + \Psi_2 + \Psi_{12}$ includes the energy Ψ_1 of the substrate interaction with layer 1 across layer 2, the energy Ψ_2 of the gas phase interaction with layer 2 across layer 1, and the energy Ψ_{12} of the interaction between the gas phase and the substrate across two layers (Israelachvili, 1985; Fisher and Golovin, 2005),

$$\Psi(h_1, h_2) = -\frac{A_{sg} - A_{s1} - A_{g2}}{12\pi h_1^2} - \frac{A_{s1}}{12\pi h_2^2} - \frac{A_{g2}}{12\pi (h_1 - h_2)^2}, \tag{6.22}$$

where A_{sg}, A_{s1} and A_{g2} are Hamaker constants characterizing the interactions between the solid substrate and the gas across the two layers, between the solid substrate and liquid 1 across liquid 2, and between the gas phase and liquid 2 across liquid 1, correspondingly. The disjoining pressures are computed as

$$\Pi_j(h_1, h_2) = \frac{\partial \Psi}{\partial h_j}, \; j = 1, 2. \tag{6.23}$$

Finally, the dimensionless equations governing the evolution of layers thicknesses can be written as (Fisher and Golovin, 2005)

$$\frac{\partial h_1}{\partial t} + \nabla \cdot (f_{11} \nabla p_1 + f_{12} \nabla p_2) = 0, \tag{6.24}$$

$$\frac{\partial h_2}{\partial t} + \nabla \cdot (f_{21} \nabla p_1 + f_{22} \nabla p_2) = 0, \tag{6.25}$$

where the mobility functions are:

$$f_{11} = (h_1 - h_2)\left[h_2^2\left(\frac{1}{2} - \frac{1}{3\eta}\right) + h_1 h_2 \left(-1 + \frac{2}{3\eta}\right) - \frac{1}{3\eta} h_1^2\right],$$

$$f_{12} = \frac{1}{6} h_2^3 - \frac{1}{2} h_2^2 h_1, \; f_{21} = -\frac{1}{2} h_2^2 (h_1 - h_2), \; f_{22} = -\frac{1}{3} h_2^3,$$

$\eta = \eta_1/\eta_2$, the pressures are

$$p_1 = -\sigma \nabla^2 h_1 + w_1(h_1, h_2), \tag{6.26}$$

$$p_2 = -\sigma \nabla^2 h_1 - \nabla^2 h_2 + w_2(h_1, h_2), \; \sigma = \sigma_1/\sigma_2, \tag{6.27}$$

and the potentials are

$$w_1 = \frac{a_0 - a_1 - a_2}{h_1^3} + \frac{a_2}{(h_1 - h_2)^3}, \tag{6.28}$$

$$w_2 = \frac{a_0 - a_1 - a_2}{h_1^3} + \frac{a_1}{h_2^3}. \tag{6.29}$$

Here $a_0 = sign(A_{sg})$, $a_1 = A_{s1}/|A_{sg}|$, $a_2 = A_{g2}/|A_{sg}|$. It is remarkable (Pototsky *et al.*, 2004, 2005) that the system has a *variational structure*,

$$\frac{\partial \mathbf{h}}{\partial t} = \nabla \cdot \left(\mathbf{Q}(\mathbf{h}) \nabla \frac{\delta F}{\delta \mathbf{h}}\right), \tag{6.30}$$

where $\mathbf{h} = (h_1, h_2)$, $\mathbf{Q}(\mathbf{h})$ is a symmetric mobility matrix, and F is the Lyapunov functional.

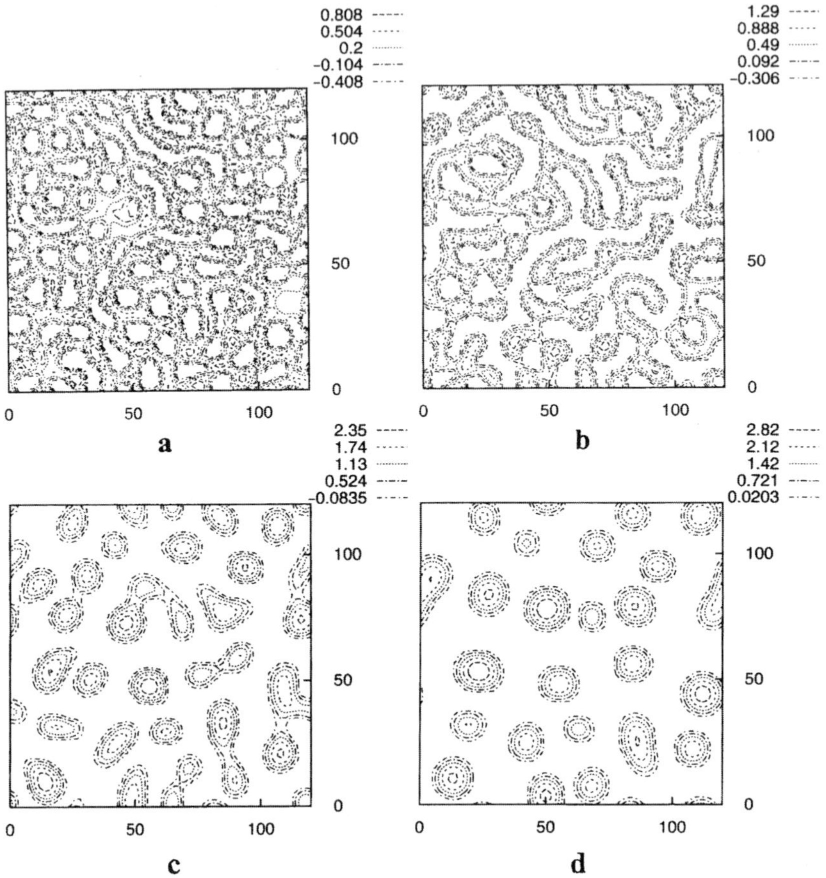

FIGURE 6.1. Isolines of h_1 for the two-dimensional numerical solution of the system (6.24), (6.25) at different moments of time for $a_0 = 1$, $a_1 = -0.4$, $a_2 = -0.1$, $\eta = 0.833$, $\sigma = 0.8$: (a) $t = 240$; (b) $t = 330$; (c) $t = 500$; (d) $t = 800$. The mean values $\langle h_2 \rangle = 1$, $\langle h_1 \rangle = 1.2$.

The linear stability theory of a two-layer film reveals bending modes (in-phase deformations of interfaces), and squeezing modes (out-of-phase deformations of interfaces). Unexpectedly, gravity effects can be relevant for the dynamics of a two-liquid system (Fisher and Golovin, 2005). Depending on parameters, the time evolution may lead either to a blow-up film rupture ("swiss-cheese" dewetting pattern), or to the formation of coarsening drops (see Fig. 6.1; Simanovskii and Nepomnyashchy, unpublished).

Merkt *et al.* (2005) considered a two-layer system sandwitched between two rigid plates, with one liquid-liquid interface. It turns out that the dynamics of such systems cannot be described in a local way by the interface profile $h(\mathbf{r})$, but needs for its full description also the stream function $f(\mathbf{r})$ of the horizontal mean flow. The variational structure of the Cahn-Hilliard-like evolution equation for h is broken by the coupling to the elliptic (Poisson-like) equation governing f.

6.3.5 *Evaporating Films*

Evaporation of Pure Liquids

Evaporation-induced instabilities in pure liquids were investigated in the framework of the one-layer approach during last decades (Palmer, 1976; Burelbach *et al.*, 1988; Oron *et al.*, 1997; Oron, 2000; Colinet *et al.*, 2003). The physical effects, which are of the greatest importance for the development of the evaporative convection, are the thermocapillary effect, the interface deformation, the vapor recoil, and the intermolecular forces.

The applicability of the one-layer approach to the evaporation problem is however rather restricted. The two-layer analysis shows that the processes in the gas phase are crucial for the development of film instabilities. For instance, it is necessary to distinguish between the evaporation of liquid into its own vapor and into an inert gas (e.g., air). In the former case (Neogi and Berryman, 1982; Margerit *et al.*, 2003) the interfacial fluctuations of temperature are limited by the Clausius-Clapeyron law, though they can be produced by the dependence of the boiling point on the film thickness due to the action of intermolecular forces. In the latter case (Moussy *et al.*, 2004; Dondlinger *et al.*, 2005; Haut and Colinet, 2005) the temperature disturbances are linked only to the partial pressure of the volatile component. The reduction to a one-layer problem by means of appropriate boundary conditions is possible in some cases, but these effective boundary conditions can be derived only through the full two-layer analysis. Also, a "1.5-sided model" which consists of the full liquid layer equations and the diffusion equation for the vapor in the gas has been suggested (Margerit *et al.*, 2005). Experiments on the evaporative convection are still rather rare (Berg *et al.*, 1966; Colinet *et al.*, 2003; Mancini and Maza, 2004).

The development of instabilities in ultra-thin evaporating (and condensating) films is influenced by intermolecular forces (Oron and Bankoff, 1999, 2001). Experiments by Elbaum and Lipson (1994, 1995) have revealed a complex behavior in the course of a thin film evaporation. One observes the formation and destruction of ridges, and coexistence of droplets with different sizes (Leizerson *et al.*, 2003). These phenomena are described by means of a potential combining van-der-Waals and polar interactions, and a similar to the phenomena in first-order phase transitions (Padmakar *et al.*, 1999; Kargupta *et al.*, 2001; Pismen, 2004). Specifically, the film rupture can be caused by heterogeneous nucleation or by surface instability (Thiele *et al.*, 1998). The description of evaporation of a finite droplet is connected with the different aspects of the contact line dynamics, including stability of the contact line (Lyushnin *et al.*, 2002; Poulard *et al.*, 2003) and patterning of small particles (Deegan, 2000; Nguyen and Stebe, 2002). The evaporation near the contact line influences the spreading of droplets (Redon *et al.*, 1992; Dussaud and Troian, 1998). Stability of an evaporating film falling down an inclined plane has been studied by Miladinova and Lebon (2005).

Evaporation of Solutions

Evaporation of solutions has such important applications as the evaporation-induced self-assembly (Brinker *et al.*, 1999), pattern formation in drying polymer solutions (Weh, 1999, 2001; Weh and Venthur, 2004; Bormashenko *et al.*, 2005a, 2005b), phase-separation induced patterns (Yamamura *et al.*, 2002), and phenomena leading to formation of micro-porous membranes (Widawski *et al.*, 1994; Pitois and François, 1999a, 1999b; Peng *et al.*, 2004). However, the diverse physical phenomena which manifest themselves in the course of the solutions evaporation, are still hardly explored.

In the series of papers by Danov *et al.* (1998a, 1998b, 1998c) and Paunov *et al.* (1998), the linear and nonlinear stability of evaporating one-layer and two-layer films in the presence of a dissolved surfactant have been studied in the long-wave approximation. The Marangoni effects due to thermal and concentration gradients, surfactant adsorption, interfacial viscous fricton, and van-der-Waals forces have been taken into account. It should be noted that due to the slowness of the diffusion process, the concentration gradient in the evaporating liquid can be essentially non-uniform. This circumstance has been taken into account in the linear analysis of the solutocapillary instability in a layer by Kaminskii and Dil'man (2003), as well as in several papers on the instabilities in evaporating droplets (Aharon and Shaw, 1996; Ha and Lai, 2002). The problem of the stability of an evaporating binary-liquid film was discussed in the context of polymer films by de Gennes (2001) and Nepomnyashchy *et al.* (2005).

Dewetting in drying two-component liquid films with the viscosity dependent on the concentration and with disjoining/conjoining pressure taken into account has been studied by Schwartz *et al.* (2001). Diverse complex patterns, which are developed by the evaporation of a solvent drop on a soluble layer of polymer, have been investigated by Gonuguntla and Sharma (2004).

6.4 Diffuse-Interface Models

There exists an alternative approach for the description of systems with an interface (for a review, see Anderson *et al.*, 1998). The sharp interface between fluids is replaced by a diffuse one. The different fluids are marked by different equilibrium values of a new variable usually called the *order parameter* (this term is related to the phase transitions which have been a primary target for the application of the diffuse-interface approach). For instance, the order parameter $\phi(\mathbf{r})$ is assumed to be equal to $+1$ in the bulk of fluid 1 (e.g., in the gas phase), and to -1 in the bulk of the fluid 2 (e.g., in the liquid phase). The model is based on the *free-energy functional*

$$F = \int_V L(\phi, \nabla\phi) dV, \tag{6.31}$$

$$L = f(\phi) + \frac{K(T)}{2}(\nabla\phi)^2, \tag{6.32}$$

where $f(\phi)$, free-energy density of a homogeneous fluid, is a continuous function with the minima at $\phi = \pm 1$, e.g.,

$$f(\phi) = C\frac{(\phi^2 - 1)^2}{4}, \quad C > 0. \tag{6.33}$$

The minimization of the functional (6.31) determines the equilibrium (tanh-like) shape $\phi_0(z)$ of the stationary phase-field distribution across the (plane) diffuse interface, and the interfacial energy (which coincides with the interfacial tension),

$$\sigma = \int_{-\infty}^{\infty} K(T) \left(\frac{d\phi_0}{dz} \right)^2 dz \tag{6.34}$$

(Rowlinson and Widom, 1982; Jasnow and Viñals, 1996).

It is assumed that order parameter ϕ is a density of a *conserved quantity*, therefore the time evolution of the order parameter is governed by the equation

$$\frac{d\phi}{dt} = -\nabla \cdot \mathbf{J}, \tag{6.35}$$

where

$$\frac{d}{dt} = \frac{\partial}{\partial t} + \mathbf{v} \cdot \nabla$$

is the Lagrangian derivative, and the order parameter flux is

$$\mathbf{J} = -m\nabla \frac{\delta F}{\delta \phi}, \tag{6.36}$$

where m is the mobility (a positive constant), and $\delta F/\delta\phi$ is the functional derivative of (6.31),

$$\frac{\delta F}{\delta \phi} = \frac{\partial f}{\partial \phi} - \nabla \cdot (K(T)\nabla\phi). \tag{6.37}$$

Substituting (6.36) and (6.37) into equation (6.35), we obtain a modification of the Cahn-Hilliard equation, similar to that used for the description of phase transitions with a conserved order parameter.

In the framework of the diffuse interface approach, the Navier-Stokes equation is valid in the whole space, but in the region of the diffuse interface, where $\nabla\phi$ does not vanish, one has include the contribution of the order parameter inhomogeneity into the stress tensor (Anderson *et al.*, 2000):

$$\rho \frac{d\mathbf{v}}{dt} = \nabla(-p + L) - \nabla \cdot (K\nabla\phi \otimes \nabla\phi) + \nabla \cdot (\eta\nabla\mathbf{v}). \tag{6.38}$$

The physical meaning of the term which contains $\nabla\phi$ is the capillary force.

This approach, originally suggested by Hohenberg and Halperin (1977) for modelling the phenomena near the critical point, was modified by Antanovskii (1995) in order to compute some simple two-phase thermocapillary flows. Jasnow and Viñals (1996) used a diffuse-interface model for the simulation of the thermocapillary drift of drops and for the investigation of the spinodal decomposition of a binary fluid subject to a temperature gradient. Verschueren *et al.* (2001) applied a diffuse-interface model for the computation of a two-phase thermocapillary motion in a Hele-Shaw cell. They simulated the creation of a droplet due to the deformational Marangoni instability of a plane interface. Borcia and Bestehorn (2003) used the phase-field approach for the description of the Marangoni instability in a two-layer liquid-gas system.

Another version of the diffuse-interface approach, which is valid in the case of a liquid-gas interface, is based on using density ρ as the order parameter in the free-energy functional corresponding to the van der Waals equation of state (Anderson *et al.*, 1998). The starting point is the free energy functional

$$F = \int_V \left[\rho f(\rho, T) + \frac{1}{2} K |\nabla \rho|^2 \right] dV, \tag{6.39}$$

which gives the (reversible) stress-tensor

$$\mathbf{T} = \left[-p + K\rho \nabla^2 \rho + \frac{1}{2} K |\nabla \rho|^2 \right] \mathbf{I} - K \nabla \rho \otimes \nabla \rho, \tag{6.40}$$

to be added to the viscous stress tensor in the Navier-Stokes equation.

Denniston and Yeomans (1999) took into account the latent heat of the phase transformation. Let us note also the application of the diffuse-interface theory for solving the paradoxes connected with the contact line propagation (Pismen and Pomeau, 2000; Pismen, 2001).

In conclusion to this section, we will mention the so-called "mesoscopic" methods for the description of the interfacial phenomena. The simulation of the fluid motion is based on the consideration of the dynamics of some artificial "mesoscopic particles" rather than on the solution of partial differential equations. The most widespread mesoscopic method for modelling the fluid flows is *the lattice Boltzmann method* (LBM) suggested by McNamara and Zanetti (1988) as a modification of the lattice-gas approach (Frisch *et al.*, 1986); for a review, see Chen and Doolen, 1998. The model is formulated for distribution functions which are defined on a fixed discrete spatial lattice and evolvein discrete time. The LBM was used for the description of strong interfacial deformations and multiphase flows (Chen and Doolen, 1998), as well as for the simulation of non-isothermal flows (Alexander *et al.*, 1993). This method may have good perspectives in modelling of various interfacial phenomena, including the multiphase flows in the presence of thermocapillary and solutocapillary convection.

Another method of the mesoscopic modelling is the *dissipative particle dynamics* (DVD); see Hoogerbrugge and Koelman, 1992; Coveney and Novik, 1996; Español *et al.*, 1997. It deals with interacting discrete Brownian particles, which represent large-size clusters of molecules moving in the continuous space. The DVD method, which provides a bridge between the macroscopic models and the microscopic molecular dynamics, has been applied for modelling multiscale interfacial instabilities (Dzwinel and Yuen, 2001), nonisothermic flows (Mackie *et al.*, 1999) and other physical phenomena. This approach seems to be especially appropriate for simulations of phenomena that include both the microstructural dynamics and the large-scale flows.

7

Outlook

The theory of interfacial convection contains many branches which are beyond the scope of the present book. In this chapter, we shall present an outlook of the most important scientific directions and new tools of the investigation.

7.1 Extension of the Linear Stability Theory. Influence of Lateral Boundaries

The linear stability approach described in Sec. 2.1 is based on the investigation of normal modes (2.2) with a *real* wave vector **k** and a *complex* growth rate $\lambda = \lambda_r + i\lambda_i$. According to the standard instability criterion, which is developed in the case of an *infinite* region, the equilibrium state or the flow is unstable if $\lambda_r > 0$ for some **k**. However, in a large but *finite* region, this criterion may be insufficient.

In the case $\lambda_i \neq 0$, the above-mentioned criterion determines the threshold of a *convective* instability (i.e., the instability that manifests itself in the reference frame *moving* with the group velocity of waves) (Lifshitz and Pitaevskii, 1981). The local growth of a disturbance in the motionless reference frame takes place in the case of an *absolute* instability. If the instability is convective but not absolute, the disturbance *decays* in each point although it grows *downstream*. In the latter case, the influence of the *lateral boundaries* of the region becomes crucial. The lateral boundary can act as a source of a finite-amplitude disturbance. The latter is localized near the lateral boundary (in the case of the "nontransparency") or fills the whole region (in the case of the "amplification"); see Lifshitz and Pitaevskii (1981). The *reflection conditions* for the disturbance at the lateral boundaries determine whether a *global* instability is developed in a closed cavity. The mathematical analysis of the above-mentioned types of instability is based on the calculation of the growth rate as the function of a *complex* wave vector, which corresponds to disturbances exponentially growing and decaying in space.

As an impressive example that demonstrates the importance of the extended linear stability theory, we will discuss the instability of a parallel flow in a long but finite horizontal layer under the combined action of the thermocapillarity and the buoyancy. The standard linear stability theory (Parmentier *et al.*, 1993) predicts the transition from oblique hydrothermal waves moving upstream (see Sec. 5.1) to transverse waves moving downstream with the growth of the dynamic Bond number $Bo_{dyn} = R/M$. Surprisingly, many experiments (Schwabe *et al.*, 1992; De Saedeleer *et al.*, 1996; Riley and Neitzel, 1998; Burguete *et al.*, 2001) revealed the appearance of one more flow pattern, *steady* transverse multicellular pattern ("cat's-eye flow"), in an apparent disagreement with the theory. The explanation of this paradox was given by Priede and Gerbeth (1997). They developed an extended linear stability theory for $P = 13.9$. For relatively small values of the dynamical Bond number, $Bo_{dyn} < 0.2$, the boundary of the absolute instability is rather close to that of the convective instability; therefore, the standard linear stability theory gives a rather good prediction. In an intermediate interval, $0.2 < Bo_{dyn} < 0.3$, the boundary of the absolute instability is much higher than the boundary of the convective instability. For larger values of Bo_{dyn}, $Bo_{dyn} > 0.3$, the parallel flow is replaced by the steady multicellular flow produced by the lateral boundary. The predictions of the extended linear stability theory are in a good agreement with experimental observations (Riley and Neitzel, 1998; Burguete *et al.*, 2001) and nonlinear simulations (Shevtsova *et al.*, 2003b).

Note that even in the case of a monotonic instability, the influence of lateral walls is important in many aspects. In the case of infinite layers, the disturbances with the wave-vector modulus $|\mathbf{k}| = k_c$ have the same excitation threshold M_c (or G_c) independently of the orientation of the vector $\mathbf{k}$; the flow pattern is selected due to the nonlinear interaction of disturbances with different directions of $\mathbf{k}$. In a confined container, the set of eigenmodes is discrete, and the flow pattern near the threshold is determined by the "most dangerous" eigenmode. The components of the wave vector of this mode (in the case of a circular container, the azimuthal and radial wave numbers) depend on the aspect ratios of the container in a rather complicated way (Dauby *et al.*, 1993; Rosenblat *et al.*, 1982a, 1982b). The dependence of patterns on the aspect ratios is especially strong near the instability threshold (Koschmieder and Prahl, 1990; Dijkstra, 1995; Dauby and Lebon, 1996; Johnson and Narayanan, 1996; Zaman and Narayanan, 1996; Dauby *et al.*, 1997; see also the review paper by Johnson and Narayanan, 1999 and references therein). With the growth of the Marangoni (or Grashof) number, the patterns selected by the lateral boundaries (e.g., those with the definite value of the azimuthal wave number m in circular containers) are replaced by polygonal patterns (Ramon *et al.*, 1999, 2001). The most interesting dynamics, including nonlinear oscillations, has been found near the codimension-2 points, where the critical Marangoni (Grashof) numbers for two different eigenmodes coincide (Rosenblat *et al.*, 1982a; Johnson and Narayanan, 1996).

In two-layer systems, the aspect ratio can influence (through the wave number selection) the type of the coupling between the flows in both layers. The intervals of the aspect-ratio values where the viscous coupling (through the tangential viscous stresses) prevails interchange with the intervals where the thermal coupling (through the disturbances of the interfacial temperature) prevails (Johnson and Narayanan, 1997, 1998). Also, the type of the correlation between the interfacial deformation and the interfacial temperature disturbance depends on the wave number; hence, it is influenced by the aspect ratio (Johnson and Narayanan, 1997).

7.2 Three-Dimensional Convective Flows

The convective flows are typically three-dimensional. This circumstance stimulates the development of special numerical methods intended for the computation of three-dimensional interfacial convective flows.

In the case of an infinitely extended fluid layer, it is natural to use the Fourier expansion for functions of the horizontal coordinates, in order to transform the differential operators to algebraic operations. At the same time, due to the locality of the nonlinear terms in the governing equations, it is more convenient to calculate these terms in the real space. Switching between the Fourier space and the real space is the basis of the *pseudospectral method* (Gottlieb and Orszag, 1977; Canuto *et al.*, 1988) which is widely used for the treatment of three-dimensional convective flows, starting with the works of Bestehorn (1993) and Thess and Orszag (1995). In the former paper, a finite-difference method was used for the approximation of the derivatives with respect to the vertical coordinate. Thess and Orszag (1995) used the Chebyshev expansion across the layer. The latter approach, which makes it possible to use the fast discrete Fourier transform and the fast Chebyshev transform, was applied to one-layer systems (Boeck and Thess, 1999a) and to two-fluid systems (Boeck and Thess, 1999b; Boeck *et al.*, 2002).

In the case of a finite cavity, the finite-difference methods are often used. The high precision of solutions is achieved by using nonuniform grids and a multigrid approach: During the course of the computation, the solution is approximated on grids with different mesh sizes (see, e.g., Sass *et al.*, 1996). The finite-volume method was used in three-dimensional simulations of the oscillatory thermocapillary convection by Xu and Zebib (1998).

Three-dimensional simulations are especially important for the description of the thermocapillary flows in cylindrical liquid bridges beyond the instability threshold of the axisymmetric flow (see Fig. 7.1). In 1989, Rupp *et al.* first carried out three-dimensional computations of steady and unsteady Marangoni convection using a semi-implicit finite-difference method for equations with the pressure-velocity linkage developed by Patankar (1980). Modern simulations of the Marangoni convection in cylindrical liquid bridges are performed typically on a staggered mesh (Savino and Monti, 1996), on the basis of a finite-volume method (Leypoldt *et al.*, 2000; Shevtsova *et al.*, 2001; Melnikov, 2004), using parallel computers (Lappa and Savino, 1999). Numerical simulations of three-dimensional flows in cylindrical liquid bridges have been carried out also using the finite-element method (Levenstam and Amberg, 1995; Levenstam *et al.*, 2001). Let us mention also the three-dimensional simulation of Rayleigh and Marangoni convection in a vertical circular cylinder with rigid lateral walls

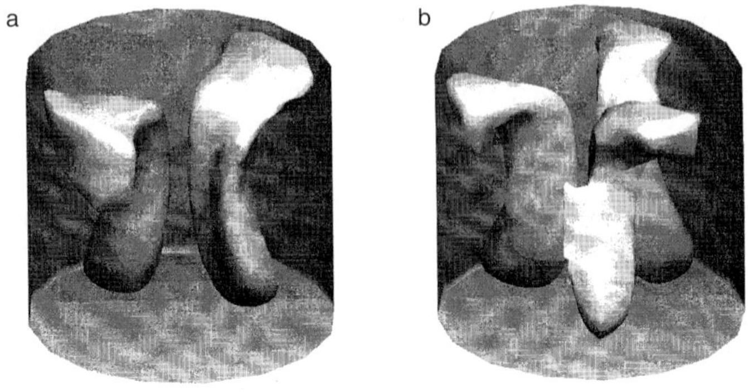

FIGURE 7.1. Surfaces of constant temperature-disturbance field for different symmetry patterns; $P = 4$, (a) $m = 2$ solution and (b) $m = 3$ solution. The axisymmetric part is subtracted from the total temperature distribution. For both solutions, the snapshots display the surfaces of the same disturbance temperature $\theta = \theta_{min} + 0.65(\theta_{max} - \theta_{min})$, where θ_{max} and θ_{min} are the maximal and the minimal values of θ in the bulk (Shevtsova *et al.*, 2003a)

(Wagner *et al.*, 1994). A further description of the three-dimensional thermocapillary convection in cylindrical and rectangular geometry is presented by Kuhlmann (1999).

7.3 Deformation of the Interface

Deformation of the interface is especially important under the microgravity conditions, when the natural shape of the interface is spherical rather than flat. The computation of the finite-amplitude interfacial convection in the case of a deformable interface is a difficult numerical problem, because the shape of the interface is influenced by the convective flow. Typically, the physical region with the curved (unknown) boundary is transformed into a computational region with a simple flat or cylindrical interface. The transformation can be applied to only one of the coordinates (Nepomnyashchy and Simanovskii, 1984; Chen and Floryan, 1994), or an orthogonal mapping can be used (Cliffe and Tavener, 1998; Tavener and Cliffe, 2002). If the flow is steady, the transformation is obtained simultaneously with the solution. In the nonsteady case, the shape of the interface is recalculated on each time step. The influence of the interfacial deformations turned out to be important in the development of the oscillations in a liquid bridge (Kazarinoff and Wilkowski, 1989, 1990) and in a rectangular cavity (Chen and Hwu, 1993; Hamed and Floryan, 2000a).

In some cases, the interfacial deformation is so strong that it leads to the rupture of the liquid layer and to the formation of a dry spot. The disappearance of the solution that describes a steady flow with a motionless interface have been observed for Marangoni flows in a cavity with differentially heated side walls

(Hamed and Floryan, 2000a) and in a cavity subject to point heating (Hamed and Floryan, 2000b; Jiang and Floryan, 2003).

As was mentioned in Chapter 2, the rupture of the liquid layer can be caused by the development of the long-wave Marangoni instability by heating across the layer. The initial stage of the long-wave disturbances growth was studied by Czechowski and Floryan (2001). The late stages of the instability development are characterized by the interesting phenomenon of the "cascade of structures". Boos and Thess (1999) explored this phenomenon in the framework of the Stokes equation by means of the boundary integral method. Krishnamoorthy *et al.* (1995) studied the liquid-film rupture by integration of the complete system of nonlinear equations governing the Marangoni convection. The authors used an unsteady finite-element method with the arbitrary Lagrangian/Eulerian description of moving boundary. The latter approach was applied also for the three-dimensional simulation of the rivulet formation in heated falling films by Ramaswamy *et al.* (1997).

In the case of a very strong deformation of the interface and especially under conditions of an interfacial breakup, the adaptation of the computation grid to the moving interface becomes a difficult task. In the latter case, it may be more convenient to carry out the simulation of flows with a fixed grid, and the position of the interface is defined by means of volume or surface markers (Schramm and Reineke, 1978; Simanovskii, 1983). A review of modern applications of the marker methods is presented by Scardovelli and Zaleski (1999).

Another way to track the position of the interface is implemented in the volume of fluid (VOF) method. The volume fraction of the definite phase is characterized by the *color function* determined in each computational cell. The VOF approach was applied for the three-dimensional simulation of the shear-flow instability leading to a breakup of the interface and to the droplet formation (for details, see Scardovelli and Zaleski, 1999). Gueyffier and Zaleski (1998) used the VOF method for a three-dimensional simulation of the finger formation during droplet impact on a liquid film; a two-fluid approach was applied.

The level set methods, introduced by Osher and Sethian (1988), track the position of the interface as the zero-level set of an auxiliary function (the level set function) that is transported by the field of the extension velocity. A detailed description of that group of algorithms is given by Sethian and Smereka (2003). A combination of the level set method with the VOF method was developed by Sussman and Puckett (2000), who applied it for the simulation of three-dimensional bubbles interaction.

7.4 Transition to Chaos and Interfacial Turbulence

In the previous chapters, we gave some examples of transitions to the interfacial convection regimes that were nonperiodic in time (see, e.g., Secs. 3.3.1, 3.3.2) or both in time and space (Sec. 4.3.3). The transition to disordered patterns ("interfacial turbulence") is characteristic for intensive convective motions observed in a variety of experiments (Linde *et al.*, 1979) and engineering processes like metallurgy (Hammerschmid, 1987), evaporation processes (Palmer and Maheshri, 1981), electrochemistry (Mityushev and Krylov, 1986), and crystal growth (Regel, 1987). In various separation processes like distillation, absorption, chemisorption, and extraction (Sawistowski, 1971; Thornton, 1987; Golovin, 1992) spontaneous interfacial turbulence leads to a sharp increase of mass transfer rates and serves as an effective mean of their intensification.

There is some progress in the exploration of the *temporal chaos* in interfacial flows (Frank and Schwabe, 1997; Schwabe and Frank, 1999; Konig and Neumann, 2001; Boeck and Vitanov, 2002; Ueno *et al.*, 2003). The development of the interfacial *spatio-temporal chaos* in systems with a large aspect ratio is still hardly understood. It is found that *defects* can play an important role in the appearance of disordered Marangoni patterns (Cerisier *et al.*, 1996; Young and Riecke, 2002, 2003). The development of the spatio-temporal interfacial chaos took place through different scenarios that were studied by Pontes *et al.* (1996), Roze and Gouesbet (1997), Boeck and Thess (1999a), Sain and Riecke (2000), and Mancho *et al.* (2002).

Still, the phenomenon of the strong interfacial turbulence remains a challenge for theoreticians. In contradistinction to the Kolmogorov *inertial turbulence*, which is characterized by an energy transfer from long scales to short scales without any dissipation (within the inertial interval), the interfacial turbulence is *strongly dissipative*; that is, it is characterized by the generation/dissipation of energy at any relevant scales. Nevertheless, one can expect that such a strongly dissipative turbulence may have some universal scaling properties. Scaling laws for a turbulent Marangoni convection were suggested by Pumir and Blumenfeld (1996) and by Karcher *et al.* (2000). They were studied by means of direct numerical simulations by Boeck and Thess (2001) and Boeck (2000). The future theory of interfacial turbulence should take into account the diversity of instability mechanisms characteristic for fluid systems with an interface that was demonstrated above.

7.5 Multicomponent Fluids

Various transport processes encountered in technology and nature are due to or affected by simultaneous action of temperature and solute concentration gradients. Two main physical situations are possible: (i) the temperature gradient and the concentration gradient have independent sources (double-diffusive convection) and (ii) the temperature gradient is imposed, whereas the concentration gradient is generated spontaneously due to the Soret effect. In a binary mixture, the interfacial tension depends on both the temperature and the solute concentration. In a majority of mixtures, the interfacial tension decreases with temperature and increases (decreases) with concentration of an inorganic (organic) solute. Therefore, if a layer of a binary mixture is subjected to both temperature and concentration gradients, nonuniformities of those at the interface may lead to the development of a Marangoni instability.

First, let us discuss the nondeformational instability modes. As mentioned above, the concentration gradient across the layer can be either independent of the temperature gradient or generated by the Soret effect. The linear stability analysis of the quiescent equilibrium in a layer with a free surface under the action of independent temperature and concentration gradients across the layer was carried out by McTaggart (1983), Chen and Su (1992), and Skarda *et al.* (1998). The case where the solute concentration was produced by the Soret effect has been considered by Castillo and Velarde (1978, 1980), Chen and Chen (1994), Skarda *et al.* (1998), and Slavtchev *et al.* (1999). The Dufour effect (the contribution of the concentration gradient into the heat flux) was taken into account by Skarda *et al.* (1998). The influence of the surface adsorption of the solute on the free surface was studied by Castillo and Velarde (1978). In all cases, it was found that when the thermal and solutal Marangoni numbers are both positive (i.e., the shear stresses induced separately by thermal and concentration components enhance each other), the quiescent state can lose its stability monotonically. However, when the Marangoni numbers have opposite signs (i.e., when the shear stresses induced separately by thermal and solutal components counteract), the instability is mostly oscillatory.

Investigations of nonlinear aspects of the Marangoni convection in binary liquids are much more rare. Ho and Chang (1988) analyzed, by means of amplitude equations, the flow dynamics in the neighborhood of the double-zero point (i.e., a point in the space of parameters where monotonic and oscillatory instabilities compete). Bergeon *et al.* (1998) studied numerically the two-dimensional Marangoni convection in binary mixtures in a container. Three-dimensional oscillatory convective regimes were studied by Bestehorn and Colinet (2000) by direct numerical simulations of the hydrodynamic equations, as well as on the basis of the model complex Swift-Hohenberg equation. Let us mention also the experimental work of Tanny *et al.* (1995) devoted to the combined Rayleigh-Marangoni convection in a binary solution.

Recently, Oron and Nepomnyashchy (2004) considered the case of convection in a liquid layer with *poorly conducting boundaries*, where the quiescent state is unstable with respect to *long-wave* disturbances. The weakly nonlinear evolution equations for the both monotonic and oscillatory Marangoni instabilities in a binary liquid in the presence of the Soret effect were derived. The obtained equations were used for the analysis of the pattern selection.

The surface deformation influences the instability modes described above. This influence was studied by Castillo and Velarde (1982, 1985).

At the same time, the surface deformation can lead to the appearance of a *new* long-wave stationary instability mode. In both the cases of the independent temperature and concentration gradients (Char and Chiang, 1996) and of the concentration gradient induced by the Soret effect (Bhattacharjee, 1994), it was found that for sufficiently small values of the Bond number (i.e., in very thin layers or under the reduced gravity), the neutral stability curve has an additional minimum at the value of the wave number $k = 0$, with the critical Marangoni number proportional to the Bond number. A similar result was obtained in the general case of a multicomponent fluid by Skarda and McCaughan (1999).

A liquid layer with a deformable free surface when heated *from above* is subject to a specific *transverse* oscillatory instability mode. The physical mechanism of this kind of oscillation is the excitation of capillary-gravity waves by the thermocapillary surface stresses. This kind of instability was studied for both pure liquids and binary liquids by Garcia-Ybarra and Velarde (1987).

The nonlinear evolution of deformational modes was never explored in the case of binary fluids.

7.6 Chemical Reactions

The interface is often subject to a heterogeneous chemical reaction that produces an interfacial convection due to the dependence of surface tension on the concentrations of reagents and due to the heat effect of the reaction (Sherwood and Wei, 1957; Dupeyrat *et al.*, 1984; Kai *et al.*, 1991). The chemical reaction may take place also in the bulk phase (Sherwood and Wei, 1957; Eckert and Grahn, 1999; Ermakov *et al.*, 2001).

Ruckenstein and Berbente (1964) first theoretically considered the case where the reaction takes place near the interface in one of phases. The linear stability theory for different types of heterogeneous reaction was developed by Sanfeld and Steinchen (1975), Dalle Vedove and Sanfeld (1981), and Buyevich *et al.* (1993a). Buyevich *et al.* (1993b) and Rabinovich *et al.* (1995) investigated also the nonlinear development of the Marangoni instability generated by the interfacial chemical reaction.

The development of interfacial instabilities is characteristic for some metal extraction processes (Hughes, 1978; Nakache *et al.*, 1986; Calves *et al.*, 1989; Shioi *et al.*, 2002). The theoretical analysis of the Marangoni instability in liquid/liquid systems with a reversible interfacial reaction typical for some metal extraction processes was studied by Mendes-Tatsis and Perez de Ortiz (1996).

In addition to the Marangoni instability, the chemical patterns can be generated also by the buoyancy convection (Avnir and Kagan, 1995) and by the Rayleigh-Taylor instability (Ibañez and Velarde, 1977). One observes also plumes and fingers (Eckert and Grahn, 1999), which are probably produced by double-diffusion instabilities. Recently, Bratsun and De Wit (2004) applied the two-layer approach for a theoretical analysis of the development of Marangoni pattern driven by an exothermic chemical reaction where thermal and solutal Marangoni effect were taken into account.

Let us note also the work of Texier-Picard *et al.* (2000) in which the development of the surface-tension-driven instability due to the heat production by a polymerization reaction was studied.

The Marangoni instability may significantly influence the propagation of reaction fronts (Martincigh *et al.*, 1995, 1997; Matthiessen *et al.*, 1996; Pismen, 1997). The reaction fronts can be also subject to a buoyancy-driven instabilities (Carey *et al.*, 1996; Böckmann and Müller, 2000; De Wit, 2001) and to double-diffusion instabilities (Pojman *et al.*, 1992).

7.7 Porous Layers

An interesting modification of the problem on the convection in a two-layer system appears in the case when one of the layers is porous. This problem has numerous geophysical and engineering applications, including formation of geothermal reservoirs, nuclear technology, solar energy engineering, etc.

First, let us discuss the basic models for the description of flows in a porous medium (for details, see Bear (1988) and Prasad (1991)). The simplest model (Darcy model) for the description of flows in a porous medium postulates the proportionality of the so-called filtration (or seepage) velocity $\mathbf{v}$ and the force acting on the fluid element:

$$0 = -\frac{1}{\rho}\nabla p - \frac{\nu}{K}\mathbf{v} + g\beta T\boldsymbol{\gamma}, \tag{7.1}$$

where K is called permeability. The actual mean velocity of the fluid in pores can be found as $\mathbf{v}/\phi$, where ϕ is the porosity of the medium. The inertial effects can be incorporated by adding a term $(1/\phi)\partial\mathbf{v}/\partial t$ on the left-hand side of equation (7.1). Because the order of equation (7.1) is lower than that of equation (1.6), the boundary conditions (1.1) and (1.2) cannot be used on the interface between the pure fluid layer and the porous medium saturated by fluid. One assumes that only *normal* components of equations (1.1) and (1.2) are kept. If the boundary between the fluid layer and the porous medium is horizontal, $z = 0$, one gets the continuity condition for the normal stress,

$$-p_1 + 2\eta_1\frac{\partial v_{1,z}}{\partial z} = -p_2, \tag{7.2}$$

and the continuity condition for the normal velocity,

$$v_{1,z} = v_{2,z}. \tag{7.3}$$

The tangential components of velocity in the fluid layer are determined by the phenomenological conditions of Beavers and Joseph (1967):

$$\frac{\partial v_{1,x}}{\partial z} = \frac{\alpha}{\sqrt{K}}(v_{1,x} - v_{2,x}), \quad \frac{\partial v_{1,y}}{\partial z} = \frac{\alpha}{\sqrt{K}}(v_{1,y} - v_{2,y}), \tag{7.4}$$

where $\alpha \approx 0.1$ is called the Beavers-Joseph constant. On the boundary between the porous medium and a solid, only the normal component of velocity vanishes. An invariant form of the model and the boundary conditions were obtained by Nir (1976).

A more sophisticated Brinkman model incorporates a viscous term on the right-hand side of equation (7.1) with a certain modified, apparent viscosity η' *in addition* to the Darcy term, and it restores the viscous boundary conditions on the fluid/porous interface. In the case of a large filtration velocity, the model is improved by Forchheimer's modification, which replaces ν in Darcy's law by $\nu + B|\mathbf{v}|$, where B is a new constant determined by the properties of the porous medium.

In the heat transfer equation, it is necessary to take into account that a mean value of any thermophysical parameter of the saturated medium f_* is different from the values of this parameter for the pure liquid f_1 and for the solid matrix f_2. Usually, one assumes that $f_* = \phi f_1 + (1-\phi) f_2$. The equation for temperature in the saturated porous medium is written as

$$(\rho c_p)_* \frac{\partial T}{\partial t} + \rho_1 c_{p,1} \mathbf{v} \cdot \nabla T = \kappa_* \nabla^2 T. \tag{7.5}$$

The problem of the convective stability of the system that consists of a viscous-fluid layer and a saturated fluid layer, was first studied by Nield (1977) in the framework of the Darcy model. Nield solved the problem analytically in the case of zero Biot number (i.e., for a constant heat flux), where a long-wave instability takes place ($k_c = 0$). The upper boundary of the liquid layer was assumed free, and the thermocapillary effect was taken into account. Among the early works on the convection in a fluid/porous layer, let us mention the investigation of the stability of superposed fluid and liquid layers in the presence of the internal heat generation by Somerton and Catton (1982), the work of Beckermann *et al.* (1988), where the authors considered the sideways heating for different variations of combined fluid, and the numerical simulations of Poulikakos (1986) in the framework of the extended model that includes both Brinkman's and Forchheimer's terms.

For a two-layer system of a viscous fluid and a saturated porous medium sandwiched between two rigid boundaries, the linear stability theory was developed by Chen and Chen (1988). Both thermal and solutal buoyancy effects were taken into account. It was found that depending on the ratio of layer thichnesses $a = a_2/a_1$, the convection is either confined within the fluid layer (if $a < a_* \approx 7.7$) or forms large cells in the porous layer (if $a > a_*$). The predictions of the linear theory were verified in the experiment by Chen and Chen (1989) and in numerical simulations by Chen and Chen (1992).

The investigation of the mixed buoyancy and Marangoni instability started by Nield (1977) was completed by Desaive *et al.* (2001), for which the authors used the Brinkman model for the description of the porous layer.

Let us mention also the works devoted to the investigation of the Marangoni convection generated by tangential stresses on the surface of the porous medium itself. Such a type of convection is expected in a sparsely packed porous medium of high permeability. Hennenberg *et al.* (1997) considered the appearance of the Marangoni instability in a liquid-saturated porous layer with a rigid lower boundary and with the upper boundary in contact with air. They included the Marangoni stress into the tangential boundary condition on the upper boundary of the porous medium (in the framework of the Brinkman model), defined the relevant dimensionless parameters, and solved the linear stability problem analytically. Rudraiah and Prasad (1998) investigated the stability of a sparsely packed liquid-saturated porous layer with two adiabatic free boundaries with respect to the development of the Marangoni convection. The analysis was done by two techniques: the single-term Galerkin approximation and the method of matched asymptotic expansions.

7.8 Contact Line Dynamics

It is well known that in the case of a static three-phase contact line between media l, g, and s (liquid, gas and solid), the balance of stresses determines the *static contact angle* θ_s (Landau and Lifshitz, 1987):

$$\sigma_{ls} + \sigma_{lg} \cos\theta_s = \sigma_{sg}, \tag{7.6}$$

where σ_{ls}, σ_{lg} and σ_{sg} are corresponding interfacial tensions. Let us consider now the situation when the three-phase contact line is moving. This situation, which is of a general interest, may appear in various physical situations, e.g., (i) due to the film rupture; (ii) in the course of the film evaporation; (iii) when a droplet is sliding on an inclined plane or moves because of the Marangoni effect; (iv) when a droplet spreads due to the own weight, complete wetting order under the action of surfactant.

Obviously, there is no reason to use the formula (7.6) for determining the *dynamic contact angle* which is observed when the *macroscopically visible* contact line is moving on the solid surface. Indeed, measurements show that the "macroscopic" dynamic contact angle depends on the direction of the contact line motion (one can distinguish between *advancing* and *receding* contact angles), as well as of the velocity of the contact line (Dussan V, 1979).

The physical origin of the contact angle dependence on the direction of motion ("contact angle hysteresis) is not fully clear. The *roughness* of the interface (Richardson, 1973; Huh and Mason, 1977; Cox, 1983; Cazabat and Cohen Stuart, 1986) and its *chemical heterogeneity* (Joanny and de Gennes, 1984; Pomeau and Vannimenus, 1985; Raphaël and de Gennes, 1989; Alexeyev and Vinogradova, 1996) definitely contribute to this effect (see also Leger and Joanny, 1992). However, contact angle hysteresis is observed also on molecularly smooth homogeneous substrates (Starov, 1992).

The dependence of the contact angle on the velocity is especially pronounced in thin capillaries (Sobolev *et al.*, 2000), and it is strongly influenced by the radius of the capillary. It was shown that the processes connected with the formation of an adsorption film in front of the moving meniscus were crucial for this phenomenon.

The attempt to use the standard no-slip condition for the fluid velocity near the contact angle leads to a singularity (Dussan V and Davis, 1974), which is usually eliminated by an artificial slip condition on the liquid-solid interface near the contact line (Hocking, 1977). An up-to-date review of applications of this phenomenological approach, which is combined with the lubrication approximation (Hocking, 1990) and includes thermocapillary effect (Ehrhard and Davis, 1991), evaporation (Anderson and Davis, 1995) and gravitational flow (Lopez *et al.*, 1996), can be found in the review by Oron *et al.* (1997). The phenomenological approach described above provides a good coincidence with experiments.

Several physical mechanisms have been suggested for the explanation of the slippage in the case of a liquid flowing over a *lyophobic* solid surface (for review, see Vinogradova, 1999), which include a real molecular slippage caused by the fact that the strength of attraction between the liquid molecules is greater than that between the liquid and the solid ones (Tolstoi, 1952a, 1952b; Blake, 1984), and the hypothesis of the existence of a gas gap between the solid and the liquid (Ruckenstein and Rajora, 1983). The most convincing explanation links the apparent slippage effect with a *decrease* in liquid's viscosity in a transition region between the liquid and the lyophobic surface (Churaev *et al.*, 1984). It should be noted however that on the *lyophilic* solid surface the liquid's viscosity *increases* near the boundary (Churaev *et al.*, 1971; Kiseleva *et al.*, 1979). In that case the physical nature of the discussed phenomena should be different.

There are some attempts to go beyond standard phenomenological theory based on the consideration of the physical phenomena which take place on the contact line.

Pismen and Pomeau (2000) suggested a diffuse-interface model for the description of a film flowing on a solid substrate, which takes into account the *disjoining potential* caused by intermolecular forces (see, e.g., Williams and Davis, 1982; Sharma and Ruckenstein, 1986a). It was found that the condensation and evaporation phenomena strongly affected the dynamics near the contact line, which has been recently justified by direct molecular-dynamics simulations (Freund, 2003). A nonlocal generalization of this model has been developed by Pismen (2001). The diffuse-interface approach turns out to be an efficient method for elimination of singularities characteristic for the sharp-interface theory (Pismen, 2002). The model of Pismen and Pomeau was used by Thiele *et al.* (2001b, 2001c, 2002a, 2002b) for investigation of the dewetting phenomena, film rupture and drop motion (see also Thiele, 2003, and Pismen and Pomeau, 2004).

In conclusion, let us discuss briefly the problem of *stability* of the contact line. Even in the absence of thermal and concentrational inhomogeneities, the contact line is subject to *fingering instabilities*, both in the case of a gravitationally driven motion (Huppert, 1982; Silvi and Dussan V, 1985; de Bruyn, 1992; Thiele and Knobloch, 2003) and in the case of wetting (R. Williams, 1977; Troian *et al.*, 1989a). A bibliography on this subject can be found in the recent papers by Ye and Chang (1999) and Johnson *et al.* (1999). The fingering instabilities are found also in presence of heat and mass transfer (Troian *et al.*, 1989b; Cazabat *et al.*, 1990). A specific instability leading to pattern formation was found in experiments on the evaporation of thin wetting films (Elbaum and Lipson, 1994, 1995).

7.9 Feedback Control of Interfacial Instabilities

In the present book, we have considered interfacial instabilities that appear and develop in "a natural way". The engineering applications need, however, *controlling* the instabilities. In some cases any instabilities should be eliminated, and in other cases some definite patterns should be selected and controlled. The abovementioned problem was studied mainly in the context of the influence of the interfacial convection on the crystal growth, specifically under microgravity conditions. In order to control the interfacial convection, one has investigated the influence of a magnetic field (Series and Hurle, 1991; Wilson, 1993a, 1993b; Priede and Gerbeth, 2000; Munakata *et al.*, 2002), vibrations (Zharikov *et al.*, 1990; Lyubimov *et al.*, 1997), liquid encapsulation (Johnson *et al.*, 1975; Géoris *et al.*, 1999, Someya *et al.*, 2002), and other factors. The evaporative interfacial convection can be influenced, e.g., by surfactants (Edwards *et al.*, 1991) and by an internal heating (Oron, 2000). All the approaches mentioned above provide a *passive* suppression of instability and can be considered as "open-loop" control strategies.

Recently, another approach based on the *feedback (online) control* has been suggested. The feedback control of the collective behaviour of nonlinear systems can be much more efficient, because in many cases it is not invasive and nearly vanishes when the desired state is achieved. In order to explain the idea of the feedback control, let us consider a system described by a certain finite-component vector $\mathbf{x}(t)$ (*state vector*) and governed by a system of linear ordinary differential equations (*"open-loop"* system)

$$\frac{d\mathbf{x}}{dt} = \mathbf{A}\mathbf{x}. \tag{7.7}$$

If at least one eigenvalue of the matrix $\mathbf{A}$ has a positive real part, the stationary state $\mathbf{x} = \mathbf{0}$ is unstable.

In order to stabilize the stationary state, one introduces the *control parameter vector* $\mathbf{u}$,

$$\mathbf{u} = \mathbf{K}\mathbf{x} \tag{7.8}$$

(the matrix $\mathbf{K}$ is the *gain matrix*) and replaces the system (7.7) by the *"closed-loop"* system)

$$\frac{d\mathbf{x}}{dt} = \mathbf{A}\mathbf{x} + \mathbf{B}\mathbf{y}, \tag{7.9}$$

where $\mathbf{B}$ is the *input matrix*. The closed-loop system can be written as

$$\frac{d\mathbf{x}}{dt} = (\mathbf{A} + \mathbf{B}\mathbf{K})\mathbf{x}. \tag{7.10}$$

The goal of the feedback control is to find such matrices $\mathbf{B}$ and $\mathbf{K}$ that all the eigenvalues of the matrix $(\mathbf{A} + \mathbf{B}\mathbf{K})$ have negative real parts, so that the stationary solution is stable.

The construction of the control parameter vector $\mathbf{u}$ can be restricted by the fact that the state vector as the whole is unknown (unmeasurable), and the scheme has to be based on the partial information provided by the *measurement vector* $\mathbf{y}$,

$$\mathbf{y} = \mathbf{C}\mathbf{x} \tag{7.11}$$

(the dimension of the vector $\mathbf{y}$ can be much smaller than that of $\mathbf{x}$). The control parameter vector $\mathbf{u}$ is defined on the background of measurements as

$$\mathbf{u} = \mathbf{K}\mathbf{y}. \tag{7.12}$$

The idea described above can be generalized in many aspects. First, it can be applied to nonlinear systems of ordinary differential equations,

$$\frac{d\mathbf{x}}{dt} = \mathbf{F}(\mathbf{x}), \tag{7.13}$$

with the goal of the stabilization of definite stationary or periodic solutions. In the case of the linear control (7.9), the stability analysis can be reduced to the investigation of a linearized problem. A *nonlinear control*

$$\mathbf{u} = \mathbf{K}(\mathbf{x}) \tag{7.14}$$

can be applied in order to bring the nonlinear system from an initial state to a definite final state.

A more difficult problem is the feedback control of infinite-dimensional systems governed by partial differential equations. Note that the fluid dynamics problems generally belong to the latter class of problems. However, in some cases the spatiotemporal dynamics of the infinite-dimensional system is effectively low-dimensional. In the case of convection, the low-dimensional dynamics is typical for flows in cavities with the aspect ratios $O(1)$. In the latter case, one can expect that a control based on a small number of measurements can be efficient.

As an example of a successful application of the feedback control one can mention controlling the thermocapillary convection in *small-aspect ratio* cylindrical containers. Petrov *et al.* (1996) applied a nonlinear control algorithm based on the local temperature measurements ("variable x" has been chosen as the difference between adjacent local temperature maxima) and the local heating of the surface ("variable u" was the voltage at the heating element) in order to stabilize regular oscillations in a cylindrical liquid bridge between plane boundaries kept at different temperatures ("half zone"). Both variables have been defined as functions of the discrete time i. The algorithm consisted of two stages: *identification* and control. During the identification stage, the responses of the measured variable x to random perturbations of u have been accumulated. At the control stage, the time series of measurements of x and u have been used for the construction of a finite-dimensional state vector $\mathbf{y}$(i). The nonlinear control algorithm has been developed in order to bring the system into the target state characterized by a state vector independent of time (i.e., a periodic, one-frequency, flow). An extended version of the approach described above has been used by Petrov *et al.* (1998) for the complete suppression of the time-dependent behavior, i.e., the stabilization of an unstable steady state.

Another approach, which needs no reference data constructed beforehand, has been used by Shiomi *et al.* (2001) in order to suppress the oscillations on the background of a thermocapillary flow generated in an open cylindrical container by a radial temperature gradient on a free surface. In order to control non-axisymmetric waves with the azimuthal number $m = 3$, two sensors were placed with 60° difference in the azimuthal direction. A simple proportional feedback control was realized by locally heating the surface at a single position. Although a significant attenuation of the oscillations was achieved, this approach did not lead to a global stabilization. The latter goal has been achieved by means of two sensor/heater pairs by Shiomi and Amberg (2002). In addition to the above-mentioned experiments, Shiomi and Amberg (2005a) have performed 3D numerical simulations of the thermocapillary convection that allowed to clarify the picture of the spatial structures of the controlled oscillations, to investigate the role of the positioning of sensors and heaters, and to analyze the influence of the heaters' properties. The influence of a linear feedback control on the thermocapillary oscillations has been also studied by Shiomi and Amberg (2005b) by means of a "toy model".

Shiomi *et al.* (2003) applied a similar approach for the suppression of the thrmocapillary oscillations in a half-zone liquid bridge. With optimal placement of sensors and heaters, the proportional control could significantly raise the critical Marangoni number and reduce the amplitude of oscillations. A weakly nonlinear (cubic) control was applied to improve the performance of the control and alter the bifurcation characteristics. Let us mention also the work of Thiessen *et al.* (2002) who applied an active electrostatic stabilization for suppression of the Rayleigh-Plateau instability of a liquid bridge.

The feedback control of the thermocapillary convection in *large-aspect ratio* containers is a much more difficult problem, because in that case the system has many degrees of freedom. The use of the feedback control for the stabilization of large-aspect ratio systems was demonstrated first in the case of the Rayleigh-Bénard convection, both theoretically (Tang and Bau, 1993, 1994) and in experiments (Howle, 1997; Tang and Bau, 1998). Still, there are only few works on the feedback of the interfacial convection in large-aspect ratio containers.

Benz *et al.* (1998) suppressed the development of hydrothermal waves in a thin layer of silicone oil subject to a horizontal temperature gradient. A non-intrusive sensing of free-surface temperature perturbation has been carried out by an infrared camera. The signal from the camera has been converted into a video signal and used for the real-time determination of the hydrothermal-wave phase velocity. The control was achieved by heating the free surface with a sheet of infrared radiation along lines parallel to the crests of the hydrothermal waves corresponding to low-temperature disturbances.

Several works are devoted to a theoretical analysis of the feasibility of a feedback control for the suppression of Rayleigh-Bénard-Marangoni instability in a fluid layer heated from below. Bau (1999) applied the linear proportional control,

$$\theta(x, y, 0, t) = -K\theta(x, y, 1, t), \tag{7.15}$$

where θ is the deviation of the fluid's temperature from its conductive value. By an appropriate choice of the controller gain, K, both the Pearson's instability mode (non-deforming surface) and the Sternling-Scriven's instability mode (deforming surface) is suppressed. However, the control strategy (7.15) turns out to be non-efficient for longwave disturbances. At sufficiently large values of K, an oscillatory instability is generated. More general feedback control laws,

$$\theta(x, y, 0, t) = -K_p\theta(x, y, 1, t) - K_q\theta^2(x, y, 1, t) - K_c\theta^3(x, y, 1, t) \quad (7.16)$$

and

$$\partial\theta(x, y, 0, t)/\partial z = -K_p\theta(x, y, 1, t) - K_q\theta^2(x, y, 1, t) - K_c\theta^3(x, y, 1, t), \quad (7.17)$$

have been used by Or and Kelly (2001) for the nonlinear stabilization of the Pearson-Nield instability in the presence of both thermocapillarity and buoyancy (but in the absence of the surface deformation). Besides the linear stabilization of the quiescent state, the nonlinear control altered a wide range of weakly nonlinear properties of the subcritical hexagonal convection.

A suppression of the longwave deformational Marangoni mode has been achieved by Or *et al.* (1999) who applied a nonlinear feedback control strategy,

$$\theta(x, y, 0, t) = K_1 h + K_2 h^2 + K_3 h^3, \quad (7.18)$$

where h is the deformation of the free surface. The appropriate values of the gain constants, K_1, K_2 and K_3 have been determined by means of a longwave weakly nonlinear analysis. The nonlinear feedback control is efficient for both suppressing the linear instability and eliminating non-linear subcritical instability.

Grigoriev (2002) considered the problem of the stabilization of instabilities which are developed by the evaporation of thin liquid films. He derived the nonlinear evolution equation for the film surface deformation in the presence of a volumetric absorption of the electromagnetic radiation and an absorption at the bottom interface. Several feedback control strategies for the suppression of instabilities in pure liquids and binary solutions have been suggested and analyzed.

The feedback control has been successfully applied for the suppression of the fingering contact-line instability in a thermocapillary-driven fluid film. Grigoriev (2003) applied a slip model for the description of the spreading of a liquid film under the action of the thermocapillary effect, and suggested a control algorithm which makes the dynamics asymptotically stable without increasing the transient amplification of disturbances characteristic for the uncontrolled contact line. The proposed control algorithm has been verified experimentally by Garnier *et al.* (2003). The shape of the contact line has been monitored by computer-controlled digitalization and processing of video images from high-resolution CCD cameras. The control has been achieved by suitable spatial and temporal perturbation of the light intensity.

With few exceptions, the feedback-control approach was applied for a total linear suppression of the interfacial instabilities of a mechanical equilibrium state or a uniform flow, and some relatively simple algorithms of a local proportional control have been used until now. Controlling in order to sustain a definite type of motion, unstable in the absence of the control, is a worthwhile direction of investigation.

7.10 Biological Surface-Tension-Driven Flows

As noticed above, the interfacial phenomena become increasingly important when the characteristic scale of the flow decreases. Hence these phenomena play the major role in many microscale biological processes related to cell biology, tissue biology, blood circulation, as well as in bioengineering. In this section we present some impressive examples.

Flows in the Respiratory System

Airway reopening. The lung is composed of many bifurcating flexible airways, surrounded by an elastic network of parenchymal tissue. At birth, all airways are filled with fluid. With a newborn's first breath, the air penetrates into the lung opening the airways. Once open, the interior surface of the airways are coated with a thin lining fluid.

The problem of the penetration of an inviscid fluid finger into a viscous liquid in a *rigid* channel or tube was studied analytically by Bretherton (1961) and Park and Homsy (1984) for small values of the capillary number, $Ca = \eta U/\sigma$, where U is the finger velocity, σ is the surface tension and η is the liquid's viscosity, and numerically by Reinelt and Saffman (1985), Martinez and Udell (1989), Halpern and Gaver III (1994) and others for finite Ca.

The penetration of an air finger into a *flexible* liquid-filled tube was studied first in the experiments of Gaver III *et al.* (1990). The airway was mimicked by a thin-walled polyethylene tube. The authors obtained the dimensionless relation between the meniscus velocity U and the applied fixed bubble pressure P,

$$PR/\sigma = 8.3 + 7.7Ca^{0.82},$$

where R is the tube radius. The threshold value of the pressure that leads to the airway reopening ("yield pressure") was later justified in direct physiological experiments (Naureckas *et al.*, 1994). Perun and Gaver III (1995a) used a two-dimensional model of the airway, and they found a different value of the yield pressure. In the experiments where the meniscus velocity U was fixed, a non-monotonic evolution of the bubble pressure was observed (Perun and Gaver III, 1995a, 1995b). Similar experiments have been carried out for non-Newtonian liquids (Hsu *et al.*, 1994, 1996; Low *et al.*, 1997).

Gaver III *et al.* (1996) performed a theoretical investigation of the problem described above, using the combination of the lubrication approximation and the boundary element method. The analysis revealed two distinct flow regimes: (i) "pushing" regime for relatively small values of the capillary number Ca, which is characterized by a recirculation region ahead of the air finger; (ii) "peeling" regime for relatively large values of Ca dominated by the wall tension and fluid viscous forces (the latter regime was studied also by Jensen *et al.* (2002)). Asymptotic approach for studying unsteady effect was suggested by Naire and Jensen (2003). The existence of the pushing and peeling flow regimes was confirmed by direct 2D (Heil, 2000) and 3D (Hazel and Heil, 2003a, 2003b) simulations.

A similar problem on the propagation of a liquid plug in an elastic tube with a preexisting film was studied by Howell *et al.* (2000).

Marangoni flows. In the above studies, the surface tension at the air-liquid interface is assumed to be constant. In reality, the lining fluid coating the internal surfaces of the airways and lung alveoli, contains a *pulmonary surfactant* which is important for their stability. The transport of the surfactant may cause surface tension gradients influencing the flow behavior.

Several investigations have been devoted to the analysis of the influence of a surfactant-induced Marangoni stresses on the bubble flow in rigid tubes and channels (Ginley and Radke, 1989; Ratulowski and Chang, 1989; Wassmuth *et al.*, 1993; Stebe and Bartes-Biesel, 1995).

The theoretical investigation of the influence of a surfactant on the pulmonary airway reopening was carried out by Yap and Gaver III (1998). The airway is modelled by a flexible-walled channel with the stress-free separation distance $2H$. The equilibrium relationship between the surface concentration of the surfactant Γ and its bulk concentration are based upon Henry's isotherm ($\Gamma = KC$), and the kinetic transfer of the surfactant between the bulk and interface occurs with a rate k. Two new dimensionless parameters are found to be significant: the surface elasticity number $El = -(d\sigma/d\Gamma)(\Gamma/\sigma)$, and the modified Stanton number $St = (k/K)/(U/H)$ which characterizes the relative rate of surfactant adsorption. An increase in $d\sigma/d\Gamma$ leads to the increase of the dimensional airway reopening pressure P_*, while an increase in Γ reduces P_*. Fast adsorption ($St > 10$) eliminated the Marangoni stresses.

Marangoni flows caused by the surfactant redistribution are developed also by breathing in airways (Espinosa and Kamm, 1997; Bull and Grotberg, 2003) and alveoli (Gradon and Podgorski, 1989; Podgorski and Gradon, 1993; Zelig and Haber, 2002; Wei *et al.*, 2003).

Blood Flows

Blood flows induced by the ultrasound. A widespread method for imaging the blood vessels used in medicine is based on the injection of encapsulated (e.g., by a thin albumin coating) gas bubbles with diameters of few μm. These microbubbles are called ultrasound contrast agents (UCA). Krasovitski *et al.* (2004) studied, both experimentally and theoretically, the behavior of a cloud of encapsulated microbubbles under the action of ambient pressure changes. A detailed analysis of gas bubble pulsations in a blood capillary caused by an ultrasonic field has been done by Sassaroli and Hynynen (2004) and Krasovitski and Kimmel (2004).

Blood flow into a capillary tube. In Section 7.8 we have discussed the propagation of a triple gas/liquid/solid line. Huang *et al.* (2001) considered the propagation of such a triple point when the liquid is blood coming into contact with a certain artificial solid biomaterial forming a capillary tube. Their theoretical analysis was based on the equation of motion obtained by the integration of the Navier-Stokes equations across the capillary tube. They predicted the existence of several flow patterns (entry regime, Poiseuille regime, and surface traction regime).

Let us mention also the application of surface-tension-driven flows for dispensing of enzymes using for fast blood sensing (Tseng *et al.*, 2004).

References

1. Agble D., Mendes-Tatsis M. A., 2000, The effect of surfactants on interfacial mass transfer in binary liquid-liquid systems, Int. J. Heat Mass Transfer **43**, 1025–1034.
2. Agble D., Mendes-Tatsis M. A., 2001, The prediction of Marangoni convection in binary liquid-liquid systems with added surfactants, Int. J. Heat Mass Transfer **44**, 1439–1449.
3. Aharon I., Shaw B. D., 1996, Marangoni instability of bi-component droplet gasification, Phys. Fluids **8**, 1820–1827.
4. Alexander F. J., Chen S., Sternling J. D., 1993, Lattice Boltzmann thermohydrodynamics, Phys. Rev. E **47**, R2249–R2252.
5. Alexeev V. L., Vinogradova L. V., 1996, Flow of a liquid in a nonuniformly hydrophobized capillary, Colloids Surfaces A **108**, 173–179.
6. Anderson D. M., Davis S. H., 1995, The spreading of volatile liquid droplets on heated surfaces, Phys. Fluids **7**, 248–265.
7. Anderson D. M., McFadden G. B., Wheeler A. A., 1998, Diffuse-interface methods in fluids mechanics, Annu. Rev. Fluid Mech. **30**, 139–165.
8. Anderson D. M., McFadden G. B., Wheeler A. A., 2000, A phase-field model of solidification with convection, Physica D **135**, 175–194.
9. Antanovskii L. K., 1995, A phase field model of capillarity, Phys. Fluids **7**, 747–753.
10. Avnir D., Kagan M. L., 1995, The evolution of chemical patterns in reactive liquids, driven by hydrodynamic instabilities, Chaos **5**, 589–601.
11. Badratinova L. G., 1985, Nonlinear equations of long waves in the problem of thermocapillary convection in a two-layer fluid, in *Continuous Media Dynamics*, Novosibirsk, **69**, 3–18. In Russian.
12. Balasubramaniam R., Subramanian R. S., 2000, The migration of a drop in a uniform temperature gradient at large Marangoni numbers, Phys. Fluids **12**, 733–743.
13. Bandyopadhyay D., Gulabani R., Sharma A., 2005, Instability and dynamics of thin liquid bilayers, Ind. Eng. Chem. Res. **44**, 1259–1272.
14. Bar D. E., Nepomnyashchy A. A., 1995, Stability of periodic waves governed by the modified Kawahara equation, Physica D **86**, 90–95.
15. Bar D. E., Nepomnyashchy A. A., 1999, Stability of periodic waves generated by long-wavelength instabilities in isotropic and anisotropic systems, Physica D **132**, 411–427.
16. Barrat J., Bocquet L., 1999, Large slip effect at a nonwetting fluid-solid interface, Phys. Rev. Lett. **82**, 4671–4674.

17. Bau H. H., 1999, Control of Marangoni-Bénard convection, Int. J. Heat Mass Transfer **42**, 1327–1341.
18. Bear Y., 1988, *Dynamics of Fluids in Porous Media*, Dover, New York.
19. Beavers G. S., Joseph D. D., 1967, Boundary conditions at a naturally permeable wall, J. Fluid Mech. **30**, 197–207.
20. Beckermann C., Viskanta R., Ramadhyani S., 1988, Natural convection in vertical enclosures with fluid and porous layers, J. Fluid Mech. **186**, 257–284.
21. Bekki S., Vignes-Adler M., Nakache E., Adler P. M., 1990, Solutal Marangoni effect, I. Pure interfacial transfer, J. Colloid Interface Sci. **140**, 492–506.
22. Bekki S., Vignes-Adler M., Nakache E., 1992, Solutal Marangoni effect, II. Dissolution, J. Colloid Interface Sci. **152**, 314–324.
23. Benz S., Hintz P., Riley R. J., Neitzel G. P., 1998, Instability of thermocapillary-buoyancy convection in shallow layers, Part 2. Suppression of hydrothermal waves, J. Fluid Mech. **359**, 165–180.
24. Berejnov V., Leshansky A. M., Lavrenteva O. M., Nir A., 2002, Spontaneous thermocapillary interaction of drops: Effect of surface deformation at nonzero capillary number, Phys. Fluids **14**, 1326–1339.
25. Berg J. C., Acrivos A. A., 1965, The effect of surface active agents on convection cells induced by surface tension, Chem. Eng. Sci. **20**, 737–745.
26. Berg J. C., Acrivos A. A., Boudart M., 1966, Evaporative convection, Adv. Chem. Eng. **6**, 61–123.
27. Bergeon A., Henry D., Benhadid H., Tuckerman L. S., 1998, Marangoni convection in binary mixtures with Soret effect, J. Fluid Mech. **375**, 143–177.
28. Bernoff A. J., Bertozzi A. L., 1995, Singularities in a modified Kuramoto-Sivashinsky equation describing interface motion for phase transition, Physica D **85**, 375–404.
29. Bestehorn M., 1993, Phase and amplitude instabilities for Bénard-Marangoni convection in fluid layers with large aspect ratio, Phys. Rev. E **48**, 3622–3634.
30. Bestehorn M., Colinet P., 2000, Bénard-Marangoni convection of a binary mixtures as an example of an oscillatory bifurcation under strong symmetry-breaking effects, Physica D **145**, 84–109.
31. Bestehorn M., Neuffer K., 2001, Surface patterns of laterally extended thin liquid films in three dimensions, Phys. Rev. Lett. **87**, 046101 (1–4).
32. Bestehorn M., Pototsky A., Thiele U., 2003, 3D large scale Marangoni convection in liquid films, Eur. Phys. J. B **33**, 457–467.
33. Bhattacharjee J. K., 1994, Marangoni convection in binary liquids, Phys. Rev. E **50**, 1198–1205.
34. Birikh R. V., 1966, Thermocapillary convection in horizontal layer of liquid, J. Appl. Mech. Tech. Phys. **7**, 43–49.
35. Blake T. D., 1984, Wetting, in *Surfactants*, edited by Academic, London, 221–275.
36. Böckmann M., Müller S. C., 2000, Growth rates of the buoyancy-driven instability of an autocatalytic reaction front in a narrow cell, Phys. Rev. Lett. **85**, 2506–2509.
37. Boeck Th., 2000, Bénard-Marangoni convection at low Prandtl numbers, PhD thesis, Ilmenau Technical University.
38. Boeck Th., Nepomnyashchy A., Simanovskii I., Golovin A., Braverman L., Thess A., 2002, Three-dimensional convection in a two-layer system with anomalous thermocapillary effect, Phys. Fluids **14**, 3899–3911.
39. Boeck Th., Thess A., 1999a, Bénard-Marangoni convection at low Prandtl number, J. Fluid Mech. **399**, 251–275.

40. Boeck Th., Thess A., 1999b, Direct numerical simulation of Marangoni convection in a two-fluid system, in *Scientific Computing in Chemical Engineering II*, edited by F. Keil, W. Mackens, H. Voss, and J. Werther, Springer-Verlag, Berlin, 352–359.
41. Boeck Th., Thess A., 2001, Power-law scaling in Benard-Marangoni convection at large Prandtl numbers, Phys. Rev. E **64**, 027303 (1–4).
42. Boeck Th., Vitanov N.K., 2002, Low-dimensional chaos in zero-Prandtl-number Bénard-Marangoni convection, Phys. Rev. E **65**, 037203 (1–4).
43. Boos W., Thess A., 1999, Cascade of structures in long-wavelength Marangoni instability, Phys. Fluids **11**, 1484–1494.
44. Borcia R., Bestehorn M., 2003, Phase-field model for Marangoni convection in liquid-gas systems with a deformable interface, Phys. Rev. E **67**, 066307 (1–10).
45. Bormashenko E., Pogreb R., Stanevsky O., Bormashenko Y., Stein T., Gaisin V.-Z., Cohen R., Gendelman O. V., 2005a, Mesoscopic patterning in thin polymer films formed under the fast dip-coating process, Macromol. Mater. Eng. **290**, 114–121.
46. Bormashenko E., Pogreb R., Stanevsky O., Bormashenko Y., Stein T., Cohen R., Nunberg M., Gaisin V.-Z., Gorelik M., Gendelman O. V., 2005b, Mesoscopic and submicroscopic patterning in thin polymer films: Impact of the solvent, Mater. Lett. **59**, 2461–2464.
47. Bratsun D. A., De Wit A., 2004, On Marangoni convective patterns driven by exothermic chemical reaction in two-layer systems, Phys. Fluids **16**, 1082–1096.
48. Braverman L. M., Eckert K., Nepomnyashchy A. A., Simanovskii I. B., Thess A., 2000, Convection in two-layer systems with anomalous thermocapillary effect, Phys. Rev. E **62**, 3619–3631.
49. Bretherton F. P., 1961, The motion of long bubbles in tubes, J. Fluid Mech. **10**, 166–188.
50. Brinker C. J., Lu Y. F., Sellinger A., Fan N. Y., 1999, Evaporation-induced self-assembly: nanostructures made easy, Adv. Mater. **11**, 579–585.
51. Bull J. L., Grotberg J. B., 2003, Surfactant spreading on thin viscous films: film thickness evolution and periodic wall stretch, Exp. Fluids **34**, 1–15.
52. Burelbach J. P., Bankoff S. G., Davis S. H., 1988, Nonlinear stability of evaporating/condensating liquid films, J. Fluid Mech. **195**, 463–494.
53. Burguete J., Mukolobwiez N., Daviaud F., Garnier N., Chiffaudel A., 2001, Buoyant-thermocapillary instabilities in extended liquid layers subjected to a horizontal temperature gradient, Phys. Fluids **13**, 2773–2787.
54. Buyevich Y. A., Rabinovich L. M., Vyazmin A. V., 1993a, Chemo-Marangoni convection. I. Linear analysis and criteria of instability, J. Colloid Interface Sci. **157**, 202–210.
55. Buyevich Y. A., Rabinovich L. M., Vyazmin A. V., 1993b, Chemo-Marangoni convection. II. Nonlinear stability analysis, J. Colloid Interface Sci. **157**, 211–218.
56. Calves J.-Y., Danes F. E., Gentric E., Lijour Y., Sanfeld A., Saumagne P., 1989, Kinetics and instabilities in the extraction of Co(II) by sodium di-(2-ethylhexyl) phosphate at the water/toluene interface. II. Spontaneous turbulences induced by interfacial reactions, J. Colloid Interface Sci. **129**, 130–138.
57. Canright D., 1994, Thermocapillary flow near a cold wall, Phys. Fluids **6**, 1415–1424.
58. Canuto C., Hussaini M. Y., Quarteroni A., Zang T., 1988, *Spectral Methods in Fluid Dynamics*, Springer-Verlag, Berlin.
59. Cardin Ph., Nataf H.-C., 1991, Nonlinear dynamical coupling observed near the threshold of convection in a two-layer system, Europhys. Lett. **14**, 655–660.

60. Cardin Ph., Nataf H.-C., Dewost Ph., 1991, Thermal coupling in layered convection: evidence for an interface viscosity control from mechanical experiments and marginal stability analysis, J. Phys. II **1**, 599–622.
61. Carey M. R., Morris S. W., Kolodner P., 1996, Convective fingering of an autocatalytic reaction front, Phys. Rev. E **53**, 6012–6015.
62. Castillo J. L., Velarde M. G., 1978, Thermal diffusion and the Marangoni-Bénard instability of a two-component fluid layer heated from below, Phys. Lett. A **66**, 489–491.
63. Castillo J. L., Velarde M. G., 1980, Microgravity and the thermoconvective stability of a binary liquid layer open to the ambient air, J. Nonequilib. Thermodyn. **5**, 111–124.
64. Castillo J. L., Velarde M. G., 1982, Buoyancy-thermocapillary instability: the role of interfacial deformation in one- and two-component fluid layers heated from below or above, J. Fluid Mech. **125**, 463–474.
65. Castillo J. L., Velarde M. G., 1985, Marangoni convection in liquid films with a deformable open surface, J. Colloid Interface Sci. **108**, 264–270.
66. Cazabat A. M., Cohen Stuart M. A., 1986, Dynamics of wetting: effects of surface roughness, J. Phys. Chem. **90**, 5845–5849.
67. Cazabat A. M., Heslot F., Troian S. M., Carles P., 1990, Fingering instability of thin spreading films driven by temperature gradients, Nature **346**, 824–826.
68. Cerisier P., Rahal S., Billia B., 1996, Extrinsic effects of the disorder dynamics of Bénard-Marangoni patterns, Phys. Rev. E **54**, 3508–3517.
69. Chandrasekhar S., 1961, *Hydrodynamic and Hydromagnetic Stability*, Oxford University Press, Oxford.
70. Char M. I., Chiang K. T., 1996, Effect of interfacial deformation on the onset of convective instability in a doubly diffusive fluid layer, Int. J. Heat Mass Transfer **39**, 407–418.
71. Chen C. F., Chen C. C., 1994, Effect of surface tension on the stability of a binary fluid layer under reduced gravity, Phys. Fluids **6**, 1482–1490.
72. Chen C. F., Floryan J. M., 1994, Numerical simulation of nonisothermal capillary interfaces, J. Comp. Phys. **111**, 183–193.
73. Chen C. F., Su T. F., 1992, Effect of surface tension on the onset of convection in a double-diffusive layer, Phys. Fluids A **4**, 2360–2367.
74. Chen F., Chen C. F., 1988, Onset of finger convection in a horizontal porous layer underlying a fluid layer, J. Heat Transfer **110**, 403–409.
75. Chen F., Chen C. F., 1989, Experimental investigation of convective stability in a superposed fluid and porous layer when heated from below, J. Fluid Mech. **207**, 311–321.
76. Chen F., Chen C. F., 1992, Convection in superposed fluid and porous layers, J. Fluid Mech. **234**, 97–119.
77. Chen J.-C., Hwu F.-S., 1993, Oscillatory thermocapillary flow in a rectangular cavity, Int. J. Heat Mass Transfer **36**, 3743–3749.
78. Chen S., Doolen G. D., 1998, Lattice Boltzmann method for fluid flows, Annu. Rev. Fluid Mech. **30**, 329–364.
79. Christov C. I., Velarde M. G., 1995, Dissipative solitons, Physica D **86**, 323–347.
80. Chu X.-L., Velarde M. G., 1988, Sustained transverse and longitudinal waves at the open surface of a liquid, Physicochem. Hydrodyn. **10**, 727–737.
81. Chu X.-L., Velarde M. G., 1989, Transverse and longitudinal waves induced and sustained by surfactant gradients at liquid-liquid interfaces, J. Colloid Interface Sci. **131**, 471–484.

82. Chu X.-L., Velarde M. G., 1991, Korteweg-de Vries soliton excitation in Bénard-Marangoni convection, Phys. Rev. A **43**, 1094–1096.
83. Churaev N. V., 2003, Surface forces in wetting films, Adv. Colloid Interface Sci. **103**, 197–218.
84. Churaev N. V., Sobolev V. D., Somov A. N., 1984, Slippage of liquids over lyophobic solid surfaces, J. Colloid Interface Sci. **97**, 574–581.
85. Churaev N. V., Sobolev V. D., Zorin Z. M., 1971, Measurement of viscosity of liquids in quartz capillaries, Trans. Faraday Soc. **52**, 213–220.
86. Cliffe K. A., Tavener S. J., 1998, Marangoni-Bénard convection with a deformable free surface, J. Comput. Phys. **145**, 193–227.
87. Colinet P., Géoris Ph., Legros J.-C., Lebon G., 1996, Spatially quasiperiodic convection and temporal chaos in two-layer thermocapillary instabilities, Phys. Rev. E **54**, 514–524.
88. Colinet P., Joannes L., Iorio C. S., Hant B., Bestehorn M., Lebon G., Legros J.-C., 2003, Interfacial turbulence in evaporating liquids: theory and preliminary results of the ITEL - Maser 9 sounding rocket experiment, Adv. Space Res. **32**, 119–137.
89. Colinet P., Legros J.-C., 1994, On the Hopf bifurcation occuring in the two-layer Rayleigh-Bénard convective instability, Phys. Fluids **6**, 2631–2639.
90. Coveney P. V., Novik K. E., 1996, Computer simulations of domain growth and phase separation in two-dimensional binary immiscible fluids using dissipative particle dynamics, Phys. Rev. E **54**, 5134–5141.
91. Cox R. G., 1983, The spreading of a liquid on a rough solid surface, J. Fluid Mech. **131**, 1–26.
92. Czechowski L., Floryan J. M., 2001, Marangoni instability in a finite container - transition between short and long wavelengths modes, J. Heat Transfer **123**, 96–104.
93. Dalle Vedove W., Sanfeld A., 1981, Hydrodynamic and chemical stability of fluid-fluid reacting interfaces. I. General theory for aperiodic regimes, J. Colloid Interface Sci. **84**, 318–327.
94. Danov K. D., Alleborn N., Raszillier H., Durst F., 1998a, The stability of evaporating thin liquid films in the presence of surfactant. I. Lubrication approximation and linear analysis, Phys. Fluids **10**, 131–143.
95. Danov K. D., Paunov V. M., Alleborn N., Raszillier H., Durst F., 1998b, Stability of evaporating two-layered liquid film in the presence of surfactant - I. The equations of lubrication approximation, Chem. Eng. Sci. **53**, 2809–2822.
96. Danov K. D., Paunov V. M., Stoyanov S. D., Alleborn N., Raszillier H., Durst F., 1998c, Stability of evaporating two-layered liquid film in the presence of surfactant - II. Linear analysis, Chem. Eng. Sci. **53**, 2823–2837.
97. Darhuber A. A., Davis J. M., Troian S. M., Reisner W. W., 2003, Thermocapillary actuation of liquid flow on chemically patterned surfaces, Phys. Fluids **15**, 1295–1304.
98. Dauby P. C., Lebon G., 1996, Bénard-Marangoni instability in rigid rectangular containers, J. Fluid Mech. **120**, 91–138.
99. Dauby P. C., Lebon G., Bouhy E., 1997, Linear Benard-Marangoni instability in rigid circular containers, Phys. Rev. E **56**, 520–530.
100. Dauby P. C., Lebon G., Colinet P., Legros J. C., 1993, Hexagonal Marangoni convection in a rectangular box with slippery walls, Q. J. Mech. Appl. Math. **46**, 683–707.
101. Davis S. H., 1987, Thermocapillary instabilities, Annu. Rev. Fluid Mech. **19**, 403–435.
102. De Bruyn J. R., 1992, Growth of fingers at a driven three-phase contact line, Phys. Rev. A **46**, R4500–R4503.

103. De Gennes P. G., 1985, Wetting: statistics and dynamics, Rev. Mod. Phys. **57**, 827–863.
104. De Gennes P. G., 2001, Instabilities during the evaporation of a film: Non-glassy polymer plus volatile solvent, Eur. Phys. J. E **6**, 421–424.
105. De Gennes P. G., Brochard-Wyart F., Quéré D., 2004, *Capillary and Wetting Phenomena*, Springer, New York.
106. De Saedeleer C., Garcimartin A., Chapeveyer G., Platten J. K., Lebon G., 1996, The instability of a liquid layer heated from the side when the upper surface is open to air, Phys. Fluids **8**, 670–676.
107. De Wit A., 2001, Fingering of chemical fronts in porous media, Phys. Rev. Lett. **87**, 054502 (1–4).
108. Deegan R. D., 2000, Pattern formation in drying drops, Phys. Rev. E **61**, 475–485.
109. Degen M. M., Colovas P. W., Andereck C. D., 1998, Time-dependent patterns in the two-layer Rayleigh-Bénard system, Phys. Rev. E **57**, 6647–6659.
110. Denniston C., Yeomans J. M., 1999, Diffuse interface simulation of Marangoni convection, PCCP **1**, 2157–2161.
111. Derjaguin B. V., Churaev N. V., Muller V. M., 1987, *Surface Forces*, Consultants Bureau, New York.
112. Desaive Th., Lebon G., Hennenberg M., 2001, Coupled capillary and gravity-driven instability in a liquid film overlying a porous layer, Phys. Rev. E **64**, 066304 (1–8).
113. Dijkstra H. A., 1995, Surface-tension-driven cellular patterns in 3-dimensional boxes - linear stability, Micrograv. Sci. Technol. **VII**, 307–312.
114. Doi T., Koster J. N., 1993, Thermocapillary convection in two immiscible liquid layers with free surface, Phys. Fluids **5**, 1914–1927.
115. Dondlinger M., Margerit J., Dauby P. C., 2005, Weakly nonlinear study of Marangoni instabilities in an evaporating liquid layer, J. Colloid Interface Sci. **283**, 522–532.
116. Dos Santos F. D., Ondarçuhu T., 1995, Free-running droplets, Phys. Rev. Lett. **75**, 2972–2975.
117. Dowden J., 2002, Interaction of the keyhole and weld pool in laser keyhole welding, J. Laser Appl. **14**, 204–209.
118. Dupeyrat M., Nakache E., Vignes-Adler M., 1984, Chemically driven interfacial instabilities, in *Chemical Instabilities*, edited by G. Nicolis and F. Baras, Reidel, Dordrecht, 233–245.
119. Dussan V E. B., 1979, On the spreading of liquids on solid surfaces: static and dynamic contact lines, Annu. Rev. Fluid Mech., **11**, 371–400.
120. Dussan V E. B., Davis S. H., 1974, On the motion of fluid-fluid interface along a solid surface, J. Fluid Mech., **65**, 71–95.
121. Dussaud A. D., Troian S. M., 1998, Dynamics of spontaneous spreading with evaporation on a deep fluid layer, Phys. Fluids **10**, 23–38.
122. Dzwinel W., Yuen D. A., 2001, Mixing driven by Rayleigh-Taylor instability in the mesoscale modeled with dissipative particle dynamics, Int. J. Mod. Phys. C **12**, 91–118.
123. Dzyaloshinskii I. E., Lifshitz E. M., Pitaevskii L. P., 1960, Van der Waals forces in liquid films, Sov. Phys. JETP **10**, 161–170.
124. Dzyaloshinskii I. E., Pitaevskii L. P., 1959, Van der Waals forces in an inhomogeneous dielectric, Sov. Phys. JETP **9**, 1282–1287.
125. Eckert K., Grahn A., 1999, Plume and finger regimes driven by an exothermic interfacial reaction, Phys. Rev. Lett. **82**, 4436–4439.
126. Eckert K., Thess A., 2001, unpublished.

127. Edwards D. A., Brenner H., Wasan D. T., 1991, *Interfacial Transport Processes and Rheology*, Butterworth-Heinemann, Boston.
128. Ehrhard P., Davis S. H., 1991, Non-isothermal spreading of liquid drops on horizontal plates, J. Fluid Mech. **229**, 365–388.
129. Elbaum M., Lipson S. G., 1994, How does a thin wetted film dry up, Phys. Rev. Lett. **72**, 3562–3565.
130. Elbaum, M., Lipson S. G., 1995, Pattern formation in the evaporation of thin liquid films, Israel J. Chem. **35**, 27–32.
131. Engel A., Swift J. B., 2000, Planform selection in two-layer Bénard-Marangoni convection, Phys. Rev. E **62**, 6540–6553.
132. Ermakov S. A., Ermakov A. A., Chupakhin O. N., Vaissov D. V., 2001, Mass transfer with chemical reaction in conditions of spontaneous interfacial convection in processes of liquid extraction, Chem. Eng. J. **84**, 321–324.
133. Erneux Th., Davis S. H., 1993, Nonlinear rupture of free films, Phys. Fluids A **5**, 1117–1122.
134. Español P., Serrano N., Zuniga I., 1997, Coarse-graining of a fluid and its relation with dissipative particle dynamics and smoothed particle dynamics, Int. J. Mod. Phys. C **8**, 899–908.
135. Espinosa F. F., Kamm R. D., 1997, Thin layer flows due to surface tension gradients over a membrane undergoing nonuniform, periodic strain, Annu. Biomed. Eng. **25**, 913–925.
136. Eyer A., Leiste H., 1985, Striation-free silicon crystal by float-zoning with surface-coated melt, J. Crystal Growth **71**, 249–252.
137. Eyer A., Leiste H., Nitsche R., 1985, Float zone growth of silicon under microgravity in a sounding rocket, J. Crystal Growth **71**, 173–182.
138. Fan Z., Ji S., Zhang J., 2001, Processing of immiscible metallic alloys by rheomixing process, Mater. Sci. Techn. **17**, 837–842.
139. Ferm E. N., Wollkind D. J., 1982, Onset of the Rayleigh - Bénard - Marangoni instability; comparison between theory and experiment, J. Non-Equilib. Thermodyn. **7**, 170–190.
140. Fisher L. S., Golovin A. A., 2005, Nonlinear stability analysis of a two-layer thin liquid film: Dewetting and autophobic behavior, J. Colloid Interface Sci. **291**, 515–528.
141. Fletcher C. A. J., 1988, *Computational Techniques for Fluid Dynamics*, Springer-Verlag, Berlin.
142. Frank S., Schwabe D., 1997, Temporal and spatial elements of thermocapillary convection in floating zones, Exp. Fluids **23**, 234–251.
143. Frenkel A. L., Halpern D., 2002, Stokes-flow instability due to interfacial surfactant, Phys. Fluids **14**, L45–L48.
144. Freund J. B., 2003, The atomic detail of a wetting/de-wetting flow, Phys. Fluids **15**, L33–L36.
145. Frisch U., Hasslacher B., Pomeau Y., 1986, Lattice-gas automata for the Navier-Stokes equation, Phys. Rev. Lett. **56**, 1505–1508.
146. Frisch U., She Z. S., Sulem P. L., 1987, Large-scale flow driven by the anisotropic kinetic alpha effect, Physica D **28**, 382–392.
147. Fuhrich T., Berger P., Hugel H., 2001, Marangoni effect in laser deep penetration welding of steel, J. Laser Appl. **13**, 178–186.
148. Funada T., 1987, Nonlinear surface waves driven by the Marangoni instability in a heat transfer system, J. Phys. Soc. Japan **56**, 2031–2038.

149. Garazo A. N., Velarde M. G., 1991, Dissipative Korteweg-de Vries description of Marangoni-Bénard oscillatory convection, Phys. Fluids A **3**, 2295–2300.
150. Garazo A. N., Velarde M. G., 1992, Marangoni-driven solitary waves, in *Proceedings of the VIII European Symposium on Materials and Fluid Sciences in Microgravity*, European Space Agency, Paris, ESA SP-333, 711–715.
151. Garcia-Ybarra P. L., Castillo J. L., Velarde M. G., 1987a, A nonlinear evolution equation for Bénard-Marangoni convection with deformable boundary, Phys. Lett. A **122**, 107–110.
152. Garcia-Ybarra P. L., Castillo J. L., Velarde M. G., 1987b, Bénard-Marangoni convection with a deformable interface and poorly conducting boundaries, Phys. Fluids **30**, 2655–2661.
153. Garcia-Ybarra P. L., Velarde M. G., 1987, Oscillatory Marangoni-Bénard interfacial instability and capillary-gravity waves in single- and two-component liquid layers with or without Soret thermal diffusion, Phys. Fluids **30**, 1649–1655.
154. Garnier N., Grigoriev R. O., Schatz M. F., 2003, Optical manipulation of microscale fluid flow, Phys. Rev. Lett. **91**, 054501 (1–4).
155. Gaver III D. P., Halpern D., Jensen O. E., Grotberg J. B., 1996, The steady motion of a semi-infinte bubble through a flexible-walled channel, J. Fluid Mech. **319**, 25–65.
156. Gaver III D. P., Samsel R. W., Solway J., 1990, Effect of surface tension and viscosity on airway reopening, J. Appl. Physiol. **69**, 74–85.
157. Géoris Ph., 1994, Contribution à l'étude des instabilités de Marangoni-Bénard et Rayleigh-Bénard pour les systèmes multicouches, PhD thesis, Université Libre de Bruxelles.
158. Géoris Ph., Hennenberg M., Simanovskii I., Nepomnyashchy A., Wertgeim I., Legros J.-C., 1993, Thermocapillary convection in a multilayer system, Phys. Fluids A **5**, 1575–1582.
159. Géoris Ph., Legros J.-C., 1996, Pure thermocapillary convection in a multilayer system: first results from the IML-2 mission, in *Materials and Fluids under Low Gravity*, edited by L. Ratke, H. Walter, and B. Feuerbacher, Springer-Verlag, Berlin, 299–311.
160. Géoris Ph., Legros J.-C., Nepomnyashchy A., Simanovskii I., Viviani A., 1997, Numerical simulation of convection flows in multilayer fluid systems, Micrograv. Sci. Techn. **10**, 13–20.
161. Géoris Ph., Hennenberg M., Lebon G., Legros J.-C., 1999, Thermocapillary convection in a three-liquid-layer system, J. Fluid Mech. **389**, 209–228.
162. Gershuni G. Z., Zhukhovitsky E. M., 1976, *Convective Stability of Incompressible Fluid*, Keter, Jerusalem.
163. Gershuni G. Z., Zhukhovitsky E. M., 1980, On the instability of a horizontal layers system of immiscible fluids when heating from above, Izv. AN SSSR, Mekh. Zhidk. Gaza No. 6, 28–34. In Russian.
164. Gershuni G. Z., Zhukhovitsky E. M., 1982, On monotonic and oscillatory instability of a two-layer immiscible fluids system heated from below, Sov. Phys. Dokl. **27**, 531–533.
165. Gershuni G. Z., Zhukhovitsky E. M., Simanovskii I. B., 1981, On equilibrium stability and finite-amplitude motions in a two-layer system when heating from above, in *Convective Flows* , Perm Pedagogical Institute, Perm, 3–11. In Russian.
166. Gertsberg V. L., Sivashinsky G. I., 1981, Large cells in nonlinear Rayleigh-Bénard convection, Progr. Theor. Phys. **66**, 1219–1229.
167. Gilev A. Yu., Nepomnyashchy A. A., Simanovskii I. B., 1986, Onset of the thermogravitational convection in a two-layer system in the presence of a surfactant on the interface, Zh. Prikl. Mekh. Tekhn. Fiz. *5*, 76–81. In Russian.

168. Gilev A. Yu., Nepomnyashchy A. A., Simanovskii I. B., 1987a, Onset of convection in two-layer system under a combined action of Rayleigh and thermocapillary mechanisms of instability, Fluid Dyn. **22**, 142–145.
169. Gilev A. Yu., Nepomnyashchy A. A., Simanovskii I. B., 1987b, Generation of thermocapillary and thermogravitational convection in an air-water system, Fluid Mech. - Sov. Res. **16**, 44–48.
170. Gilev A. Yu., Nepomnyashchy A. A., Simanovskii I. B., 1987c, Onset of oscillatory thermogravitational convection in a two-layer system when heating from below, in *Dynamics of Viscous Fluid*, Sverdlovsk, USC AS USSR, 36–37.
171. Gilev A. Yu., Nepomnyashchy A. A., Simanovskii I. B., 1991, Thermocapillary convection with the interfacial heat release depending on temperature, Zh. Prikl. Mekh. Tekh. Fiz. *3*, 115–119.
172. Ginley G. M., Radke C. J., 1989, Influence of soluble surfactants on the flow of long bubbles through a cylindrical capillary, ACS Symp. Series **396**, 480–501.
173. Golovin A. A., 1992, Mass transfer under interfacial turbulence, Chem. Eng. Sci. **47**, 2069–2080.
174. Golovin A. A., Nepomnyashchy A. A., Davis S. H., Zaks M. A., 2001, Convective Cahn-Hilliard models: From coarsening to roughening, Phys. Rev. Lett. **86**, 1550–1553.
175. Golovin A. A., Nepomnyashchy A. A., Pismen L. M., 1994, Interaction between short-scale Marangoni convection and long-scale deformational instability, Phys. Fluids **6**, 34–48.
176. Golovin A. A., Nepomnyashchy A. A., Pismen L. M., 1995, Pattern formation in large-scale Marangoni convection with deformable interface, Physica D **81**, 117–147.
177. Golovin A. A., Nepomnyashchy A. A., Pismen L. M., 1997a, Nonlinear evolution and secondary instabilities of Marangoni convection in a liquid-gas system with deformable interface, J. Fluid Mech. **341**, 317–341.
178. Golovin A. A., Nepomnyashchy A. A., Pismen L. M., 2003, Non-potential effects in nonlinear dynamics of Marangoni convection, Int. J. Bifurc. Chaos **12**, 2487–2500.
179. Golovin A. A., Nepomnyashchy A. A., Pismen L. M., Riecke H., 1997b, Steady and oscillatory side-band instabilities in Marangoni convection with deformable interface, Physica D **106**, 131–147.
180. Gonuguntla M., Sharma A., 2004, Polymer patterns in evaporating droplets on dissolving substrates, Langmuir **20**, 3456–3463.
181. Goodrich F. C., 1981, The theory of capillary excess viscosities, Proc. Roy. Soc. London A **374**, 341–370.
182. Goodwin R. T., Schowalter W. R., 1995, Arbitrarily oriented capillary-viscous planar jets in the presence of gravity, Phys. Fluids **7**, 954–963.
183. Gottlib D., Orszag S. A., 1977, in *Numerical Analysis of Spectral Methods*, CBMS-NSF Regional Conference Series in Applied Mathematics, Philadelphia.
184. Gradon L., Podgorski, A., 1989, Hydrodynamical model of pulmonary clearance, Chem. Eng. Sci. **44**, 741–749.
185. Grigoriev R. O., 2002, Control of evaporatively driven instabilities of thin liquid films, Phys. Fluids **14**, 1895–1909.
186. Grigoriev R. O., 2003, Contact line instability and pattern selection in thermally driven liquid films, Phys. Fluids **15**, 1363–1374.
187. Groothuis H., Zuiderweg F. G., 1960, Influence of mass transfer on coalescence of drops, Chem. Eng. Sci. **12**, 288–289.

188. Grotberg J. B., Jensen O. E., 2004, Biofluid mechanics in flexible tubes, Annu. Rev. Fluid Mech. **36**, 121–147.
189. Gueyffier D., Zaleski S., 1998, Formation de digitations lors de l'impact d'une goutte sur un film liquide, C. R. Acad. Sci. Paris, Série II b **326**, 839–844.
190. Gumerman R. J., Homsy G. M., 1974a, Convective instabilities in concurrent two-phase flow. Part I. Linear stability, AIChE J. **20**, 981–988.
191. Gumerman R. J., Homsy G. M., 1974b, Convective instabilities in concurrent two-phase flow. Part II. Global stability, AIChE J. **20**, 1161–1167.
192. Gumerman R. J., Homsy G. M., 1975, The stability of radially bounded thin films, Chem. Eng. Commun. **2**, 27.
193. Ha V.-M., Lai C.-L., 2002, Onset of Marangoni instability of a two-component evaporating droplet, Int. J. Heat Mass Transf. **45**, 5143–5158.
194. Halpern D., Gaver III D. P., 1994, Boundary element analysis of the time-dependent motion of a semi-infinite bubble in a channel, J. Comput. Phys. **115**, 366–375.
195. Halpern D., Frenkel A. L., 2003, Destabilization of a creeping flow by interfacial surfactant: linear theory extended to all wavenumbers, J. Fluid Mech. **485**, 191–220.
196. Hamaker H. C., 1937, The London-van der Waals attraction between spherical particles, Physica **4**, 1058–1072.
197. Hamed M., Floryan J. M., 2000a, Marangoni convection. Part 1. A cavity with differentially heated sidewalls, J. Fluid Mech. **405**, 79–110.
198. Hamed M., Floryan J. M., 2000b, Marangoni convection. Part 2. A cavity subject to point heating, J. Fluid Mech. **405**, 111–129.
199. Hammerschmid P., 1987, Bedeutung des Marangoni-Effekts für metallurgische Vorgänge, Stahl u. Eisen **2**, 61–66.
200. Haut B., Colinet P., 2005, Surface-tension-driven instabilities of a pure liquid layer evaporating into an inert gas, J. Colloid Interface Sci. **285**, 296–305.
201. Hazel A. L., Heil M., 2003a, Three-dimensional airway reopening: the steady propagation of a semi-infinite bubble into a buckled elastic tube, J. Fluid Mech. **478**, 47–70.
202. Hazel A. L., Heil M., 2003b, Steady finite-Reynolds-number flows in three-dimensional collapsible tubes, J. Fluid Mech. **486**, 79–103.
203. Heil M., 2000, Finite Reynolds number effects in the propagation of an air finger into a liquid-filled flexible-wall channel, J. Fluid Mech. **424**, 21–44.
204. Hennenberg M., Bisch P. M., Vignes-Adler M., Sanfeld A., 1979, Mass transfer, Marangoni effect, and instability of interfacial longitudinal waves, I. Diffusional exchanges, J. Colloid Interface Sci. **69**, 128–137.
205. Hennenberg M., Bisch P. M., Vignes-Adler M., Sanfeld A., 1980, Mass transfer, Marangoni effect, and instability of interfacial longitudinal waves, II. Diffusional exchanges and adsorption-desorption processes, J. Colloid Interface Sci. **74**, 495 508.
206. Hennenberg M., Chu X.-L., Sanfeld A., Velarde M. G., 1992, Transverse and longitudinal waves at the air-liquid interface in the presence of an adsorption barrier, J. Colloid Interface Sci. **150**, 7–21.
207. Hennenberg M., Saghir M. Z., Rednikov A., Legros J.-C., 1997, Porous media and the Bénard-Marangoni problem, Transport Porous Media **27**, 327–355.
208. Higgins A. M., Jones R. A. L., 2000, Anisotropic spinodal dewetting as a route to self-assembly of patterned surfaces, Nature **404**, 476–478.
209. Ho K.-L., Chang H.-C., 1988, On nonlinear doubly-diffusive Marangoni instability, AIChE J. **34**, 705–722.

210. Hocking L. M., 1977, A moving fluid interface. II. The removal of the force singularity by a slip flow, J. Fluid Mech. **79**, 209–229.
211. Hocking L. M., 1990, Spreading and instability of a viscosity fluid sheet, J. Fluid Mech. **211**, 373–392.
212. Hocking L. M., 1993, The influence of intermolecular forces on thin fluid layers, Phys. Fluids **5**, 793–799.
213. Hohenberg P. C., Halperin B. I., 1977, Theory of dynamic critical phenomena, Rev. Mod. Phys. **49**, 435–479.
214. Hondros E. D., McLean M., Mills K. C. (editors), 1998, *Marangoni and Interfacial Phenomena in Materials Processing: Originating from Contributions to a Discussion of the Royal Society of London*, The University Press, Cambridge.
215. Hoogerbrugge P. J., Koelman J. M. V. A., 1992, Simulating microscopic hydrodynamic phenomena with dissipative particle dynamics, Europhys. Lett. **19**, 155–160.
216. Howell P. D., Waters S. L., Grotberg J. B., 2000, The propagation of a liquid bolus along a liquid-lined flexible tube, J. Fluid Mech. **406**, 309–335.
217. Howle L. E., 1997, Active control of Rayleigh-Bénard convection, Phys. Fluids **9**, 1861–1863.
218. Hsu S. H., Strohl K. P., Haxhiu M. A., Jamieson A. M., 1996, Role of viscoelasticity in tube model of airway reopening. 2. Non-Newtonian gels and airway simulation, J. Appl. Physiol. **80**, 1649–1659.
219. Hsu S. H., Strohl K. P., Jamieson A. M., 1994, Role of viscoelasticity in tube model of airway reopening. 1. Non-Newtonian sols, J. Appl. Physiol. **76**, 2481–2489.
220. Huang W., Bhullar R. S., Fung Y. C., 2001, The surface-tension-driven flow of blood from a droplet into a capillary tube, J. Biomech. Eng. - Trans. ASME **123**, 446–454.
221. Hughes M. A., 1978, On the direct observation of films formed at a liquid-liquid interface during the extraction of metals, Hydrometallurgy **3**, 85–90.
222. Huh C., Mason S. G., 1977, Effects of surface roughness on wetting (theoretical), J. Colloid Interface Sci. **60**, 11–38.
223. Huppert H. E., 1982, Flow and instability of a viscous gravity current down a slope, Nature **300**, 427–429.
224. Ibañez J. L., Velarde M. G., 1977, Hydrochemical stability of an interface between 2 immiscible liquids: the role of Langmuir-Hinshelwood saturation law, J. Phys. (France) **38**, 1479–1483.
225. Imaishi N., Fujinawa K., 1974a, Theoretical study of the stability of two-fluid layers, J. Chem. Eng. Japan **7**, 81–86.
226. Imaishi N., Fujinawa K., 1974b, Thermal instability in two-fluid layers, J. Chem. Eng. Japan **7**, 87–92.
227. Israelashvili J. H., 1985, *Intermolecular and Surface Forces,* Academic, New York.
228. Jasnow D., Viñals J., 1996, Coarse-grain description of thermo-capillary flow, Phys. Fluids **8**, 660–669.
229. Jensen O. E., Horsburgh M. K., Halpern D., Gaver III D. P., 2002, The steady propagation of a bubble in a flexible-wall channel: Asymptotic and computational models, Phys. Fluids **14**, 443–457.
230. Jiang Y., Floryan J. M., 2003, Influence of heat transfer at the interface on the thermocapillary convection in the adjacent phase, J. Heat Transfer **125**, 190–194.
231. Joanny J. F., de Gennes P. G., 1984, A model for contact angle hysteresis, J. Chem. Phys. **81**, 552–562.

232. Johnson D., Narayanan R., 1996, Experimental observation of dynamic mode switching in interfacial-tension-driven convection near a codimension-two point, Phys. Rev. E **54**, R3102–R3104.
233. Johnson D., Narayanan R., 1997, Geometric effects on convective coupling and interfacial structures in bilayer convection, Phys. Rev. E **56**, 5462–5472.
234. Johnson D., Narayanan R., 1998, Marangoni convection in multiple bounded fluid layers and its application to materials processing, Phil. Trans. R. Soc. Lond. A **356**, 885–897.
235. Johnson D., Narayanan R., 1999, A tutorial on the Rayleigh-Marangoni-Benard problem with multiple layers and side wall effects, Chaos **9**, 124–140.
236. Johnson E. S., 1975, Liquid encapsulated float zone melting for GaAs, J. Crystal Growth **30**, 249–256.
237. Johnson M. F. G., Schluter R. A., Miksis M. J., Bankoff S. G., 1999, Experimental study of rivulet formation on an inclined plate by fluorescent imaging, J. Fluid Mech. 394, 339–354.
238. Joo S. W., Davis S. H., Bankoff S. G., 1991, Longwave instabilities of heated falling films: two-dimensional theory of uniform layers, J. Fluid Mech. **230**, 117–146.
239. Joseph D. D., Renardy Y. Y., 1993, *Fundamentals of Two-Fluid Dynamics*, Springer-Verlag, New York.
240. Juel A., Burgess J. M., McCormick W. D., Swift J. B., Swinney H. L., 2000, Surface tension-driven convection patterns in two liquid layers, Physica D **143**, 169–186.
241. Kai S., Muller S. C., Mori T., Miki M., 1991, Chemically driven nonlinear waves and oscillations at an oil-water interface, Physica D **50**, 412–428.
242. Kaminskii V. A., Dil'man V. V., 2003, Marangoni instability in evaporation of binary mixtures, Theor. Found. Chem. Eng. **37**, 533–538.
243. Karcher C., Schaller R., Boeck T., Metzner C., Thess A., 2000, Turbulent heat transfer in liquid iron during electron beam evaporation, Int. J. Heat Mass Transfer **43**, 1759–1766.
244. Kargupta K., Konnur R., Sharma A., 2001, Spontaneous dewetting and ordered patterns in evaporating thin liquid films on homogeneous and heterogeneous substrates, Langmuir **17**, 1294–1305.
245. Kataoka D. E., Troian S. M., 1999, Patterning liquid flow on the microscopic scale, Nature **402**, 794–797.
246. Kats-Demyanets V., 1996, Marangoni instability in liquid three layer systems, MSc thesis, Technion, Haifa.
247. Kats-Demyanets V., Oron A., Nepomnyashchy A. A., 1997, Marangoni instability in tri-layer systems, Eur. J. Mech. B/Fluids **16**, 49–74.
248. Kawahara T., Toh S., 1985, Nonlinear dispersive periodic waves in the presence of instability and damping, Phys. Fluids **28**, 1636–1638.
249. Kawasaki K., Ohta T., 1982, Kink dynamics in one-dimensional nonlinear systems, Physica A **116**, 573–593.
250. Kazarinoff N. D., Wilkowski J. S., 1989, A numerical study of Marangoni flows in zone-refined silicon crystals, Phys. Fluids A **1**, 625–627.
251. Kazarinoff N. D., Wilkowski J. S., 1990, Bifurcations of numerically simulated thermocapillary float zones, Phys. Fluids A **2**, 1797–1807.
252. Khomenko G. A., Moiseev S. S., Tur A. V., 1991, The hydrodynamical alpha-effect in a compressible medium, J. Fluid Mech. **225**, 335–369.
253. Kiseleva O. A., Sobolev V. D., Starov V. M., Churaev N. V., 1979, Changes in viscosity of water close to quartz surface, Colloid J. USSR **41**, 192–197.

254. Kliakhandler I. L., Nepomnyashchy A. A., 1999, Kinetic instabilities of three-layer thermocapillary creeping flows, Phys. Fluids **11**, 2139–2142.
255. Kliakhandler I. L., Nepomnyashchy A. A., Simanovskii I. B., Zaks M. A., 1998, Nonlinear Marangoni waves in multilayer systems, Phys. Rev. E **58**, 5765–5775.
256. Kliakhandler I. L., Sivashinsky G. I., 1995, Kinetic alpha effect in viscosity stratified creeping flows, Phys. Fluids **7**, 1866–1871.
257. Kliakhandler I. L., Sivashinsky G. I., 1997, Viscous damping and instabilities in stratified liquid film flowing down a slightly inclined plane, Phys. Fluids **9**, 23–30.
258. Knobloch E., 1990, Pattern selection in long-wavelength convection, Physica D **41**, 450–479.
259. Konig F., Neumann W., 2001, On the spatio-temporal flow phenomena in low Prandtl number melts, Part 1. The case of stationary vertical molten zones, Crystal Res. Techn. **36**, 239–254.
260. Koplik J., Banavar J. R., Willemsen J. F., 1989, Molecular dynamics of fluid flow at solid surfaces, Phys. Fluids A **1**, 781–794.
261. Koschmieder E. L., 1992, *Bénard Cells and Taylor Vortices*, Cambridge University Press, Cambridge.
262. Koschmieder E. L., Biggerstaff M. I., 1986, Onset of surface-tension-driven convection, J. Fluid Mech. **167**, 49–64.
263. Koschmieder E. L., Prahl S., 1990, Surface-tension-driven Bénard convection in small containers, J. Fluid Mech. **215**, 571–583.
264. Kovalchuk N. M., Vollhardt D., 2000, Autooscillations of surface tension in water-alcohol systems, J. Phys. Chem. B **104**, 7987–7992.
265. Kovalchuk N. M., Vollhardt D., 2002, Theoretical description of repeated surface-tension auto-oscillations, Phys. Rev. E **66**, 026302 (1–9).
266. Kovalchuk N. M., Vollhardt D., 2004, Effect of substance properties on the appearance and characteristics of repeated surface tension auto-oscillation driven by Marangoni force, Phys. Rev. E **69**, 016307 (1–13).
267. Krasovitski B., Kimmel E., 2004, Shear stress induced by a gas bubble pulsating in an ultrasonic field near the wall, IEEE Trans. on Ultrasonics, Ferroelectrics, and Frequency Control **51**, 973–979.
268. Krasovitski B., Kimmel E., Sapunar M., Adam D., 2004, Ultrasound attenuation by encapsulated microbubbles: time and pressure effects, Ultrasound in Med. and Biol. **30**, 793–802.
269. Krishnamoorthy S., Ramaswamy B., Joo S. W., 1995, Spontaneous rupture of thin liquid films due to thermocapillarity: a full-scale direct numerical simulation, Phys. Fluids **7**, 2291–2293.
270. Kuhlmann H. C., 1999, *Thermocapillary Convection in Models of Crystal Growth*, Springer-Verlag, Berlin.
271. Kuskova T. V., Chudov L. A., 1968, On approximate boundary conditions for vortex at calculation of the flows of viscous incompressible fluid, Comp. Meth. Programming **11**, 27–31. In Russian.
272. Landau L. D., Lifshitz E. M., 1987, *Fluid Mechanics*, Pergamon Press, Oxford.
273. Langer J. S., 1980, Instabilities and pattern formation in crystal growth, Rev. Mod. Phys. **52**, 1–28.
274. Lappa M., Savino R., 1999, Parallel solution of three-dimensional Marangoni flows in liquid bridges, Int. J. Numer. Meth. Fluids **31**, 911–935.
275. Leger L., Joanny J. F., 1992, Liquid spreading, Rep. Prog. Phys. **55**, 431–486.
276. Legros J.-C., 1986, Problems related to non-linear variations of surface tension, Acta Astr. **13**, 697–703.

277. Legros J.-C., Géoris Ph., 1997, Oscillatory thermocapillary instability, report, Microgravity Research Center, Université Libre de Bruxelles.
278. Leizerson I., Lipson S. G., Lyushnin A. V., 2003, When larger drops evaporate faster?, Nature **422**, 395–396.
279. Leshansky A. M., 2001, On the influence of mass transfer on coalescence of bubbles, Int. J. Multiphase Flow **27**, 189–196.
280. Levchenko E. B., Chernyakov A. L., 1981, Instability of surface waves in a nonuniformly heated liquid, Sov. Phys. JETP **54**, 102–105.
281. Levenstam M., Amberg G., 1995, Hydrodynamical instabilities of thermocapillary flow in a half-zone, J. Fluid Mech. **297**, 357–372.
282. Levenstam M., Amberg G., Winkler C., 2001, Instabilities of thermocapillary convection in a half-zone at intermediate Prandtl numbers, Phys. Fluids **13**, 807–816.
283. Levich V. G., 1962, *Physicochemical hydrodynamics*, Prentice Hall, Englewood Cliffs, N. J.
284. Levich V. G., 1981, The influence of surface-active substances on the motion of liquids, Physicochem. Hydr. **2**, 85–94.
285. Levich V. G., Krylov V. S., 1969, Surface-tension-driven phenomena, Annu. Rev. Fluid Mech. **1**, 293–316.
286. Leypoldt J., Kuhlmann H. C., Rath H. J., 2000, Three-dimensional numerical simulation of thermocapillary flows in cylindrical liquid bridges, J. Fluid Mech. **414**, 285–314.
287. Li C. H., 1969, Instability of 3-layer viscous stratified fluid, Phys. Fluids **7**, 2473–2481.
288. Lifshitz E. M., 1956, The theory of molecular attractive forces between solids, Sov. Phys. JETP **2**, 73–83.
289. Lifshitz E. M., Pitaevskii L. P., 1981, *Physical Kinetics*, Pergamon Press, Oxford.
290. Lin Z. Q., Kerle T., Baker S. M., Hoagland D. A., Schaffer E., Steiner U., Russel T. P., 2001, Electric field induced instabilities at liquid-liquid interfaces, J. Chem. Phys. **114**, 2377–2381.
291. Lin Z. Q., Kerle T., Russel T. P., Schaffer E., Steiner U., 2002a, Structure formation at the interface of liquid-liquid bilayer in electric field, Macromolecules **35**, 3971–3976.
292. Lin Z. Q., Kerle T., Russel T. P., Schaffer E., Steiner U., 2002b, Electric field induced dewetting at polymer/polymer interfaces, Macromolecules **35**, 6255–6262.
293. Linde H., Schwarz P., Wilke H., 1979, Dissipative structures and nonlinear kinetics of the Marangoni instability, in *Dynamics and Instability of Fluid Interfaces*, edited by T. S. Sorensen, Springer-Verlag, Berlin, 75–119.
294. Lopez J. C., Bankoff S. G., Miksis M. J., 1996, Non-isothermal spreading of a thin liquid film on an inclined plane, J. Fluid Mech. **324**, 261–286.
295. Low H. T., Chew Y. T., Zhou C. W., 1997, Pulmonary airway reopening: effects of non-Newtonian fluid viscosity, J. Biomech. Eng. - Trans. ASME **119**, 298–308.
296. Lu X. Y., Cao C. D., Wei B., 2001, Microstructure evolution of undercooled iron-copper hypoperitectic alloy, Mater. Sci. Eng. A **313**, 198–206.
297. Lucassen J., 1968, Longitudinal capillary waves, Part 1. Theory, Trans. Faraday Soc. **64**, 2221–2229; Part 2. Experiments, Trans. Faraday Soc. **64**, 2230–2235.
298. Lucassen-Reynders E. H., Lucassen J., 1969, Properties of capillary waves, Adv. Colloid Interface Sci. **2**, 347–395.
299. Lyubimov D. V., Lyubimova T. P., Meradji S., Roux B., 1997, Vibrational control of crystal growth from liquid phase, J. Crystal Growth **180**, 648–659.

300. Lyubimov D. V., Zaks M. A., 1983, Two mechanisms of the transition to chaos in finite-dimensional models of convection, Physica D **9**, 52–64.
301. Lyushnin A. V., Golovin A. A., Pismen L. M., 2002, Fingering instability of thin evaporating liquid films, Phys. Rev. E **65**, 021602 (1–7).
302. Mackie A. D., Avalos J. B., Navas V., 1999, Dissipative particle dynamics with energy conservation: modelling of heat flow, Phys. Chem. Chem. Phys. **1**, 2039–2049.
303. Madruga S., Perez-Garcia C., Lebon G., 2003, Convective instabilities in two superposed horizontal liquid layers heated laterally, Phys. Rev. E **68**, 041607 (1–12).
304. Madruga S., Perez-Garcia C., Lebon G., 2004, Instabilities in two-liquid layers subject to a horizontal temperature gradient, Theoret. Comput. Fluid Dynamics **18**, 277–284.
305. Maillard M., Motte L., Ngo A. T., Pileni M. P., 2000, Rings and hexagons made of nanocrystals: a Marangoni effect, J. Phys. Chem. B **104**, 11,871–11,877.
306. Majda A., Pego R. L., 1985, Stable viscosity matrices for systems of conservation laws, J. Diff. Eq. **56**, 229–262.
307. Mancho A. M., Riecke H., Sain F., 2002, Instabilities and spatio-temporal chaos of long-wave hexagon patterns in rotating Marangoni convection, Chaos **12**, 706–718.
308. Mancini H., Maza D., 2004, Pattern formation without heating in an evaporative convection experiment, Europhys. Lett. **66**, 812–818.
309. Margerit J., Colinet P., Lebon G., Iorio C. S., Legros J. C., 2003, Interfacial nonequilibrium and Bénard-Marangoni instability of a liquid-vapour system, Phys. Rev. E **68**, 041601 (1–14).
310. Margerit J., Dondlinger M., Dauby P. C., 2005, Improved 1.5 - sided model for the weakly nonlinear study of Bénard - Marangoni instabilities in an evaporating liquid layer, J. Colloid Interface Sci. **290**, 220–230.
311. Martincigh B. S., Chinake C. R., Howes T., Simoyi R. H., 1997, Self-organization with traveling waves: a case for a convective torus, Phys. Rev. E **55**, 7299–7303.
312. Martincigh B. S., Hauser M. J. B., Simoyi R. H., 1995, Formation of thermal plumes in an autocatalytic exothermic chemical reaction, Phys. Rev. E **52**, 6146–6153.
313. Martinez M. J., Udell K. S., 1989, Boundary integral analysis of creeping flow of long bubbles in capillaries, J. Appl. Mech. - Trans. ASME **56**, 211–217.
314. Maruyama N., Koito T., Nishida J., Sawadaishi T., Cieren X., Ijiro K., Karthaus O., Shimomura M., 1998, Mesoscopic patterns of molecular agregates on solid substrates, Thin Solid Films **327 - 329**, 854–856.
315. Matthiessen K., Wilke H., Müller S. C., 1996, Influence of surface tension changes on hydrodynamic flow induced by traveling chemical waves, Phys. Rev. E **53**, 6056–6060.
316. McNamara G., Zanetti G., 1988, Use of the Boltzmann equation to simulate lattice gas automata, Phys. Rev. Lett. **61**, 2332–2335.
317. McTaggart C. L., 1983, Convection driven by concentration- and temperature-dependent surface tension, J. Fluid Mech. **134**, 301–310.
318. Melnikov D., 2004, Development of numerical code for the study of Marangoni convection, PhD thesis, Université Libre de Bruxelles.
319. Mendes-Tatsis M. A., Agble D., 2000, The effect of surfactants on Marangoni convection in the isobutanol/water system, J. Non-Equilib. Thermodyn. **25**, 239–249.
320. Mendes-Tatsis M. A., Perez de Ortiz E. S., 1996, Marangoni instabilities in systems with an interfacial chemical reaction, Chem. Eng. Sci. **51**, 3755–3761.
321. Merkt D., Bestehorn M., 2003, Benard-Marangoni convection in a strongly evaporating fluid, Physica D **185**, 196–208.

322. Merkt D., Pototsky A., Bestehorn M., Thiele U., 2005, Long-wave theory of bounded two-layer films with a free liquid-liquid interface: Short- and long-time evolution, Phys. Fluids **17**, 064104 (1–20).
323. Miladinova S., Lebon G., 2005, Effect of nonuniform heating and thermocapillarity in evaporating films falling down an inclined plate, Acta Mech. **174**, 33–49.
324. Mityushev P. V., Krylov V. S., 1986, Hydrodynamic instabilities caused by electric forces acting upon electrolyte solution surfaces, Sov. Electrochem. **22**, 520–524.
325. Moffat H. K., 1964, Viscous and resistive eddies near a sharp corner, J. Fluid Mech. **18**, 1–18.
326. Moiseev S. S., Sagdeev R. Z., Tur A. V., Khomenko G. A., Yanovskii V. V., 1983, Theory of the origin of large-scale structures in hydrodynamical turbulence, Sov. Phys. JETP **58**, 1149–1153.
327. Moussy C., Lebon G., Margerit J., 2004, Influence of evaporation on Bénard-Marangoni instability in a gas-liquid bilayer with a deformable interface, Eur. Phys. J. B **40**, 327–335.
328. Mukai K., Lei Z. X., Segawa H., Wang Z., Izu D., 2003, Behaviour of bubbles in liquid caused by Marangoni effect in relation to steel refining process, High Temp. Mat. Proc. **22**, 309–317.
329. Munakata T., Someya S., Tanasawa I., 2002, Suppression of Marangoni convection in the FZ melt by high-frequency magnetic field, J. Crystal Growth **235**, 167–172.
330. Nagai K., Sumino Y., Kitahata H., Yoshikawa K., 2005, Mode selection in the spontaneous motion of an alcohol droplet, Phys. Rev. E **71**, 065301 (1–6) (R).
331. Nagayama M., Nakata S., Doi Y., Hayashima Y., 2004, A theoretical and experimental study on the unidirectional motion of a camphor disk, Physica D **194**, 151–165.
332. Naire S., Jensen O. E. 2003, An asymptotic model of unsteady airway reopening, J. Biomech. Eng. - Trans. ASME **125**, 823–831.
333. Nakache E., Dupeyrat M., Lemaire J., 1986, Enhanced recovery of metals by liquid-liquid extraction: influence of the interfacial properties of the extractant, J. Chem. Phys. **83**, 339–342.
334. Nakache E., Dupeyrat M., Vignes-Adler M., 1983, Experimental and theoretical study of an interfacial instability at some oil-water interfaces involving a surface-active agent: I. Physicochemical description and outlines for a theoretical approach, J. Colloid Interface Sci. **94**, 120–127.
335. Nakache E., Raharimalala S., 1988, Interfacial convection driven by surfactant compounds at liquid interfaces: Characterization by a solutal Marangoni number, in *Physicochemical Hydrodynamics: Interfacial Phenomena*, NATO ASI Series, Series B, edited by M. G. Velarde, Plenum Press, New York, 359–366.
336. Nakata S., Iguchi Y., Ose S., Kuboyama M., Ishii T., Yoshikawa K., 1997, Self-rotation of a camphor scraping on water: new insight into the old problem, Langmuir **13**, 4454–4458.
337. Napolitano L. G., 1978, Thermodynamics and dynamics of pure interfaces, Acta Astronaut. **5**, 655–670.
338. Nataf H. C., Moreno S., Cardin Ph., 1988, What is responsible for thermal coupling in layered convection?, J. Phys. (France) **49**, 1707–1714.
339. Naureckas E. T., Dawson C. A., Gerber B. S., Gaver III D. P., Gerber H. L., Linehan J. H., Solway J., Samsel R. W., 1994, Airway reopening pressure in isolated rat lungs, J. Appl. Physiol. **76**, 1372–1377.
340. Nekorkin V. I., Velarde M. G., 1994, Solitary waves of a dissipative Korteweg-de Vries equation describing Marangoni-Bénard convection and other thermoconvective instabilities, Int. J. Bifurc. Chaos **4**, 1135–1146.

341. Neogi P., Berryman J., 1982, Stability of thin liquid films evaporating into saturated vapor, J. Colloid Interface Sci. **88**, 100–110.
342. Nepomnyashchy A. A., 1976, Wavy motions in the layer of viscous fluid flowing down the inclined plane, in *Fluid Dynamics, Part 8*, Proc. of Perm State University **362**, 114–124. In Russian.
343. Nepomnyashchy A., Golovin A., Volpert V., Tikhomirova A., 2005, Interfacial phenomena in evaporation of binary mixtures, in Proc. of 10th National Congress on Theoretical and Applied Mechanics, Varna, Bulgaria, in press.
344. Nepomnyashchy A. A., Simanovskii I. B., 1983a, Thermocapillary convection in a two-layer system, Sov. Phys. Doklady **28**, 838–839.
345. Nepomnyashchy A. A., Simanovskii I. B., 1983b, Thermocapillary convection in a two-layer system, Fluid Dyn. **18**, 629–633.
346. Nepomnyashchy A. A., Simanovskii I. B., 1984, Thermocapillary and thermogravitational convection in a two-layer system with a distorted interface, Fluid Dyn. **19**, 494–499.
347. Nepomnyashchy A. A., Simanovskii I. B., 1986, On the onset of the thermocapillary convection in the presence of surfactants, Fluid Dyn. **21**, 169–174.
348. Nepomnyashchy A. A., Simanovskii I. B., 1988, Generation of the thermocapillary convection in a two-layer system with the soluble surfactant, Fluid Dyn. **23**, 302–306.
349. Nepomnyashchy A. A., Simanovskii I. B., 1989a, Generation of the thermogravitational convection with the soluble surfactant on the interface, Zh. Prikl. Mekh. Tekhn. Fiz. *1*, 146–149. In Russian.
350. Nepomnyashchy A. A., Simanovskii I. B., 1989b, Generation of the oscillatory convection in two-layer systems due to the surfactant on the interface, Sov. Phys. Doklady **34**, 420–422.
351. Nepomnyashchy A. A., Simanovskii I. B., 1990a, Generation of convection when heating from above and heat release at the interface, Fluid Dyn. **25**, 340–343.
352. Nepomnyashchy A. A., Simanovskii I. B., 1990b, Long-wave thermocapillary convection in layers with a deformable interface, J. Appl. Math. Mech. **54**, 490–496.
353. Nepomnyashchy A. A., Simanovskii I. B., 1990c, Generation of the thermocapillary convection in a two-layer system with heat release at the interface, Zh. Prikl. Mekh. Tekhn. Fiz. *1*, 69–73. In Russian.
354. Nepomnyashchy A. A., Simanovskii I. B., 1991, Onset of oscillatory thermocapillary convection in systems with a deformable interface, Fluid Dyn. **26**, 484–488.
355. Nepomnyashchy A. A., Simanovskii I. B., 1992, Onset of thermocapillary convection in systems with surfactant at the deformable interface, Zh. Prikl. Mekh. Tekhn. Fiz. *3*, 73–79. In Russian.
356. Nepomnyashchy A. A., Simanovskii I. B., 1995, Oscillatory convection instabilities in systems with an interface, Int. J. Multiphase Flow **21**, 129–139.
357. Nepomnyashchy A. A., Simanovskii I. B., 1997, New types of longwave oscillatory Marangoni instabilities in multilayer systems, Q. J. Mech. Appl. Math. **50**, 149–163.
358. Nepomnyashchy A. A., Simanovskii I. B., 1999, Combined action of different mechanisms of convective instability in multilayer systems, Phys. Rev. E **59**, 6672–6686.
359. Nepomnyashchy A. A., Simanovskii I. B., 2001, Nonlinear investigation of anticonvection and Rayleigh-Bénard convection in systems with heat release on the interface, Eur. J. Mech. B/Fluids **20**, 75–86.
360. Nepomnyashchy A. A., Simanovskii I. B., 2002, Combined action of anticonvective and thermocapillary mechanisms of instability, Phys. Fluids **14**, 3855–3867.
361. Nepomnyashchy A. A., Simanovskii I. B., 2003, Influence of buoyancy on thermocapillary oscillations in a two-layer system, Phys. Rev. E **68**, 026301 (1–8).

362. Nepomnyashchy A. A., Simanovskii I. B., 2004, Influence of thermocapillary effect and interfacial heat release on convective oscillations in a two-layer system, Phys. Fluids **16**, 1127–1139.
363. Nepomnyashchy A. A., Simanovskii I. B., 2006, Convective flows in a two-layar system with a temperature gradient along the interface, Phys. Fluids **18**, 032105 (1–7)
364. Nepomnyashchy A. A., Simanovskii I. B., Boeck Th., Golovin A. A., Braverman L. M., Thess A., 2003, Convective instabilities in layered systems, in *Lecture Notes in Physics*, Springer-Verlag, Berlin, 21–44.
365. Nepomnyashchy A. A., Simanovskii I. B., Braverman L. M., 2000, Anticonvection in systems with heat release on the interface, Phys. Fluids **12**, 1129–1132.
366. Nepomnyashchy A. A., Simanovskii I. B., Braverman L. M., 2001, Stability of thermocapillary flows with inclined temperature gradient, J. Fluid Mech. **442**, 141–155.
367. Nepomnyashchy A. A., Simanovskii I. B., Braverman L. M., 2005, Convective oscillations in multilayer systems under the combined action of buoyancy and thermocapillary effect, Phys. Fluids **17**, 022103 (1–9).
368. Nepomnyashchy A. A., Velarde M. G., 1994, A three-dimensional description of solitary waves and their interaction in Marangoni-Bénard layers, Phys. Fluids **6**, 187–198.
369. Nguyen V. X., Stebe K. J., 2002, Patterning of small particles by a surfactant-enhanced Marangoni-Bénard instability, Phys. Rev. Lett. **88**, 164501 (1–4).
370. Nield D. A., 1977, Onset of convection in a fluid layer overlying a layer of porous medium, J. Fluid Mech. **81**, 513–522.
371. Nir A., 1976, Linear shear flow past a porous particle, Appl. Sci. Res. **32**, 313–325.
372. Noble R. D., Douglas Way J., 1987, Liquid membrane technology: an overview, in *Liquid membranes: theory and applications*, ACS Symp. Series, American Chemical Society, Washington, DC.
373. Or A. C., Kelly R. E., 2001, Feedback control of weakly nonlinear Rayleigh-Bénard-Marangoni convection, J. Fluid Mech. **440**, 27–47.
374. Or A. C., Kelly R. E., Cortelezzi L., Speyer J. L., 1999, Control of long-wavelength Marangoni-Bénard convection, J. Fluid Mech. **387**, 321–341.
375. Oron A., 2000, Nonlinear dynamics of irradiated thin volatile liquid films, Phys. Fluids **12**, 29–41.
376. Oron A., Bankoff S. G., 1999, Dewetting of a heated surface by an evaporating liquid film under conjoining/disjoining pressures, J. Colloid Interface Sci. **218**, 152–166.
377. Oron A., Bankoff S. G., 2001, Dynamics of a condensing liquid film under conjoining/disjoining pressures, Phys. Fluids **13**, 1107–1117.
378. Oron A., Davis S. H., Bankoff S. G., 1997, Long-scale evolution of thin liquid films, Rev. Mod. Phys. **69**, 931–980.
379. Oron A., Nepomnyashchy A. A., 2004, Long-wavelength thermocapillary instability with the Soret effect, Phys. Rev. E **69**, 016313 (1–17).
380. Oron A., Rosenau Ph., 1997, Evolution and formation of dispersive-dissipative patterns, Phys. Rev. E **55**, R1267–R1270.
381. Osher S. J., Sethian J. A., 1988, Fronts propagating with curvature dependent speed: algorithms based on Hamilton-Jacobi formulations, J. Comput. Phys. **79**, 12–49.
382. Ostrach S., 1982, Low-gravity fluid flows, Annu. Rev. Fluid Mech. **14**, 313–345.
383. Padmakar A., Kargupta K., Sharma A., 1999, Instability and dewetting of evaporating thin water films on partially and completely wettable substrates, J. Chem. Phys. **110**, 1735–1744.

384. Palmer H. J., 1976, The hydrodynamic stability of rapidly evaporating liquids at reduced pressure, J. Fluid Mech. **75**, 487–511.
385. Palmer H. J., Berg J. C., 1972, Hydrodynamic stability of surfactant solutions, heated from below, J. Fluid Mech. **51**, 385–402.
386. Palmer H. J., Maheshri J. C., 1981, Enhanced interfacial heat transfer by differential vapor recoil instabilities, Int. J. Heat Mass Transfer **24**, 117–123.
387. Park C.-W., Homsy G. M., 1984, Two-phase displacement in Hele-Shaw cells: Theory, J. Fluid Mech. **139**, 291–308.
388. Parmentier P. M., Regnier V. C., Lebon G., 1993, Buoyant-thermocapillary instabilities in medium-Prandtl-number fluid layers subject to a horizontal temperature gradient, Int. J. Heat Mass Transfer **36**, 2417–2427.
389. Parmentier P. M., Regnier V. C., Lebon G., Legros J.-C., 1996, Nonlinear analysis of coupled gravitational and capillary thermoconvection in thin fluid layers, Phys. Rev. E **54**, 411–423.
390. Patankar S. V., 1980, *Numerical Heat Transfer and Fluid Flow*, Hemisphere, Washington, DC.
391. Paunov V. M., Danov K. D., Alleborn N., Raszillier H., Durst F., 1998, Stability of evaporating two-layered liquid film in the presence of surfactant - III. Non-linear stability analysis, Chem. Eng. Sci. **53**, 2839–2857.
392. Pearson J. R., 1958, On convection cells induced by surface tension, J. Fluid Mech. **4**, 489–500.
393. Peng J., Han Y., Yang Y., 2004, The influencing factors on the macroporous formation in polymer films by water droplet templating, Polymer **45**, 447–452.
394. Perestenko O. V., Ingel L. Kh., 1991, On one possible type of convective instability in the ocean-atmosphere system, Izv. Akad. Nauk SSSR, Fiz. Atmosf. Okeana **27**, 408–418.
395. Perestenko O. V., Ingel L. Kh., 1995, The occurence of moist 'anticonvection' in a water-air system, J. Fluid Mech. **287**, 1–20.
396. Perez de Ortiz E. S., Sawistowski H., 1973a, Interfacial stability of binary liquid-liquid systems, I. Stability analysis, Chem. Eng. Sci. **28**, 2051–2061.
397. Perez de Ortiz E. S., Sawistowski H., 1973b, Interfacial stability of binary liquid-liquid systems, II. Stability behaviour of selected systems, Chem. Eng. Sci. **28**, 2063–2069.
398. Perry R. H., 1997, *Perry's Chemical Engineers' Handbook.*
399. Perun M. L., Gaver III D. P., 1995a, An experimental-model investigation of the opening of a collapsed untethered pulmonary airway, J. Biomech. Eng. - Trans. ASME **117**, 245–253.
400. Perun M. L., Gaver III D. P., 1995b, Interaction between airway lining fluid forces and parenchymal tethering during pulmonary airway reopening, J. Appl. Physiol. **79**, 1717–1728.
401. Petrov V., Haaning A., Muehlner K. A., Van Hook S. J., Swinney H. L., 1998, Model-independent nonlinear control algorithm with application to a liquid bridge experiment, Phys. Rev. E **58**, 427–433.
402. Petrov V., Schatz M. F., Muehlner K. A., Van Hook S. J., McCormick W. D., Swift J. B., Swinney H. L., 1996, Nonlinear control of remote unstable states in a liquid bridge convection experiment, Phys. Rev. Lett. **77**, 3779–3782.
403. Peyret R., Taylor D., 1990, *Computational Methods for Fluid Flow*, Springer-Verlag, Berlin.
404. Pismen L. M., 1986a, Inertial effects in long-scale thermal convection, Phys. Lett. A **116**, 241–244.

405. Pismen L. M., 1986b, Nonlinear dynamics of dissipative wave patterns near a bifurcation point, Dyn. Stabil. Syst. **1**, 97–113.
406. Pismen L. M., 1988, Selection of long-scale oscillatory convective patterns, Phys. Rev. A **38**, 2564–2572.
407. Pismen L. M., 1997, Interaction of reaction-diffusion fronts and Marangoni flow on the interface of a deep fluid, Phys. Rev. Lett. **78**, 382–385.
408. Pismen L. M., 2001, Nonlocal diffuse interface theory of thin films and the moving contact line, Phys. Rev. E **64**, 021603 (1–9).
409. Pismen L. M., 2002, Mesoscopic hydrodynamics of contact line motion, Colloids and Surfaces **206**, 11–30.
410. Pismen L. M., 2004, Spinodal dewetting in a volatile liquid film, Phys. Rev. E **70**, 021601 (1–9).
411. Pismen L. M., Pomeau Y., 2000, Disjoining potential and spreading of thin liquid layers in the diffuse-interface model coupled to hydrodynamics, Phys. Rev. E **62**, 2480–2492.
412. Pismen L. M., Pomeau Y., 2004, Mobility and interactions of weakly nonwetting droplets, Phys. Fluids **16**, 2604–2612.
413. Pitois O., François B., 1999a, Crystallization of condensation droplets on a liquid surface, Colloid and Polymer Sci. **277**, 574–578.
414. Pitois O., François B., 1999b, Formation of ordered micro-porous membranes, Eur. Phys. J. B **8**, 225–231.
415. Podgorski A., Gradon L., 1993, An improved mathematical model of hydrodynamic self-cleansing of pulmonary alveoli, Annu. Occup. Hyg. **37**, 347–365.
416. Pojman J. A., Craven R., Khan A., West W., 1992, Convective instabilities in traveling fronts of additional polymerization, J. Phys. Chem. **96**, 7466–7472.
417. Pomeau Y, Vannimenus J., 1984, Contact angle of heterogeneous surfaces: weak heterogeneities, J. Colloid Interface Sci. **104**, 477–488.
418. Pontes J., Christov C. I., Velarde M. G., 1996, Numerical study of patterns and the evolution in finite geometries, Int. J. Bifurc. Chaos **6**, 1883–1890.
419. Pototsky A., Bestehorn M., Merkt D., Thiele U., 2004, Alternative pathways of dewetting for a thin liquid two-layer film, Phys. Rev. E **1**, 2039–2049.
420. Pototsky A., Bestehorn M., Merkt D., Thiele U., 2005, Morphology changes in the evolution of liquid two-layer films, J. Chem. Phys. **122**, 224711 (1–13).
421. Poulard C., Benichou O., Cazabat A., 2003, Freely receding evaporating droplets, Langmuir **19**, 8828–8834.
422. Poulikakos D., 1986, Buoyancy-driven convection in a horizontal fluid layer extended over a porous substrate, Phys. Fluids **29**, 3949–3957.
423. Prakash A., Koster J. N., 1993, Natural and thermocapillary convection in three layers, Eur. J. Mech. B/Fluids **12**, 635–655.
424. Prakash A., Koster J. N., 1994a, Convection in multiple layers of immiscible liquids in a shallow cavity I. Steady natural convection, Int. J. Multiphase Flow **20**, 383–396.
425. Prakash A., Koster J. N., 1994b, Convection in multiple layers of immiscible liquids in a shallow cavity II. Steady thermocapillary convection, Int. J. Multiphase Flow **20**, 397–414.
426. Prakash A., Koster J. N., 1996, Steady Rayleigh-Bénard convection in a two-layer system of immiscible liquids, J. Heat Transfer **118**, 366–373.
427. Prasad V., 1991, Convective flow interaction and heat transfer between fluid and porous layers, in *Convective Heat and Mass Transfer in Porous Media*, edited by S. Kakaç, B. Kilkis, F. A. Kulacki, and F. Arinç, Kluwer Academic, Dordrecht, 563–615.

428. Prévost M., Gallez D., 1986, Nonlinear rupture of thin free liquid films, J. Chem. Phys. **84**, 4043–4048.
429. Priede J., Gerbeth, G., 1997, Convective, absolute, and global instabilities of thermocapillary-buoyancy convection in extended layers, Phys. Rev. E **56**, 4187–4199.
430. Priede J., Gerbeth G., 2000, Hydrothermal wave instability of thermocapillary-driven convection in a transverse magnetic field, J. Fluid Mech. **404**, 211–250.
431. Pshenichnikov A. F., Tokmenina G. A., 1983, Deformation of the free surface of a liquid by thermocapillary motion, Fluid Dyn. **18**, 463–465.
432. Pukhnachev V. V., 1987, *Thermocapillary Convection at Reduced Gravitation*, Institute of Hydrodynamics SB AS USSR, Novosibirsk.
433. Pumir A. and Blumenfeld L., 1996, Heat transport in a liquid layer locally heated on its free surface, Phys. Rev. E **54**, R4528–R4531.
434. Rabinovich L. M., Vyazmin A. V., Buyevich Y. A., 1995, Chemo-Marangoni convection. III. Interface mass transfer, J. Colloid Interface Sci. **173**, 1–7.
435. Ramaswamy B., Krishnamoorthy S., Joo S. W., 1997, Three-dimensional simulation of instabilities and rivulet formation in heated falling films, J. Comp. Phys. **131**, 70–88.
436. Ramon M. L., Maza D. M., Mancini H. L., 1999, Patterns in small aspect ratio Benard-Marangoni convection, Phys. Rev. E **60**, 4193–4198.
437. Ramon M. L., Maza D. M., Mancini H. L., Mancho A. M., Herrero H., 2001, Hexagonal structure in intermediate aspect ratio Benard-Marangoni convection, Int. J. Bifurc. Chaos **11**, 2779–2787.
438. Raphael E., de Gennes P. G., 1989, Dynamics of wetting with nonideal surfaces. The single defect problem, J. Chem. Phys. **90**, 7577–7584.
439. Rasenat S., Busse F. H., Rehberg I., 1989, A theoretical and experimental study of double-layer convection, J. Fluid Mech. **199**, 519–540.
440. Ratke L., Walter H., Feuerbacher B. (editors), 1989, *Material and Fluid under Low Gravity*, Springer-Verlag, Berlin.
441. Ratulowski J., Chang H.-C., 1989, Transport of gas bubbles in capillaries, Phys. Fluids A **1**, 1642–1655.
442. Rednikov A. Ye., Colinet P., Velarde M. G., Legros J. C., 1998, Two-layer Bénard-Marangoni instability and the limit of transverse and longitudinal waves, Phys. Rev. E **57**, 2872–2884.
443. Rednikov A. Ye., Colinet P., Velarde M. G., Legros J. C., 2000, Rayleigh-Marangoni oscillatory instability in a horizontal liquid layer heated from above: coupling and mode mixing of internal and surface dilational waves, J. Fluid Mech. **405**, 57–77.
444. Redon C., Brochart-Wyart F., Rondelez F., 1992, Festoon instabilities of slightly volatile liquids during spreading, J. Physique II **2**, 1671–1676.
445. Regel L. A., 1987, *Material Science in Space*, Halsted, New York.
446. Reichenbach J., Linde H., 1981, Linear perturbation analysis of surface-tension-driven convection at a plane interface (Marangoni instability), J. Colloid Interface Sci. **84**, 433–443.
447. Reinelt D. A., Saffman P. G., 1985, The penetration of a finger into a viscous fluid in a channel and tube, SIAM J. Sci. Stat. Comp. **6**, 542–561.
448. Reiter G., 1992, Dewetting of thin polymer films, Phys. Rev. Lett. **68**, 75–78.
449. Renardy Y. Y., 1986, Interfacial stability in a two-layer Bénard problem, Phys. Fluids **29**, 356–363.
450. Renardy Y. Y., 1996, Pattern formation for oscillatory bulk-mode competition in a two-layer Bénard problem, Z. Angew. Math. Phys. **47**, 567–590.

451. Renardy Y. Y., Joseph D. D., 1985, Oscillatory instability in a Bénard problem of two-fluids, Phys. Fluids **28**, 788–793.
452. Riley R. J., Neitzel G. P., 1998, Instability of thermocapillary-buoyancy convection in shallow layers, Part 1. Characterization of steady and oscillatory instabilities, J. Fluid Mech. **359**, 143–164.
453. Rosenblat S., Davis S. H., Homsy G. M., 1982a, Non-linear Marangoni convection in bounded layers, Part 1. Circular cylindrical containers, J. Fluid Mech. **120**, 91–122.
454. Rosenblat S., Homsy G. M., Davis S. H., 1982b, Non-linear Marangoni convection in bounded layers, Part 2. Rectangular cylindrical containers, J. Fluid Mech. **120**, 123–138.
455. Rother M. A., Davis R. H., 1999, The effect of slight deformation on thermocapillary-driven droplet coalescence and growth, J. Colloid Interface Sci. **214**, 297–318.
456. Rowlinson J. S., Widom B., 1982, *Molecular Theory of Capillarity*, Clarendon Press, Oxford.
457. Roze C., Gouesbet G., 1997, Complex 1D spatio-temporal behaviour generated by hot wire heating below an interface, Phys. Lett. A **227**, 79–86.
458. Ruckenstein E., Berbente C., 1964, The occurence of interfacial turbulence in the case of diffusion accompanied by chemical reaction, Chem. Eng. Sci. **19**, 329–347.
459. Ruckenstein E., Jain R. K., 1974, Spontaneous rupture of thin liquid films, J. Chem. Soc. Faraday Trans. II **70**, 132–147.
460. Ruckenstein E., Rajora P., 1983, On the no-slip boundary condition of hydrodynamics, J. Colloid Interface Sci. **96**, 488–491.
461. Rudraiah N., Prasad V., 1998, Effect of Brinkman boundary layer on the onset of Marangoni convection in a fluid-saturated porous layer, Acta Mech. **127**, 235–246.
462. Rupp R., Müller G., Neumann G., 1989, Three-dimensional time dependent modelling of the Marangoni convection in zone melting configurations for GaAs, J. Crystal Growth **97**, 34–41.
463. Sain F., Riecke H., 2000, Instabilities and spatio-temporal chaos in hexagon patterns with rotation, Physica D **144**, 124–141.
464. Sammarco T. S., Burns M. A., 1999, Thermocapillary pumping of discrete droplets in microfabricated analysis devices, AIChE J. **45**, 350–366.
465. Sanfeld A., Steinchen A., 1975, Coupling between a transconformation surface reaction and hydrodynamic motion, Biophys. Chem. **3**, 99–106.
466. Sass V., Kuhlmann H. C., Rath H. J., 1996, Investigation of three-dimensional thermocapillary convection in a cubic container by a multi-grid method, Int. J. Heat Mass Transfer **39**, 603–613.
467. Sassaroli E., Hynynen K., 2004, Forced linear oscillations of microbubbles in blood capillaries, J. Acoust. Soc. Am. **115**, 3235–3243.
468. Savino R., Monti R., 1996, Oscillatory Marangoni convection in cylindrical liquid bridges, Phys. Fluids **9**, 2906–2915.
469. Savino R., Paterna D., Lappa M., 2003, Marangoni flotation of liquid droplets, J. Fluid Mech. **479**, 307–326.
470. Sawistowski H., 1971, Interfacial phenomena, in *Recent Advances in Liquid-Liquid Extraction*, edited by C. Hanson, Pergamon Press, Oxford.
471. Scanlon J. W., Segel L. A., 1967, Finite-amplitude cellular convection induced by surface tension, J. Fluid Mech. **30**, 149–162.
472. Scardovelli R., Zaleski S., 1999, Direct numerical simulation of free-surface and interfacial flow, Annu. Rev. Fluid Mech. **31**, 567–603.
473. Schatz M. F., Van-Hook S. J., McCormic W. D., Swift J. B., Swinney H. L., 1995, Onset of surface-tension-driven Bénard convection, Phys. Rev. Lett. **75**, 1938–1941.

474. Schramm R., Reineke H., 1978, Numerical convection in a horizontal layer of two different fluids with internal heat sources, in *Proceedings of the 6th International Heat Transfer Conference* (Toronto), 299–304.
475. Schwabe D., Frank S., 1999, Experiments on the transition to chaotic thermocapillary flow in floating zones under microgravity, Adv. Space Res. **24**, 1391–1396.
476. Schwabe D., Möller U., Schneider J, Scharmann A., 1992, Instabilities of shallow dynamic thermocapillary liquid layers, Phys. Fluids A **4**, 2638–2381.
477. Schwartz L. W., Roy R. V., Eley R. R., Petrash S., 2001, Dewetting patterns in a drying liquid film, J. Colloid Interface Sci. **214**, 363–374.
478. Series R. W., Hurle D. T. J., 1991, The use of magnetic fields in semiconductor crystal growth, J. Crystal Growth **113**, 305–328.
479. Sethian J. A., Smereka P., 2003, Level set methods for fluid interfaces, Annu. Rev. Fluid Mech. **35**, 341–372.
480. Sharma A., 1993, Relationship of thin film stability and morphology to macroscopic parameters of wetting in the apolar and polar systems, Langmuir **9**, 861–869.
481. Sharma A., 2003, Many paths to dewetting of thin films: anatomy and physiology of surface instability, Eur. Phys. J. E **12**, 397–408.
482. Sharma A., Khanna R., 1998, Pattern formation in unstable thin liquid films, Phys. Rev. Lett. **81**, 3463–3466.
483. Sharma A., Ruckenstein E., 1986a, An analytical nonlinear theory of thin film rupture and its application to wetting films, J. Colloid Interface Sci. **113**, 456–479.
484. Sharma, A., Ruckenstein E., 1986b, Rupture of thin free films with insoluble surfactants: nonlinear aspects, AIChE Symp. Ser. **252**, 130–144.
485. Sherwood T. S., Wei J. C., 1957, Interfacial phenomena in liquid extraction, Ind. Eng. Chem. **49**, 1030–1034.
486. Shevtsova V. M., Melnikov D. E., Legros J.-C., 2001, Three-dimensional simulations of hydrodynamic instability in liquid bridges: Influence of temperature-dependent viscosity, Phys. Fluids **13**, 2851–2865.
487. Shevtsova V. M., Melnikov D. E., Legros J.-C., 2003a, Multistability of the oscillatory thermocapillary convection in liquid bridge, Phys. Rev. E **68**, 066311 (1–13).
488. Shevtsova V. M., Nepomnyashchy A. A., Legros J.-C., 2003b, Thermocapillary-buoyancy convection in shallow cavity heated from the side, Phys. Rev. E **67**, 066308 (1–14).
489. Shevtsova V. M., Simanovskii I. B., Nepomnyashchy A. A., Legros J.-C., 2005, Thermocapillary convection in a three-layer system with the temperature gradient directed along the interfaces, Compt. Rend. Mecan. **333**, 311–318.
490. Shikhmurzaev Yu. D., 1997, Moving contact lines in liquid/liquid/solid systems, J. Fluid Mech. **334**, 211–249.
491. Shioi A., Kumagai H., Sugiura Y., Kitayama Y., 2002, Oscillations of interfacial tension and spontaneous interfacial flow at a water/oil interface composed of di-(2-ethylhexyl) phosphoric acid, Langmuir **18**, 5516–5522.
492. Shiomi J., Amberg G., 2002, Active control of a global thermocapillary instability, Phys. Fluids **14**, 3039–3045.
493. Shiomi J., Amberg G., 2005a, Numerical investigation of feedback control of thermocapillary instability, Phys. Fluids **17**, 054107 (1–12).
494. Shiomi J., Amberg G., 2005b, Proportional control of oscillatory thermocapillary convection in a toy model, Eur. J. Mech. B/Fluids **24**, 296–313.
495. Shiomi J., Amberg G., Alfredsson H., 2001, Active control of oscillatory thermocapillary convection, Phys. Rev. E **64**, 031205 (1–7).

496. Shiomi J., Kudo M., Ueno I., Kawamura H., Amberg G., 2003, Feedback control of oscillatory thermocapillary convection in a half-zone liquid bridge, J. Fluid Mech. **496**, 193–211.
497. Shtilman L., Sivashinsky G., 1991, Hexagonal structure of large-scale Marangoni convection, Physica D **52**, 477–488.
498. Silber M., Knobloch E., 1991, Hopf bifurcation on a square lattice, Nonlinearity **4**, 1063–1106.
499. Silvi N., Dussan V E. B., 1985, On the rewetting of an inclined solid-surface by a liquid, Phys. Fluids **28**, 5–7.
500. Simanovskii I. B., 1979a, Finite-amplitude convection in a two-layer system, Fluid Dynamics **14**, 637–642.
501. Simanovskii I. B., 1979b, Numerical investigation of convection in the system of two immiscible fluids when heating from below, in *Convective Flows and Hydrodynamic Stability*, Sverdlovsk, USC AS USSR, 126–130. In Russian.
502. Simanovskii I. B., 1980, Convective stability of two-layer systems. PhD thesis, Leningrad State University, Leningrad. In Russian.
503. Simanovskii I. B., 1983, On stability of two-layer systems to the convective mixing, Fluid Dynamics **18**, 434–438.
504. Simanovskii I. B., 1991, Convective stabilities in systems with interface. D.Sc. thesis, Moscow Physical - Technical Institute, Moscow. In Russian.
505. Simanovskii I. B. 1996, Instability of the three layer system, in Proc. of the Second European Symposium *Fluids in Space*, edited by A. Viviani, European Space Agency, Naples, Italy, 264–271.
506. Simanovskii I. B., 1997, Anticonvective, steady and oscillatory mechanisms of instability in multi-layer systems, Physica D **102**, 313–327.
507. Simanovskii I. B., 2000, Combined action of anticonvective and thermocapillary mechanisms of instability in multilayer systems, Eur. J. Mech. B/Fluids **19**, 123–137.
508. Simanovskii I. B., 2004, Combined action of anticonvection and thermocapillarity in two- and three - layer systems, Compt. Rend. Mecan. **332**, 597–604.
509. Simanovskii I. B., 2005, Nonlinear oscillations in multilayer systems, in Proc. of 5th Euromech *Nonlinear Dynamics Conference*, Eindhoven, The Netherlands, 04-033, 654–660.
510. Simanovskii I., Colinet P., Nepomnyashchy A., Legros J.-C., 2002, Anticonvection with an inclined temperature gradient, Phys. Rev. E **66**, 056305 (1–10).
511. Simanovskii I., Georis Ph., Hennenberg M., Van Vaerenbergh S., Wertgeim I., Legros J.-C., 1992, Numerical investigation of Marangoni-Benard instability in multilayer systems, in *ESA SP-333* , European Space Agency, Paris, 729–734.
512. Simanovskii I., Georis Ph., Nepomnyashchy A., Legros J.-C., 2003, Oscillatory instability in multilayer systems, Phys. Fluids **15**, 3867–3870.
513. Simanovskii I., Georis Ph., Nepomnyashchy A., Legros J.-C., 2004, Nonlinear Marangoni oscillations in multilayer systems, Micrograv. Sci. Techn. **XV/2**, 25–34.
514. Simanovskii I. B., Nepomnyashchy A. A., 1993, *Convective Instabilities in Systems with Interface*, Gordon and Breach, London.
515. Simanovskii I. B., Nepomnyashchy A. A., 2006, Nonlinear development of oscillatory instability in a two-layer system under the combined action of buoyancy and thermocapillary effect, J. Fluid Mech. **555**, 177–202.
516. Sivashinsky G. I., 1982, Large cells in nonlinear Marangoni convection, Physica D **4**, 227–235.
517. Sivashinsky G. I., 1983, On cellular instability in the solidification of a dilute binary alloy, Physica D **8**, 243–248.

518. Skarda J. R. L., Jacqmin D., McCaughan F. E., 1998, Exact and approximate solutions to the double diffusive Marangoni-Bénard problem with cross-diffusive terms, J. Fluid Mech. **366**, 109–133.
519. Skarda J. R. L., McCaughan F. E., 1999, Exact solution to stationary onset of convection due to surface tension variation in a multicomponent fluid layer with interfacial deformation, Int. J. Heat Mass Transfer **42**, 2387–2398.
520. Slavtchev S., Hennenberg M., Legros J.-C., Lebon G., 1998, Stationary solutal Marangoni instability in a two-layer system, J. Colloid Interface Sci. **203**, 354–368.
521. Slavtchev S., Mendes M. A., 2004, Marangoni instability in binary liquid-liquid systems, Int. J. Heat Mass Transfer **47**, 3269–3278.
522. Slavtchev S., Simeonov G., Van Vaerenbergh S., Legros J.-C., 1999, Marangoni instability of a layer of binary liquid in the presence of nonlinear Soret effect, Int. J. Heat Mass Transfer **42**, 3007–3011.
523. Smith, K. A., 1966, On convective instability induced by surface-tension gradients, J. Fluid Mech. **24**, 401–414.
524. Smith M. K., Davis S. H., 1982, Inviscid instability of isothermal flow, J. Fluid Mech. **121**, 187–206.
525. Smith M. K., Davis S. H., 1983a, Instabilities of dynamic thermocapillary liquid layers, Part 1. Convective instabilities, J. Fluid Mech. **132**, 119–144.
526. Smith M. K., Davis S. H., 1983b, Instabilities of dynamic thermocapillary liquid layers, Part 2. Surface-wave instabilities, J. Fluid Mech. **132**, 145–162.
527. Sobolev V. D., Churaev N. V., Velarde M. G., Zorin Z. M., 2000, Surface tension and dynamic contact angle of water in thin quartz capillaries, J. Coll. Interface Sci. **222**, 51–54.
528. Somerton C. W., Catton I., 1982, On the thermal instability of superposed porous and fluid layers, J. Heat Transfer **104**, 160–165.
529. Someya S., Munakata T., Nishio M., Okamoto K., Madarame H., 2002, Flow observation in two immiscible liquid layers subject to a horizontal temperature gradient, J. Crystal Growth **235**, 626–632.
530. Srinivasao M., Collings D., Philips A., Patel S., 2001, Three-dimensionally ordered array of air bubbles in a polymer film, Science **292**, 79–83.
531. Starov V. M., 1992, Equilibrium and hysteresis contact angles, Adv. Colloid Interface Sci. **39**, 147–173.
532. Stebe K. J., Bartes-Biesel D., 1995, Marangoni effects of adsorption-desorption controlled surfactants on the leading end of an infinitely long bubble in a capillary, J. Fluid Mech. **286**, 25–48.
533. Sternling C. V., Scriven L. E., 1959, Interfacial turbulence: hydrodynamic instability and the Marangoni effect, AIChE J. **5**, 514–523.
534. Subramanian R. S., 1981, Slow migration of a gas bubble in a thermal gradient, AIChE J. **27**, 646–654.
535. Sumino Y., Kitahata H., Yoshikawa K., Nagayama M., Nomura S. M., Magome N., Mori Y., 2005a, Chemosensitive running droplet, Phys. Rev. E **72**, 041603 (1–8).
536. Sumino Y., Magome N., Hamada T., Yoshikawa K., 2005b, Self-running droplet: emergence of regular motion from nonequilibrium noise, Phys. Rev. Lett. **94**, 068301 (1–4).
537. Sussman M., Puckett E. G., 2000, A coupled level set and volume-of-fluid method for computing 3D and axisymmetric incompressible two-phase flows, J. Comput. Phys. **162**, 301–337.
538. Swift J. W., 1988, Hopf bifurcation with the symmetry of the square, Nonlinearity **1**, 333–377.

539. Szekely J., 1979, *Fluid Flow Phenomena in Metals Processing*, Academic Press, New York.
540. Takashima M., 1981, Surface tension driven instability in a horizontal liquid layer with a deformable free surface, II. Overstability, J. Phys. Soc. Japan **50**, 2751–2756.
541. Tan M. J., Bankoff S. G., Davis S. H., 1990, Steady thermocapilary flows of thin liquid layers, Phys. Fluids A **2**, 313–321.
542. Tang J., Bau H. H., 1993, Stabilization of the no-motion state in the Rayleigh-Bénard convection through the use of feedback-control, Phys. Rev. Lett. **70**, 1795–1798.
543. Tang J., Bau H. H., 1994, Stabilization of the no-motion state in the Rayleigh-Bénard problem, Proc. R. Soc. Lond. A **447**, 587–607.
544. Tang J., Bau H. H., 1998, Experiments in stabilization of the no-motion state of a fluid layer heated from below and cooled from above, J. Fluid Mech. **363**, 153–171.
545. Tanny J., Chen C. C., Chen C. F., 1995, Effects of interaction between Marangoni and double-diffusive instabilities, J. Fluid Mech. **303**, 1–21.
546. Tavener S. J., Cliffe K. A., 2002, Two-fluid Marangoni-Bénard convection with a deformable interface, J. Comput. Phys. **182**, 277–300.
547. Texier-Picard R., Pojman J. A., Volpert V. A., 2000, Effect of interfacial tension on propagating polymerization fronts, Chaos **10**, 224–230.
548. Thess A., Orzag S. A., 1995, Surface tension driven Bénard convection at infinite Prandtl number, J. Fluid Mech. **283**, 201–230.
549. Thiele U., 2003, Open questions and promising new fields in dewetting, Eur. Phys. J. E **12**, 409–414.
550. Thiele U., John K., Bär M., 2004, Dynamical model for chemically driven running droplets, Phys. Rev. Lett. **93**, 027802 (1–4).
551. Thiele U., Knobloch E., 2003, Front and back instability of a liquid film on a slightly inclined plate, Phys. Fluids **15**, 892–907.
552. Thiele U., Mertig M., Pompe W., 1998, Dewetting of an evaporating thin liquid film: Heterogeneous nucleation and surface instability, Phys. Rev. Lett. **80**, 2869–2872.
553. Thiele U., Neuffer K., Bestehorn M., Pomeau Y., Velarde M. G., 2002a, Sliding drops on an inclined plane, Colloids and Surfaces A **206**, 87–104.
554. Thiele U., Neuffer K., Pomeau Y., Velarde M. G., 2002b, On the importance of nucleation solutions for the rupture of thin liquid films, Colloids and Surfaces A **206**, 135–155.
555. Thiele U., Velarde M. G., Neuffer K., 2001a, Dewetting: Film rupture by nucleation in the spinodal regime, Phys. Rev. Lett. **87**, 016104 (1–4).
556. Thiele U., Velarde M. G., Neuffer K., Bestehorn M., Pomeau Y., 2001b, Sliding drops in the diffuse interface model coupled to hydrodynamics, Phys. Rev. E **64**, 061601 (1–12).
557. Thiele U., Velarde M. G., Neuffer K., Pomeau, Y., 2001c, Film rupture in the diffuse interface model coupled to hydrodynamics, Phys. Rev. E **64**, 031602 (1–11).
558. Thiessen D. B., Marr-Lyon M. J., Marston P. L., 2002, Active electrostatic stabilization of liquid bridges in low gravity, J. Fluid Mech. **457**, 285–294.
559. Thompson P. A., Robbins M. O., 1989, Simulations of contact-line motion: slip and the dynamic contact angle, Phys. Rev. Lett. **63**, 766–769.
560. Thornton J., 1987, Interfacial phenomena and mass transfer in liquid-liquid extraction, Chem. Ind. **6**, 193–196.
561. Tokaruk W. A., Molteno T. C. A., Morris S. W., 2000, Bénard-Marangoni convection in two-layered liquids, Phys. Rev. Lett. **84**, 3590–3593.
562. Tolstoi D. M., 1952a, Molecular theory of fluid sliding on solid surfaces, Dokl. Akad. Nauk SSSR **85**, 1089–1093 (in Russian).

563. Tolstoi D. M., 1952b, Mercury sliding on glass, Dokl. Akad. Nauk SSSR **85**, 1329–1335 (in Russian).
564. Tomita H., Abe K., 2000, Numerical simulation of pattern formation in the Bénard-Marangoni convection, Phys. Fluids **12**, 1389–1400.
565. Troian S. M., Herzbolheimer E., Safran S. A., Joanny J. F., 1989a, Fingering instabilities of driven spreading films, Europhys. Lett. **10**, 25–30.
566. Troian S. M., Wu X. L., Safran S. A., 1989b, Fingering instability in thin wetting films, Phys. Rev. Lett. **62**, 1496–1499.
567. Truskett V., Stebe K. J., 2003, Influence of surfactants on an evaporating drop: Fluorescence images and particle deposition patterns, Langmuir **19**, 8271–8279.
568. Tseng F.-G., Lin K.-H., Hsu H.-T., Chieng C.-C., 2004, A surface-tension-driven fluidic network for precise enzyme batch-dispensing and glucose detection, Sensors and Actuators A **111**, 107–117.
569. Ueno I., Tanaka S., Kawamura H., 2003, Oscillatory and chaotic thermocapillary convection in a half-zone liquid bridge, Phys. Fluids **15**, 408–416.
570. VanHook S. J., Schatz M. F., McCormick W. D., Swift J. B., Swinney H. L., 1995, Long-wavelength instability in surface-tension-driven Bénard convection, Phys. Rev. Lett. **75**, 4397–4400.
571. VanHook S. J., Schatz M. F., Swift J. B., McCormick W. D., Swinney H. L., 1997, Long-wavelength surface-tension-driven Bénard convection: experiment and theory, J. Fluid Mech. **345**, 45–78.
572. Vargaftik N. B., 1975, *Tables on the Thermophysical Properties of Liquids and Gases.*
573. Velarde M. G., Nekorkin V. I., Maksimov A. G., 1995, Further results on the evolution of solitary waves and their bound states of a dissipative Korteweg-de Vries equation, Int. J. Bifurc. Chaos **5**, 831–839.
574. Velarde M. G., Nepomnyashchy A. A., Hennenberg M., 2001, Onset of oscillatory interfacial instability and wave motion in Bénard layers, Adv. Appl. Mech. **37**, 167–238.
575. Velarde M. G., Rednikov A. Ye., Linde H., 1999, Waves generated by surface-tension gradients and instability, in *Fluid Dynamics at Interfaces*, Cambridge University Press, Cambridge, 43–56.
576. Verschueren M., Van der Vosse F. N., Meijer H. E. H., 2001, Diffuse-interface modelling of thermocapillary flow instabilities in a Hele-Shaw cell, J. Fluid Mech. **434**, 153–166.
577. Vinogradova O. I., 1999, Slippage of water over hydrophobic surfaces, Int. J. Mineral Process. **56**, 31–60.
578. Wagner C., Friedrich R., Narayanan R., 1994, Comments on the numerical investigation of Rayleigh and Marangoni convection in a vertical circular cylinder, Phys. Fluids **6**, 1425–1433.
579. Wahal S., Bose A., 1988, Rayleigh-Bénard and interfacial instabilities in two immiscible liquid layers, Phys. Fluids **31**, 3502–3510.
580. Walter H. U. (editor), 1987, *Fluid Sciences and Materials Science in Space*, Springer-Verlag, Berlin.
581. Wassmuth F., Laidlaw W. G., Coombe D. A., 1993, Calculation of interfacial flows and surfactant redistribution as a gas-liquid interface moves between two parallel plates, Phys. Fluids A **5**, 1533–1548.
582. Weh L., 1999, Self-organized structures at the surface of thin polymer films, Mat. Sci. Eng. C **8–9**, 463–467.

583. Weh L., 2001, Hyperbolic spirals as surface structures in thin layers, J. Colloid Interface Sci. **235**, 210–217.
584. Weh L., Venthur A., 2004, Crack patterns in thin polymer layers, Macromol. Mater. Eng. **289**, 227–237.
585. Wei H. H., Benintendi S. W., Halpern D., Grotberg J. B., 2003, Cycle-induced flow and transport in a model alveolar liquid lining, J. Fluid Mech. **483**, 1–36.
586. Welander P., 1964, Convective instability in a two-layer fluid heated uniformly from above, Tellus **16**, 349–358.
587. Whitham G. B., 1974, *Linear and Nonlinear Waves*, John Wiley, New York.
588. Widawski C., Rawiso M., François B., 1994, Self-organized honeycomb morphology of star-polymer polystyrene films, Nature **369**, 397–399.
589. Wierschem A., Linde H., Velarde M. G., 2000, Internal waves excited by the Marangoni effect, Phys. Rev. E **62**, 6522–6530.
590. Williams M. B., Davis S. H., 1982, Nonlinear theory of film rupture, J. Colloid Interface Sci. **90**, 220–228.
591. Williams R., 1977, The advancing front of a spreading liquid, Nature **266**, 153–154.
592. Wilson S. K., 1993a, The effect of a uniform magnetic field on the onset of steady Bénard-Marangoni convection in a layer of conducting fluid, J. Eng. Math. **27**, 161–188.
593. Wilson S. K., 1993b, The effect of a uniform magnetic field on the onset of Marangoni convection in a layer of conducting fluid, Q. J. Mech. Appl. Math. **46**, 211–248.
594. Wu Q. F., Wong H., 2004, A slope-dependent disjoining pressure for non-zero contact angles, J. Fluid Mech. **506**, 157–185.
595. Xu J., Zebib A., 1998, Oscillatory two- and three-dimensional thermocapillary convection, J. Fluid Mech. **364**, 187–209.
596. Yamamura M., Nishio T., Kajiwara T., Adachi K., 2002, Evaporation-induced pattern formation in polymer films via secondary phase separation, Chem. Eng. Sci. **57**, 2901–2905.
597. Yap D. Y. K., Gaver III D. P., 1998, The influence of surfactant on two-phase flow in a flexible-walled channel under bulk equilibrium conditions, Phys. Fluids **10**, 1846–1863.
598. Yarin A. L., 1995, Surface tension driven flows at low Reynolds number arising in opto-electronic technology, J. Fluid Mech. **286**, 273–200.
599. Ye Y., Chang H. C., 1999, A spectral theory for fingering on a prewetted plane, Phys. Fluids **11**, 2494–2515.
600. Yeo L. Y., Matar O. K., 2003, Hydrodynamic instability of a thin viscous film between two drops, J. Colloid Interface Sci. **261**, 575–579.
601. Yeo L. Y., Matar O. K., de Ortiz E. S. P., Hewitt G. F., 2001, The dynamics of Marangoni-driven local film drainage between two drops, J. Colloid Interface Sci. **241**, 233–247.
602. Yoshikawa K., Magome N., 1993, Chemomechanical transduction in an oil-water system - regulation of the macroscopic mechanical motion, Bull. Chem. Soc. Japan **6**, 3352–3357.
603. Young Y. N., Riecke H., 2002, Mean flow in hexagonal convection: stability and nonlinear dynamics, Physica D **163**, 166–183.
604. Young Y. N., Riecke H., 2003, Induced defect nucleation and side-band instabilities in hexagons with rotation and mean flow, Physica D **176**, 107–124.

605. Zaman A. A., Narayanan R., 1996, Interfacial and buoyancy-driven convection - The effect of geometry and comparison with experiment, J. Colloid Interface Sci. **179**, 151–162.
606. Zeldovich Ya. B., Ruzmikin A. A., Sokoloff D. D., 1983, *Magnetic Fields in Astrophysics*, vol. 3 of: *The Fluid Mechanics of Astrophysics and Geophysics*, edited by P. H. Roberts, Gordon and Breach, New York.
607. Zelig D., Haber S., 2002, Hydrodynamic cleansing of pulmonary alveoli, SIAM J. Appl. Math. **63**, 195–221.
608. Zeren R. W, Reynolds W. C., 1972, Thermal instabilities in two-fluid horizontal layers, J. Fluid Mech. **53**, 305–327.
609. Zeytounian R. Kh., 1998, The Bénard-Marangoni thermocapillary-instability problem, Sov. Phys. Uspekhi **41**, 241–267.
610. Zharikov E. V., Prikhod'ko L. V., Storozhev N. R., 1990, Fluid-flow formation resulting from forced vibration of a growing crystal, J. Crystal Growth **99**, 910–914.
611. Zimmerman W. B., 2001, Excitation of surface waves due to thermocapillary effects on a stable stratified fluid layer, preprint.

Index